Curvature in Mathematics and Physics

Shlomo Sternberg

Dover Publications, Inc.
Mineola, New York

Copyright

Copyright © 2012 by Shlomo Sternberg
All rights reserved.

Bibliographical Note

Curvature in Mathematics and Physics is a new work, first published by Dover Publications, Inc., in 2012.

International Standard Book Number
ISBN-13: 978-0-486-47855-5
ISBN-10: 0-486-47855-6

Manufactured in the United States by Courier Corporation
47855605   2014
www.doverpublications.com

Curvature in Mathematics and Physics

0.1 Introduction

I have taught an advanced undergraduate course on differential geometry many times over the past fifty five years. The contents varied according to the interests and the background of the students. But the core ingredients of the course consist of giving an introduction to (semi-)Riemannian geometry and its principal physical application, Einstein's theory of general relativity, using the Cartan exterior calculus as a principal tool. The background assumed is a good grounding in linear algebra and in advanced calculus, preferably in the language of differential forms.

The golden standard, against which any book must be measured, is O'Neill's *Semi-Riemannian Geometry with applications to relativity*. I rely heavily on O'Neill's treatment, but my book is aimed at a slightly less advanced readership. Also, at various points, my "philosophy" is somewhat different (and inevitably, my notation) from O'Neill's. Another source of inspirations is M. Berger's various and excellent books on the subject.

Chapter I introduces the various curvatures associated to a hypersurface embedded in Euclidean space, motivated by the formula for the volume for the region obtained by thickening the hypersurface on one side. If we thicken the hypersurface by an amount h in the normal direction, this formula is a polynomial in h whose coefficients are integrals over the hypersurface of local expressions. These local expressions are elementary symmetric polynomials in what are known as the principal curvatures. The precise definitions are given in the text. The chapter culminates with Gauss' *theorema egregium* which asserts that if we thicken a two dimensional surface evenly on *both* sides, then these integrands depend only on the intrinsic geometry of the surface, and not on how the surface is embedded. We give two proofs of this important theorem. (We give several more later in the book.) The first proof makes use of "normal coordinates" which become so important in Riemannian geometry and, as "inertial frames," in general relativity. It was this theorem of Gauss, and particularly the very notion of "intrinsic geometry", which inspired Riemann to develop his geometry. We mention such ideas as minimal surfaces, but do not go into their study. The background assumed in this chapter is the implicit function theorem and the transformation law for multiple integrals. At Harvard, students "shop around" for courses in the first few weeks of the semester, and so I do not want to make so many demands as to scare them away.

Chapter II is a rapid review of the differential and integral calculus on manifolds, including differential forms, the d operator, and Stokes' theorem. Also vector fields and Lie derivatives. In this chapter I do, eventually, require knowledge of the basics of the theory of differentiable manifolds. My favorite source is the chapter devoted to this topic in my book with Loomis *Advanced calculus* available on my Harvard web site. A different excellent source is Chapter I of O'Neill. At the end of the chapter are a series of exercises which are essential for the rest of the book. They give experience in dealing with matrices with differential forms as entries and get to the Maurer Cartan equations and their implementation in two and three dimensional differential geometry.

0.1. INTRODUCTION

Chapter III discusses the fundamental notions of linear connections and their curvatures, and as such is pretty standard. I discuss the Jacobi equations in the setting of a general linear connection, as well as the exponential map. I relate the differential of the exponential map to solutions of the Jacobi equations thus introducing what Berger call's "Cartan's philosophy" of how to do Riemannian geometry. I conclude the chapter with the notion of a "locally symmetric" connection, which, in its Riemannian setting is one of the outcomes of "Cartan's philosophy".

Chapter IV is devoted to Levi-Civita's theorem and some of its consequences. It includes a series of problems devoted to computing the geodesics in the exterior of the Scwarzschild black hole. Perhaps this is out of order, and should, logically, come in after the discussion of the Einstein field equations and the verification that the Schwartzschild metric is indeed a solution of those equations. But this series of problems are good illustrations of how to compute geodesics, so why not put them in here? The organization of these problems follows the treatment in O'Neill. We then discuss identities of the Riemann curvature tensor and Cartan's notion of locally symmetric spaces. We conclude this chapter with a discussion of the curvature of a non-degenerate submanifold with applications to the deSitter universe.

Chapter V is devoted to the computation of various geometric objects associated with a bi-invariant metric on a Lie group. At the end, I introduce some notation related to the "Standard Model" of electroweak interactions which will be used in Chapter XVII.

Chapter VI is devoted to Cartan's method of calculation, using frame fields and differential forms, in Riemannian geometry. The list of examples calculated can be seen in the table of contents.

Chapter VII discusses Gauss's lemma, and its principal consequence - that in Riemannian geometry, geodesics locally minimize arc length. The concept of a geodesic, introduced in Chapter III in the framework of a general linear connection was that of a curve which is "self-parallel". So the fact that a geodesic locally minimizes arc length is a theorem, not a definition.

Chapter VIII gives some of the basic facts about variational formulas and their implications for Riemannian geometry. The material and treatment here are fairly standard.

Chapter IX is devoted to the Hopf-Rinow theorem. This is a foundational theorem in Riemannian geometry. The proof that we give is a beautiful argument due to de Rham. The treatment here is standard.

Chapter X continues the discussion involving variational formulas, in particular discusses Myer's theorem. Once again, the material here is fairly standard. In a more advanced course I would teach Toponogov's theorem at this point, but omit it here.

Chaper XI is a review of special relativity. I assume that the student (reader) has had a previous course on the subject.

Chapter XII introduces the star operator and its applications to electromagnetic theory. In particular we emphasize that from the point of view of special relativity the London equations for superconductivity amount to adding a mass term to the photon.

Chapter XIII develops various notions and results needed for the study of the Einstein field equations. In addition, it presents a result of Einstein, Infeld and Hoffmann which gives a purely "group theoretical' characterization of geodesics. This result is not as well known as it should be.

Chapter XIV is the high point of the course from the theoretical point of view. We discuss Einstein's general theory of relativity from the point of view of the Einstein-Hilbert functional. In fact we borrow the title of Hilbert's paper for the chapter heading. We try to put the Einstein-Hilbert approach to the equations of gravity in a more general context, and show how the notions of harmonic maps and minimal embeddings can also be derived from the Einstein, Infeld, Hoffmann approach to geodesics.

Chapter XV is devoted to the theorem of Frobenius on the integrability of differential systems. This theorem really deserves to be discussed in a course on differential topology. But it is so central to understanding Ehresmann's theory of general connections, that I decided to include it.

Chapter XVI begins by discussing the bundle of frames which is the modern setting for Cartan's calculus of moving frames and also the jumping off point for the general theory of connections on principal bundles which lie at the base of such modern physical theories as Yang-Mills fields. It continues to discuss the general theory of connections on fiber bundles and then specializes to principal and associated bundles. It serves not only as a valuable subject in its own right, but also as an introduction to the study of "Gauge theories" and the "Standard model" in elementary particle physics discussed in Chapter XVII. This chapter seems to present the most difficulty conceptually for the student.

Chapter XVII discusses Cartan's theory of reduction of principal bundles and of soldering forms, and also the current "gauge theories of forces" including the "Higgs mechanism" which is related to the mathematical concept of "reduction of a principle bundle".

Chapter XVIII discusses Quillen's theory of superconnections.

Chapter XIX returns to semi- Riemannian geometry and discusses the concept of a semi- Riemannian submersion, introduced by Robert Hermann and perfected by Barrett O'Neill.

Only a portion of this material can be covered in a one semester course. When I teach the course, I try to include at least the material in Chapters I-IV, VI,VII, and XII-XV.

Throughout the book, I have included portraits that I have downloaded from the web (mostly from the St. Andrews biographical website) of some of the key players in our subject. But there are three heroes whose work cannot be confined to a single chapter, so I include their portraits here: The first is Riemann, who is the founder of our entire subject.

Georg Friedrich Bernhard Riemann

Born: 17 Sept. 1826 in Breselenz, Hanover (now Germany)
Died: 20 July 1866 in Selasca, Italy

The second is, of course, Einstein:

Albert Einstein

(photograph by Bachrach)
Born: 14 March 1879 in Ulm, Wurttemberg, Germany
Died: 18 April 1955 in Princeton, New Jersey, USA

The third is Élie Cartan, whose exterior differential calculus, method of moving frames, "philosophy" of Riemannian geometry, symmetric spaces etc. are key ingredients in this book. His spirit inhabits every nook and cranny.

Elie Joseph Cartan

Born: 9 April 1869 in Dolomieu, Savoie, Rhone, France
Died: 6 May 1951 in Paris, France

I would like to thank Chen He for proofreading the book and saving me from some serious inconsistencies in notation.

I would like to thank Oliver Knill for technical help and advice in the general preparation of this book, and also, more specifically, for help in the preparation of various diagrams.

Above all, I would like to express my deepest gratitude to Prof. Robert Low for his help with the entire book. He not only corrected typos and caught mathematical errors, but also made useful suggestions as to exposition and also as to overall organization of the material. He has been an immense help.

Needless to say, I bear full responsibility for all remaining faults.

Contents

    0.1   Introduction 2

1 Gauss's theorema egregium.   **19**
    1.1   Volume of a thickened hypersurface 20
    1.2   Defining some of our terms. 21
    1.3   The Gauss map and the Weingarten map. 26
    1.4   Proof of the volume formula. 30
    1.5   Gauss's theorema egregium. 33
        1.5.1   First proof, using inertial coordinates. 36
        1.5.2   Second proof. The Brioschi formula. 39
    1.6   Back to the area formula. 41
        1.6.1   An alternative expression for the surface "area". 41
        1.6.2   The mean curvature and minimal surfaces. 42
        1.6.3   Minimal hypersurfaces. 43
    1.7   Problem set - Surfaces of revolution. 46

2 Rules of calculus.   **51**
    2.1   Superalgebras. 51
    2.2   Differential forms. 52
        2.2.1   Linear differential forms. 52
        2.2.2 $\Omega(M)$, the algebra of exterior differential forms. 52
    2.3   The d operator. 53
    2.4   Even and odd derivations of a superalgebra. 53
    2.5   Pullback. 55
    2.6   Chain rule. 55
    2.7   Lie derivative. 55
    2.8   Weil's formula. 56
    2.9   Integration. 58
    2.10 Stokes theorem. 58
    2.11 Lie derivatives of vector fields. 59
    2.12 Jacobi's identity. 60
    2.13 Forms as multilinear functions on vector fields. 61
    2.14 Problems. 62
        2.14.1 Matrix valued differential forms. 62
        2.14.2 Actions of Lie groups on themselves. 63

      2.14.3 The Maurer-Cartan form. 64
      2.14.4 The Maurer-Cartan equation(s). 64
      2.14.5 Restriction to a subgroup of $Gl(n)$. 65
      2.14.6 Interlude, the Haar integral. 65
      2.14.7 Frames. 68
      2.14.8 Euclidean frames. 69
      2.14.9 Frames adapted to a submanifold. 70
      2.14.10 Curves and surfaces - their structure equations. 71
      2.14.11 The sphere as an example. 71
      2.14.12 Ribbons 73
      2.14.13 The induced metric and the Weingarten map determine a surface up to a Euclidean motion. 74
      2.14.14 Back to ribbons. 75
      2.14.15 Developing a ribbon. 76
      2.14.16 The general Maurer-Cartan equations. 77
      2.14.17 Back to curves in $\mathbb{R}^3$. 78
      2.14.18 Summary in more logical order, starting from Weil's formula. 81

3 Connections on the tangent bundle.   85
  3.1 Definition of a linear connection on the tangent bundle. 85
  3.2 Christoffel symbols. 86
  3.3 Parallel transport. 86
  3.4 Parallel vector fields along a curve. 88
  3.5 Geodesics. 89
      3.5.1 Making a linear change of parameter. 90
  3.6 Torsion. 90
  3.7 Curvature. 91
      3.7.1 R is a tensor. 91
      3.7.2 The first Bianchi identity. 92
  3.8 Tensors and tensor analysis. 93
      3.8.1 Multilinear functions and tensors of type (r,s). 93
      3.8.2 Tensor multiplication. 95
      3.8.3 Why a tensor over $\mathcal{V}(M)$ is a "field" of tensors over M. . 95
      3.8.4 Tensor notation. 97
      3.8.5 Contraction. 98
      3.8.6 Tensor derivations. 99
      3.8.7 A connection as a tensor derivation, covariant differential. 101
      3.8.8 Covariant differential vs. exterior derivative. 102
  3.9 Variations and the Jacobi equations. 104
      3.9.1 Two parameter maps. 104
      3.9.2 Variations of a curve. 105
      3.9.3 Geodesic variations and the Jacobi equations. 106
      3.9.4 Conjugate points. 107
  3.10 The exponential map. 107
      3.10.1 The differential of the exponential map at 0 is the identity. 108

        3.10.2 Normal neighborhoods. 108
        3.10.3 Normal coordinates. 108
        3.10.4 The exponential map and the Jacobi equation. 109
        3.10.5 Polar maps. 111
  3.11 Locally symmetric connections. 112
  3.12 Normal neighborhoods and convex open sets. 112

4 Levi-Civita's theorem.   117
  4.1 Isometric connections. 117
  4.2 Levi-Civita's theorem. 118
  4.3 The Christoffel symbols of the Levi-Civita connection. 120
  4.4 Geodesics in orthogonal coordinates. 121
  4.5 The hereditary character of the Levi-Civita connection. 122
  4.6 Back to the isometric condition $\nabla \mathbf{g} = 0$. 123
  4.7 Problems: Geodesics in the Schwarzschild exterior. 123
        4.7.1 The Schwarzschild solution. 124
        4.7.2 Massive particles. 125
        4.7.3 Orbit Types. 126
        4.7.4 Perihelion advance. 128
        4.7.5 Massless particles. 131
        4.7.6 Kerr-Schild form. 132
        4.7.7 A brief biography of Schwarzschild culled from wikipedeia
              and St. Andrews. 133
  4.8 Curvature identities. 135
  4.9 Sectional curvature. 136
        4.9.1 Degenerate and non-degenerate planes. 136
        4.9.2 Definition of the sectional curvature. 137
        4.9.3 The sectional curvature determines the Riemann curvature tensor. 137
        4.9.4 Constant curvature spaces. 138
  4.10 Ricci curvature. 139
  4.11 Locally symmetric semi-Riemannian manifolds. 140
        4.11.1 Why the word "symmetric" in locally symmetric? 143
  4.12 Curvature of the induced metric of a submanifold. 143
        4.12.1 The case of a hypersurface. 145
  4.13 The de Sitter universe and its relatives. 145
        4.13.1 The Einstein field equations with cosmological constant Λ. 148
        4.13.2 Cosmological considerations 148

5 Bi-invariant metrics on a Lie group.   151
  5.1 The Lie algebra of a Lie group. 151
  5.2 The general Maurer-Cartan form. 153
  5.3 Left invariant and bi-invariant metrics. 155
  5.4 Geodesics are cosets of one parameter subgroups. 156
  5.5 The Riemann curvature of a bi-invariant metric. 157
  5.6 Sectional curvatures. 157

- 5.7 The Ricci curvature and the Killing form. 157
- 5.8 Bi-invariant forms from representations. 158
- 5.9 The Weinberg angle. 160

6 Cartan calculations   161
- 6.1 Frame fields and coframe fields. 161
  - 6.1.1 The tautological tensor. 162
- 6.2 Connection and curvature forms in a frame field. 162
  - 6.2.1 Connection forms. 162
  - 6.2.2 Cartan's first structural equation. 163
  - 6.2.3 Symmetry properties of ω. 163
  - 6.2.4 Curvature forms in a frame field. 164
  - 6.2.5 Cartan's second structural equation. 165
  - 6.2.6 Both structural equations in compact form. 165
- 6.3 Cartan's lemma and Levi-Civita's theorem. 165
  - 6.3.1 Cartan's lemma in exterior algebra. 166
  - 6.3.2 Using Cartan's lemma to prove Levi-Civita's theorem. . . 166
- 6.4 Examples of Cartan style computations. 167
  - 6.4.1 Polar coordinates in two dimensions. 167
  - 6.4.2 Hyperbolic geometry. 168
  - 6.4.3 The Schwarzschild metric. 170
- 6.5 The second Bianchi identity. 172
- 6.6 A theorem of F. Schur. 173
- 6.7 Friedmann Robertson Walker metrics. 173
  - 6.7.1 The expanding universe and the big bang. 176
- 6.8 The rotating black hole. 178
  - 6.8.1 Killing fields and Noether's theorem. 178
  - 6.8.2 The definition of the Kerr metric and some of its elementary properties. 179
  - 6.8.3 Checking that we do have a semi-Riemannian metric. . . 180
  - 6.8.4 The domains and the signature. 182
  - 6.8.5 An orthonormal frame field and its coframe field. 184
  - 6.8.6 The connection forms. 186
  - 6.8.7 The curvature. 187

7 Gauss's lemma.   189
- 7.1 Geodesics locally minimize arc length in a Riemannian manifold. 189
- 7.2 Gauss's lemma. 190
- 7.3 Short enough geodesics give an absolute minimum for arc length. 192

8 Variational formulas.   193
- 8.1 Jacobi fields in semi-Riemannian geometry. 193
  - 8.1.1 Tangential Jacobi fields. 193
  - 8.1.2 Perpendicular Jacobi fields. 193
  - 8.1.3 Decomposition of a Jacobi field into its tangential and perpendicular components. 194

CONTENTS

    8.2   Variations of arc length. 195
          8.2.1   The first variation. 195
    8.3   Geodesics are stationary for arc length. 196
          8.3.1   Piecewise smooth variations. 196
    8.4   The second variation. 197
          8.4.1   Synge's formula for the second variation. 197
    8.5   Conjugate points and the Morse index. 199
          8.5.1   Non-positive sectional curvature means no conjugate points. 200
    8.6   Synge's theorem. 201
    8.7   Cartan on the existence of closed geodesics. 202

9 The Hopf-Rinow theorem.   **205**
    9.1   Riemannian distance. 205
          9.1.1   Some history. 207
          9.1.2   Minimizing curves. 207
    9.2   Completeness and the Hopf-Rinow theorem. 207
          9.2.1   The key proposition - de Rham's proof. 207
          9.2.2   Geodesically complete manifolds. 209
          9.2.3   The Hopf-Rinow theorem. 210
    9.3   Hadamard's theorem. 212
    9.4   Locally isometric coverings. 212
    9.5   Symmetric spaces. 214

10 Curvature, distance, and volume.   **217**
    10.1   Sectional curvature and distance, locally. 217
    10.2   Myer's theorem. 222
          10.2.1   Back to the Ricci tensor. 223
          10.2.2   Myer's theorem. 225
    10.3   Length variation of a Jacobi vector field. 225
          10.3.1   Riemann's formula for the metric in a normal neighborhood. 226
    10.4   The Ricci tensor and volume growth. 228

11 Review of special relativity.   **231**
    11.1   Two dimensional Lorentz transformations. 231
          11.1.1   Two dimensional Minkowski spaces. 231
          11.1.2   Addition law for velocities. 233
          11.1.3   Hyperbolic angle aka "rapidity". 234
          11.1.4   Proper time. 235
          11.1.5   Time dilation. 235
          11.1.6   The Lorentz-Fitzgerald contraction. 236
          11.1.7   The reverse triangle inequality. 236
          11.1.8   Physical significance of the Minkowski distance. 237
          11.1.9   Energy-momentum 238
          11.1.10 Psychological units. 239
          11.1.11 The Galilean limit. 241
    11.2   Minkowski space. 241

  11.2.1 The Compton effect. 242
  11.2.2 Natural Units. 247

12 The star operator and electromagnetism.   249
 12.1 Definition of the star operator. 249
  12.1.1 The induced scalar product on exterior powers. 249
 12.2 Does $\star : \wedge^k V \to \wedge^{n-k} V$ determine the metric? 252
 12.3 The star operator on forms. 254
  12.3.1 Some equations of mathematical physics. 254
 12.4 Electromagnetism. 258
  12.4.1 Two non-relativistic regimes. 258
  12.4.2 Maxwell's equations. 262
  12.4.3 Natural units and Maxwell's equations. 263
  12.4.4 The Maxwell equations with a source term. 264
 12.5 The London equations. 265
  12.5.1 The London equations in relativistic form. 267
  12.5.2 Comparing Maxwell and London. 269

13 Preliminaries to the Einstein equations.   271
 13.1 Preliminaries to the preliminaries. 271
  13.1.1 Densities and n-forms. 271
  13.1.2 Densities of arbitrary order. 273
  13.1.3 Pullback of a density under a diffeomorphism. 273
  13.1.4 The Lie derivative of a density. 273
  13.1.5 The divergence of a vector field relative to a density. . . . 274
 13.2 Divergence on a semi-Riemannian manifold. 275
 13.3 The Lie derivative of a semi-Riemannian metric. 278
 13.4 The divergence of a symmetric tensor field. 278
  13.4.1 The meaning of the condition div $\mathbf{T} = 0$. 279
  13.4.2 Generalizing the condition of the vanishing of the covariant divergence. 281
 13.5 Analyzing the condition $\ell(L_V \mathbf{g}) = 0$. 281
  13.5.1 What does condition (13.16) say for a tensor field concentrated along a curve? 283
 13.6 Three different characterizations of a geodesic. 285
 13.7 The space of connections as an affine space. 286
 13.8 The Levi-Civita map and its derivative. 287
  13.8.1 The Riemann curvature and the Ricci tensor as a maps. . 287
 13.9 An important integral identity. 288

14 Die Grundlagen der Physik.   291
 14.1 The structure of physical laws. 291
  14.1.1 The Legendre transformation. 291
  14.1.2 Inverting the Legendre transformation as the "source equation" of physics. 292
 14.2 The Newtonian example. 292

- 14.3 The passive equations. 294
- 14.4 The Hilbert "function". 295
- 14.5 Harmonic maps as solutions to a passive equation. 299
- 14.6 Schrodinger's equation as a passive equation. 302

15 The Frobenius theorem. 303
- 15.1 The Frobenius theorem. 304
  - 15.1.1 Differential systems. 304
  - 15.1.2 Foliations, submersions, and fibrations. 305
  - 15.1.3 The vector fields of a differential system, the Frobenius theorem. 306
  - 15.1.4 Connected and maximal leaves of an integrable system. . 308
- 15.2 Maps into a Lie group. 309
  - 15.2.1 Applying the above to the diagonal. 310
  - 15.2.2 The induced metric and the Weingarten map determine a hypersurface up to a Euclidean motion. 310
- 15.3 Another application of Frobenius: to reduction. 311
- 15.4 A dual formulation of Frobenius' theorem. 312
- 15.5 Horizontal and basic forms of a fibration. 313
- 15.6 Reduction of a closed form. 314

16 Connections on principal bundles. 315
- 16.1 Connection and curvature forms in a frame field. 315
- 16.2 Change of frame field. 316
- 16.3 The bundle of frames. 318
  - 16.3.1 The form ϑ. 320
  - 16.3.2 The form ϑ in terms of a frame field. 320
  - 16.3.3 The definition of ω. 321
- 16.4 Connection forms in a frame field as a pull-backs. 321
- 16.5 Submersions, fibrations, and connections. 324
  - 16.5.1 Submersions. 324
  - 16.5.2 Fibrations. 327
  - 16.5.3 Projection onto the vertical. 329
  - 16.5.4 Frobenius, generalized curvature, and local triviality. . . 330
- 16.6 Principal bundles and invariant connections. 331
  - 16.6.1 Principal bundles. 331
  - 16.6.2 Connections on principal bundles. 334
  - 16.6.3 Associated bundles. 335
  - 16.6.4 Sections of associated bundles. 336
  - 16.6.5 Associated vector bundles. 337
  - 16.6.6 Exterior products of vector valued forms. 340
- 16.7 Covariant differentials and derivatives. 341
  - 16.7.1 The horizontal projection of forms. 341
  - 16.7.2 The covariant differential of forms on P. 342
  - 16.7.3 A formula for the covariant differential of basic forms. . . 343

17 Reduction of principal bundles. 347
- 17.1 A brief history of gauge theories. ... 347
- 17.2 Cartan's approach to connections. ... 348
  - 17.2.1 Cartan connections. ... 351
- 17.3 Symmetry breaking and "mass acquisition". ... 352
  - 17.3.1 The Automorphism group and the Gauge group of a principal bundle. ... 353
  - 17.3.2 Mass acquisition - the Higgs mechanism. ... 356
- 17.4 The Higgs mechanism in the standard model. ... 357
  - 17.4.1 Problems of translation between mathematicians and and physicists. ... 357
  - 17.4.2 The Higgs mechanism in the standard model. ... 360

18 Superconnections. 365
- 18.1 Superbundles. ... 366
  - 18.1.1 Superspaces and superalgebras. ... 366
  - 18.1.2 The tensor product of two superalgebras. ... 366
  - 18.1.3 Lie superalgebras. ... 367
  - 18.1.4 The endomorphism algebra of a superspace. ... 367
  - 18.1.5 Superbundles. ... 368
  - 18.1.6 The endomorphism bundle of a superbundle. ... 368
  - 18.1.7 The centralizer of multiplication by differential forms. ... 368
  - 18.1.8 Bundles of Lie superalgebras. ... 369
- 18.2 Superconnections. ... 369
  - 18.2.1 Extending superconnections to the bundle of endomorphisms. ... 370
  - 18.2.2 Supercurvature. ... 370
  - 18.2.3 The tensor product of two superconnections. ... 371
  - 18.2.4 The exterior components of a superconnection. ... 371
  - 18.2.5 A local computation. ... 372
- 18.3 Superconnections and principal bundles. ... 372
  - 18.3.1 Recalling some definitions. ... 372
  - 18.3.2 Generalizing the above to superconnections. ... 374
- 18.4 Clifford Bundles and Clifford superconnections. ... 375

19 Semi-Riemannian submersions. 377
- 19.1 Submersions. ... 378
- 19.2 The fundamental tensors of a submersion. ... 380
  - 19.2.1 The tensor T. ... 380
  - 19.2.2 The tensor A. ... 381
  - 19.2.3 Covariant derivatives of T and A. ... 382
  - 19.2.4 The fundamental tensors for a warped product. ... 384
- 19.3 Curvature. ... 385
  - 19.3.1 Curvature for warped products. ... 390
  - 19.3.2 Sectional curvature. ... 393
- 19.4 Reductive homogeneous spaces. ... 394

19.4.1 Homogeneous spaces. 394
19.4.2 Normal symmetric spaces. 394
19.4.3 Orthogonal groups. 396
19.4.4 Dual Grassmannians. 397
19.5 Schwarzschild as a warped product. 399

Bibliography. 401

Index. 402

Chapter 1

The principal curvatures and Gauss's theorema egregium.

One of the greatest, if not the greatest, achievements of the human mind was Einstein's formulation of "general relativity" in 1915, which posited that the gravitational force is due to the "curvature of space time" and that particles move along geodesics in this "curved space-time".

The concept of "curvature" that Einstein used was due to Riemann in his Habilitation lecture *Über die Hypothesen welche der Geometrie zu Grunde liegen* (On the hypotheses that lie at the foundations of geometry), delivered on 10 June 1854.

Riemann's work was strongly influenced by Gauss's work on the geometry of surfaces, in particular Gauss's *theorema egregium* (Latin: "Remarkable Theorem"). The theorem says that (what we call today) the *Gaussian curvature* of a surface can be determined entirely by measuring angle and distances on the surface itself, without further reference to the particular way in which the surface is situated in the ambient 3-dimensional Euclidean space. Thus the Gaussian curvature is an **intrinsic invariant** of a surface. Gauss presented the theorem in this way (translated from Latin): Thus the formula of the preceding article leads itself to the remarkable Theorem.

Theorem 1. *If a curved surface is developed upon any other surface whatever, the measure of curvature in each point remains unchanged.*

In more recent times the other forces of nature are also conceived as being due to a certain kind of curvature (the theory of Yang-Mills fields).

I hope to explain the (increasingly abstract) notions of curvature in this book, starting with a problem which (I hope) is very intuitive. This is a book about mathematics, not physics. But I hope to be able to explain some of the connections to physics.

In this chapter I assume two "big ticket" theorems: the implicit function theorem and the change of variables formula for a multiple integral. I will state these theorems as I need them.

1.1 Volume of a thickened hypersurface

Consider the following problem: Let $Y \subset \mathbb{R}^n$ be an oriented hypersurface, so there is a well defined unit normal vector, $\nu(y)$, at each point of Y. Let Y_h denote the set of all points of the form

$$y + t\nu(y), \quad 0 \leq t \leq h.$$

We wish to compute $V_n(Y_h)$ where V_n denotes the n-dimensional volume. We will do this computation for small h, see the discussion after the examples. atis

Examples in three dimensional space.

1. Suppose that Y is a bounded region in a plane, of area A. Clearly

$$V_3(Y_h) = hA$$

in this case.

2. Suppose that Y is a right circular cylinder of radius r and height ℓ with outwardly pointing normal. Then Y_h is the region between the right circular cylinders of height ℓ and radii r and $r + h$ so

$$\begin{aligned} V_3(Y_h) &= \pi\ell[(r+h)^2 - r^2] \\ &= 2\pi\ell rh + \pi\ell h^2 \\ &= hA + h^2 \cdot \frac{1}{2r} \cdot A \\ &= A\left(h + \frac{1}{2} \cdot kh^2\right), \end{aligned}$$

where $A = 2\pi r\ell$ is the area of the cylinder and where $k = 1/r$ is the curvature of the generating circle of the cylinder. For small h, this formula is correct, in fact, whether we choose the normal vector to point out of the cylinder or into the cylinder. Of course, in the inward pointing case, the curvature has the opposite sign, $k = -1/r$.

For inward pointing normals, the formula breaks down when $h > r$, since we get multiple coverage of points in space by points of the form $y + t\nu(y)$.

3. Y is a sphere of radius R with outward normal, so Y_h is a spherical shell, and

$$\begin{aligned} V_3(Y_h) &= \frac{4}{3}\pi[(R+h)^3 - R^3] \\ &= h4\pi R^2 + h^2 4\pi R + h^3 \frac{4}{3}\pi \\ &= hA + h^2 \frac{1}{R} A + h^3 \frac{1}{3R^2} A \\ &= \frac{1}{3} \cdot A \cdot \left[3h + 3\frac{1}{R} \cdot h^2 + \frac{1}{R^2} h^3\right], \end{aligned}$$

where $A = 4\pi R^2$ is the area of the sphere.

Once again, for inward pointing normals we must change the sign of the coefficient of h^2 and the formula thus obtained is only correct for $h \leq \frac{1}{R}$.

So in general, we wish to make the assumption that h is such that the map

$$Y \times [0, h] \to \mathbb{R}^n, \quad (y, t) \mapsto y + t\nu(y)$$

is injective. For Y compact, there always exists an $h_0 > 0$ such that this condition holds for all $h < h_0$. This can be seen to be a consequence of the implicit function theorem. But as so not to interrupt the discussion, we will take the injectivity of the map as an hypothesis, for the moment.

In a moment we will define the notion of the various averaged curvatures, $H_1, \ldots, H_{n-1}$, of a hypersurface, and find for the case of the sphere with outward pointing normal, that

$$H_1 = \frac{1}{R}, \quad H_2 = \frac{1}{R^2},$$

while for the case of the cylinder with outward pointing normal that

$$H_1 = \frac{1}{2r}, \quad H_2 = 0,$$

and for the case of the planar region that

$$H_1 = H_2 = 0.$$

We can thus write all three of the above the above formulas as

$$V_3(Y_h) = \frac{1}{3} A \left[3h + 3H_1 h^2 + H_2 h^3\right].$$

1.2 Defining some of our terms.

Before going on, we should make some attempt at defining our terms. For example:

There are two ways of defining a surface embedded in $\mathbb{R}^3$:

We can define it implicitly, or via a parametrization:

1. The implicit definition of a surface.

If we look our example of the sphere of radius R, we see that it can be described as being the set of all points satisfying

$$F(x_1, x_2, x_3) = 0 \qquad (1.1)$$

where

$$F(x_1, x_2, x_3) := x_1^2 + x_2^2 + x_3^2 - R.$$

So we might want to define a surface to be the set of points satisfying (1.1) where now F is an arbitrary smooth function. But we must avoid the types of pitfalls illustrated by the following examples:

- $F \equiv 0$. Here the set of points satisiying (1.1) is the entire three dimensional space.

- $F(x_1, x_2, x_3) := x_1^2 + x_2^2 + x_3^2 + 1$. Here there are no points satisfying (1.1).

- $F(x_1, x_2, x_3) := x_1^2 + x_2^2 + x_3^2$. Here there is exactly one point satisfying (1.1), namely the origin. So the equation (1.1) defines a single point, not a "surface".

Thus we need to make some assumptions about F.

We assume that:

1. there is a point $\mathbf{p}$ such that $F(\mathbf{p}) = 0$ and

2. at least one of the three partial derivatives

$$\frac{\partial F}{\partial x_1}, \frac{\partial F}{\partial x_2}, \frac{\partial F}{\partial x_3}$$

   does **not** vanish at $\mathbf{p}$.

The implicit function theorem then guarantees that in a neighborhood of $\mathbf{p}$ the only solution of

$$F(x_1, x_2, x_3) = 0 \qquad (1.1)$$

looks like a parametrized surface. More precisely:

To fix the ideas, suppose that $\frac{\partial F}{\partial x_3}(\mathbf{p}) \neq 0$. The implicit function theorem then says that there is a smooth function $f = f(x_1, x_2)$ defined in some neighborhood of (p_1, p_2) where $\mathbf{p} = \begin{pmatrix} p_1 \\ p_2 \\ p_3 \end{pmatrix}$ such that

- $f(p_1, p_2) = p_3$

1.2. DEFINING SOME OF OUR TERMS.

- the set of points $(x_1, x_2, f(x_1, x_2))$ all satisfy

$$F(x_1, x_2, x_3) = 0 \qquad (1.1)$$

and

- in some neighborhood of **p** in $\mathbb{R}^3$ these are the only points satisfying (1.1).

Remarks.

1. If $\frac{\partial F}{\partial x_2}(\mathbf{p}) \neq 0$ then we can solve for x_2 as a function of (x_1, x_3) and if $\frac{\partial F}{\partial x_1}(\mathbf{p}) \neq 0$ we can solve for x_1 as a function of (x_2, x_3).

2. Conversely, suppose that we are told that the surface is given parametrically as $x_3 = f(x_1, x_2)$. Then it is also given implicitly as

$$F(x_1, x_2, x_3) = 0 \qquad (1.1)$$

where $F(x_1, x_2, x_3) := x_3 - f(x_1, x_2)$.

3. There is nothing special about three dimensions in all of this: If we are in n-dimensional space with coordinates $x_1, \ldots, x_n$ and we are given a point

$$\mathbf{p} = \begin{pmatrix} p_1 \\ \vdots \\ p_n \end{pmatrix} \text{ with } F(\mathbf{p}) = 0 \text{ and}$$

$$\frac{\partial F}{\partial x_n}(\mathbf{p}) \neq 0 \qquad \text{then there is a}$$

smooth function $f = f(x_1, x_2, \ldots, x_{n-1})$ defined in some neighborhood of $(p_1, \ldots, p_{n-1})$ such that

- $f(p_1, \ldots, p_{n-1}) = p_n$
- the set of points $(x_1, \ldots, x_{n-1}, f(x_1, \ldots, x_{n-1}))$ all satisfy

$$F(x_1, \ldots, x_n) = 0$$

and

- in some neighborhood of **p** these are the only points satisfying the above equation.

In other words, we have defined a curve in the plane if $n = 2$, a surface in three space if $n = 3$. In general, we talk of a **hypersurface** in n-dimensions. In fact, we will compute the volume of a thickened hypersurface.

A surface $x_3 = f(x_1, x_2)$ drawn with MATLAB.

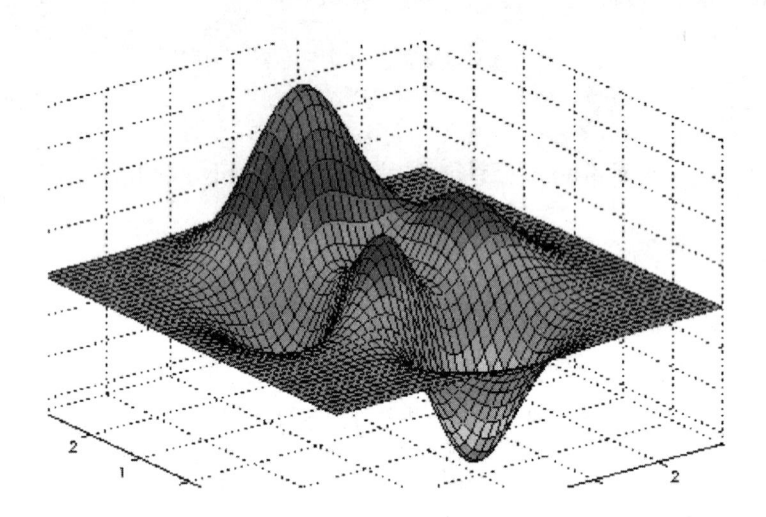

2. Parametrized surfaces.

We saw that from a (piece of) surface defined implicitly we could get a (piece of) surface parametrized by two of the three coordinates in $\mathbb{R}^3$ and also checked the converse of this fact. It is frequently convenient to use more general parametrizations.

For example, for the sphere, polar coordinates are useful.

So we assume that we are given some region M in the plane $\mathbb{R}^2$ and a map $X : M \to \mathbb{R}^3$. Let u_1, u_2 be the standard coordinates in the plane, so X is given as

$$X(u_1, u_2) = \begin{pmatrix} x_1(u_1, u_2) \\ x_2(u_1, u_2) \\ x_3(u_1, u_2) \end{pmatrix}.$$

So we are given three functions of two variables. We again have to make some assumptions:

We assume that the Jacobian matrix

$$\begin{pmatrix} \frac{\partial X}{\partial u_1} & \frac{\partial X}{\partial u_2} \end{pmatrix} = \begin{pmatrix} \frac{\partial x_1}{\partial u_1} & \frac{\partial x_1}{\partial u_2} \\ \frac{\partial x_2}{\partial u_1} & \frac{\partial x_2}{\partial u_2} \\ \frac{\partial x_3}{\partial u_1} & \frac{\partial x_3}{\partial u_2} \end{pmatrix}$$

has rank two at all points of M. This means that at any given point of M, one of the three square two by two submatrices of the Jacobian matrix is non-singular.

1.2. DEFINING SOME OF OUR TERMS.

Suppose, for example, that $\begin{pmatrix} \frac{\partial x_1}{\partial u_1} & \frac{\partial x_1}{\partial u_2} \\ \frac{\partial x_2}{\partial u_1} & \frac{\partial x_2}{\partial u_2} \end{pmatrix}$ is non-singular at a given point. The implicit function theorem guarantees that we can locally solve for u_1, u_2 as functions of x_1, x_2:

$$u_1 = u_1(x_1, x_2), \quad u_2 = u_2(x_1, x_2),$$

and so use x_1, x_2 as local parameters in the sense of the previous discussion. So we can describe this same piece of surface as $x_3 = f(x_1, x_2)$ where

$$f(x_1, x_2) = x_3(u_1(x_1, x_2), u_2(x_1, x_2)).$$

Similarly, we talk of a parametrized hypersurface Y in n dimensions. This means that we are given a map $X : M \to \mathbb{R}^n$ where M is some open subset of $\mathbb{R}^{n-1}$, and $Y = X(M)$. Let $y_1, \ldots, y_{n-1}$ be the standard coordinates on $\mathbb{R}^{n-1}$. We require that the map X be **regular** meaning that Jacobian matrix

$$J := \left(\frac{\partial X}{\partial y_1}, \ldots, \frac{\partial X}{\partial y_{n-1}} \right)$$

whose columns are the partial derivatives of the map X have rank $n-1$ everywhere. The matrix J has n rows and $n-1$ columns. So regularity means that at each point $y \in M$ the $(n-1) \times (n-1)$ sub-matrix obtained by omitting some row and evaluating at y is invertible.

The tangent space.

We now turn to a discussion of the tangent space at a point. The regularity condition in the definition of a parametrized hypersurface amounts to the assertion that for each $y \in M$ the vectors,

$$\frac{\partial X}{\partial y_1}(y), \ldots, \frac{\partial X}{\partial y_{n-1}}(y)$$

span a subspace of dimension $n-1$. If $x = X(y)$ then the **tangent space** TY_x is defined to be precisely the space spanned by the above vectors. We want to think of this as a vector space. At the moment, we may think of this as a subspace of the full n-dimensional space $\mathbb{R}^n$. Later we will give a more abstract definition. But we want to think of this vector space as labeled by x and visualize it as having its origin placed at x.

Another way of thinking about TY_x is that it consists of all tangent vectors $\gamma'(0)$ to smooth curves $t \mapsto \gamma(t)$ which lie in Y and which satisfy $\gamma(0) = x$.

Here is still another way of thinking of TY_x. If Y is given by the implicit equation $F = 0$ (satisfying the conditions above) then TY_x consists of all vectors v such that the directional derivative of F in the direction v vanishes at x. In symbols, TY_x consists of all v such that $(D_v F)(x)) = 0$.

I will let you ponder over these definitions and convince yourselves that they all define the same vector space.

Since the tangent space has dimension $n-1$ as a subspace of $\mathbb{R}^n$, its orthogonal complement is one dimensional. This is called the **line normal** to Y at x. Thus there are exactly two unit vectors in this line. We will assume that a choice has been made of one of them. By continuity, this determines such a (unit) normal vector at all nearby points.

There is one more general definition we need to make before we can get started:

The differential of a smooth map from $Y \subset \mathbb{R}^n$ to $\mathbb{R}^m$.

Suppose that G is a differentiable map from $Y \subset \mathbb{R}^n$ to $\mathbb{R}^m$. For each $x \in Y$, we can define the linear map

$$dG_x : TY_x \to \mathbb{R}^m$$

called the **differential** of G at x as follows: In terms of the "infinitesimal curve" description of the tangent space, if $v = \gamma'(0)$ then we define

$$dG_x(v) := \frac{d(G \circ \gamma)}{dt}(0).$$

(You must check that this does not depend on the choice of representing curve, γ, but this is easy.)

Alternatively, to give a linear map, it is enough to give its value at the elements of a basis. In terms of the basis coming from a parameterization, we have

$$dG_x \left(\frac{\partial X}{\partial y_i}(y) \right) = \frac{\partial (G \circ X)}{\partial y_i}(y).$$

Here $G \circ X : M \to \mathbb{R}^m$ is the composition of the map G with the map X . You must check that the map dG_x so defined does not depend on the choice of parameterization. Both of these verifications proceed by the chain rule.

One immediate consequence of either characterization is the following important property: Suppose that G takes values in a submanifold, say a hypersurface, $Z \subset \mathbb{R}^m$. Then

$$dG_x : TY_x \to TZ_{G(x)}. \tag{1.2}$$

1.3 The Gauss map and the Weingarten map.

In order to state the general formula, we make the following definitions: Let Y be an oriented hypersurface. Recall that this means that at each $x \in Y$ there is a unique (positive) unit normal vector, and hence a well defined **Gauss map**

$$\nu : Y \to S^{n-1}$$

1.3. THE GAUSS MAP AND THE WEINGARTEN MAP.

assigning to each point $x \in Y$ its unit normal vector, $\nu(x)$. Here S^{n-1} denotes the unit sphere, the set of all unit vectors in $\mathbb{R}^n$.

Examples:

1. Suppose that Y is a portion of an $(n-1)$ dimensional linear or affine subspace space sitting in $\mathbb{R}^n$. For example suppose that $Y = \mathbb{R}^{n-1}$ consisting of those points in $\mathbb{R}^n$ whose last coordinate vanishes. Then the tangent space to Y at every point is just this same subspace, and hence the normal vector is a constant. The Gauss map is thus a constant, mapping all of Y onto a single point in S^{n-1}.

2. Suppose that Y is the sphere of radius R (say centered at the origin). The Gauss map carries every point of Y into the corresponding (parallel) point of S^{n-1}. In other words, it is multiplication by $1/R$:

$$\nu(y) = \frac{1}{R}y.$$

3. Suppose that Y is a right circular cylinder in $\mathbb{R}^3$ whose base is the circle of radius r in the x^1, x^2 plane. Then the Gauss map sends Y onto the equator of the unit sphere, S^2, sending a point x into $(1/r)\pi(x)$ where $\pi : \mathbb{R}^3 \to \mathbb{R}^2$ is projection onto the x^1, x^2 plane.

Johann Carl Friedrich Gauss

Born: 30 April 1777 in Brunswick, Duchy of Brunswick (now Germany)
Died: 23 Feb. 1855 in Gottingen, Hanover (now Germany)

Let us apply (1.2) to the Gauss map, ν, which maps Y to the unit sphere, S^{n-1}. Then
$$d\nu_x : TY_x \to TS^{n-1}_{\nu(x)}.$$

But the tangent space to the unit sphere at $\nu(x)$ consists of all vectors perpendicular to $\nu(x)$ and so can be identified with TY_x. We define the **Weingarten map** to be the differential of the Gauss map, regarded as a map from TY_x to itself:
$$W_x := d\nu_x, \quad W_x : TY_x \to TY_x.$$

Julius Weingarten

Born: 2 March 1836 in Berlin, Germany
Died: 16 June 1910 in Breisgau, Germany

The **second fundamental form** is defined to be the bilinear form on TY_x given by
$$II_x(v, w) := (W_x v, w).$$
In the next section we will show, using local coordinates, that this form is symmetric, i.e. that
$$(W_x u, v) = (u, W_x v).$$

This implies, from linear algebra, that W_x is diagonalizable with real eigenvalues. These eigenvalues, $k_1 = k_1(x), \cdots, k_{n-1} = k_{n-1}(x)$, of the Weingarten map are called the **principal curvatures** of Y at the point x.

1.3. THE GAUSS MAP AND THE WEINGARTEN MAP.

Warning!

The definition that I am using here for the second fundamental form and hence for the principal curvatures (and so for the mean curvature - but not for the Gaussian curvature if $n-1$ is even) differs by a sign from that used in standard elementary texts. For example in the excellent books on curves and surfaces by do Carmo and by O'Niell, the "shape operator" is defined as the negative of what we are calling the Weingarten map, and the principal curvatures are defined as the eigenvalues of the shape operator. So, for example, the sphere of radius R has, as its principal curvatures, $-\frac{1}{R}$ according to these (and classical) authors.

I chose this convention in order to avoid a proliferation of minus signs in the computation I am about to do,

Examples:

1. For a portion of $(n-1)$ space sitting in $\mathbb{R}^n$ the Gauss map is constant so its differential is zero. Hence the Weingarten map and thus all the principal curvatures are zero.

2. For the sphere of radius R the Gauss map consists of multiplication by $1/R$ which is a linear transformation. The differential of a linear transformation is that same transformation (regarded as acting on the tangent spaces). Hence the Weingarten map is $1/R \times \mathrm{id}$ and so all the principal curvatures are equal and are equal to $1/R$.

3. For the cylinder, again the Gauss map is linear, and so the principal curvatures are 0 and $1/r$.

We let H_j denote the jth normalized elementary symmetric functions of the principal curvatures. So

$$\begin{aligned} H_0 &= 1 \\ H_1 &= \frac{1}{n-1}(k_1 + \cdots + k_{n-1}) \\ H_{n-1} &= k_1 \cdot k_2 \cdots k_{n-1} \end{aligned}$$

and, in general,

$$H_j = \binom{n-1}{j}^{-1} \sum_{1 \le i_1 < \cdots < i_j \le n-1} k_{i_1} \cdots k_{i_j}. \tag{1.3}$$

H_1 is called the **mean curvature** and H_{n-1} is called the **Gaussian curvature**. All the principal curvatures are functions of the point $x \in Y$. For notational simplicity, we will frequently suppress the dependence on x. Then the formula for the volume of the thickened hypersurface (we will call this the "volume formula" for short) is:

$$V_n(Y_h) = \frac{1}{n} \sum_{i=1}^{n} \binom{n}{i} h^i \int_Y H_{i-1} d^{n-1}A \qquad (1.4)$$

where $d^{n-1}A$ denotes the ($n-1$ dimensional) volume (area) measure on Y.

A immediate check shows that this gives the answers that we got above for the the plane, the cylinder, and the sphere.

1.4 Proof of the volume formula.

We recall that the Gauss map, ν assigns to each point $x \in Y$ its unit normal vector, and so is a map from Y to the unit sphere, S^{n-1}. The Weingarten map, W_x, is the differential of the Gauss map, $W_x = d\nu_x$, regarded as a map of the tangent space, TY_x to itself. We now describe these maps in terms of a local parameterization of Y. So let $X : M \to \mathbb{R}^n$ be a parameterization of class C^2 of a neighborhood of Y near x, where M is an open subset of $\mathbb{R}^{n-1}$. So $x = X(y)$, $y \in M$, say. Let

$$N := \nu \circ X$$

so that $N : M \to S^{n-1}$ is a map of class C^1. The map

$$dX_y : \mathbb{R}^{n-1} \to TY_x$$

gives a **frame** of TY_x. The word "frame" means an isomorphism of our "standard" $(n-1)$-dimensional space, $\mathbb{R}^{n-1}$ with our given $(n-1)$-dimensional space, TY_x. Here we have identified $T(\mathbb{R}^{n-1})_y$ with $\mathbb{R}^{n-1}$, so the frame dX_y gives us a particular isomorphism of $\mathbb{R}^{n-1}$ with TY_x.

Giving a frame of a vector space is the same as giving a basis of that vector space. We will use these two different ways of using the word "frame" interchangeably. Let $e_1, \ldots, e_{n-1}$ denote the standard basis of $\mathbb{R}^{n-1}$, and for X and N, let the subscript i denote the partial derivative with respect to the ith Cartesian coordinate. Thus

$$dX_y(e_i) = X_i(y)$$

for example, and so $X_1(y), \ldots, X_{n-1}(y)$ "is" the frame determined by dX_y (when we regard TY_x as a subspace of $\mathbb{R}^n$). For the sake of notational simplicity we will drop the argument y. Thus we have

$$\begin{aligned} dX(e_i) &= X_i, \\ dN(e_i) &= N_i, \end{aligned}$$

and so

$$W_x X_i = N_i.$$

1.4. PROOF OF THE VOLUME FORMULA.

Recall the definition, $II_x(v,w) = (W_x v, w)$, of the second fundamental form. Let (L_{ij}) denote the matrix of the second fundamental form with respect to the basis $X_1, \ldots X_{n-1}$ of TY_x. So

$$\begin{aligned} L_{ij} &= II_x(X_i, X_j) \\ &= (W_x X_i, X_j) \\ &= (N_i, X_j) \end{aligned}$$

so

$$L_{ij} = -(N, \frac{\partial^2 X}{\partial y_i \partial y_j}), \tag{1.5}$$

the last equality coming from differentiating the identity

$$(N, X_j) \equiv 0$$

in the ith direction. In particular, it follows from (1.5) and the equality of cross derivatives that

$$(W_x X_i, X_j) = (X_i, W_x X_j)$$

and hence, by linearity that

$$(W_x u, v) = (u, W_x v) \ \forall u, v \in TY_x.$$

We have proved that the second fundamental form is symmetric, and hence the Weingarten map is diagonizable with real eigenvalues.

Recall that the principal curvatures are, by definition, the eigenvalues of the Weingarten map. We will let

$$W = (W_{ij})$$

denote the matrix of the Weingarten map with respect to the basis $X_1, \ldots, X_{n-1}$. Explicitly,

$$N_i = \sum_j W_{ji} X_j.$$

If we write $N_1, \ldots, N_{n-1}, X_1, \ldots, X_{n-1}$ as column vectors of length n, we can write the preceding equation as the matrix equation

$$(N_1, \ldots, N_{n-1}) = (X_1, \ldots, X_{n-1})W. \tag{1.6}$$

The matrix multiplication on the right is that of an $n \times (n-1)$ matrix with an $(n-1) \times (n-1)$ matrix. To understand this abbreviated notation, let us write it out in the case $n=3$, so that X_1, X_2, N_1, N_2 are vectors in $\mathbb{R}^3$:

$$X_1 = \begin{pmatrix} X_{11} \\ X_{12} \\ X_{13} \end{pmatrix}, \ X_2 = \begin{pmatrix} X_{21} \\ X_{22} \\ X_{23} \end{pmatrix}, \ N_1 = \begin{pmatrix} N_{11} \\ N_{12} \\ N_{13} \end{pmatrix}, \ N_2 = \begin{pmatrix} N_{21} \\ N_{22} \\ N_{23} \end{pmatrix}.$$

Then (1.6) is the matrix equation

$$\begin{pmatrix} N_{11} & N_{21} \\ N_{12} & N_{22} \\ N_{13} & N_{23} \end{pmatrix} = \begin{pmatrix} X_{11} & X_{21} \\ X_{12} & X_{22} \\ X_{13} & X_{23} \end{pmatrix} \begin{pmatrix} W_{11} & W_{12} \\ W_{21} & W_{22} \end{pmatrix}.$$

Matrix multiplication shows that this gives

$$N_1 = W_{11}X_1 + W_{21}X_2, \quad N_2 = W_{12}X_1 + W_{22}X_2,$$

and more generally that (1.6) gives $N_i = \sum_j W_{ji} X_j$ in all dimensions.

Now consider the region Y_h, the thickened hypersurface, introduced in the preceding section except that we replace the full hypersurface Y by the portion $X(M)$. Thus the region in space that we are considering is

$$\{X(y) + \lambda N(y), y \in M, \ 0 \le \lambda \le h\}.$$

It is the image of the region $M \times [0, h] \subset \mathbb{R}^{n-1} \times \mathbb{R}$ under the map

$$(y, \lambda) \mapsto X(y) + \lambda N(y). \tag{1.7}$$

Let J denote the Jacobian of this map. By the formula for the change of variables in a multiple integral, the volume of the thickened hypersurface is given by the integral

$$V_n(h) = \int_M \int_0^h |\det J| dh dy_1 \cdots dy_{n-1}, \tag{1.8}$$

where we are assuming that the map (1.7) is injective. By (1.6), it has Jacobian matrix (differential)

$$J = (X_1 + \lambda N_1, \ldots, X_{n-1} + \lambda N_{n-1}, N) =$$

$$(X_1, \ldots, X_{n-1}, N) \begin{pmatrix} (I_{n-1} + \lambda W) & 0 \\ 0 & 1 \end{pmatrix}. \tag{1.9}$$

The right hand side of (1.9) is now the product of two n by n matrices.

Let us take the determinant of the right hand side of (1.9). The determinant of the matrix $(X_1, \ldots, X_{n-1}, N)$ is just the (oriented) n dimensional volume of the parallelepiped spanned by $X_1, \ldots, X_{n-1}, N$. Since N is of unit length and is perpendicular to the X's, this is the same as the (oriented) $n-1$ dimensional volume of the parallelepiped spanned by $X_1, \ldots, X_{n-1}$. Thus, "by definition",

$$|\det (X_1, \ldots, X_{n-1}, N)| dy_1 \cdots dy_{n-1} = d^{n-1} A. \tag{1.10}$$

(We will come back shortly to discuss why this is the right definition.) The second factor on the right hand side of (1.9) contributes

$$\det(1 + \lambda W) = (1 + \lambda k_1) \cdots (1 + \lambda k_{n-1}).$$

For sufficiently small λ, this expression is positive, so we need not worry about the absolute value sign if h small enough. Integrating with respect to λ from 0 to h gives (1.4).

We proved (1.4) *if* we define $d^{n-1}A$ to be given by (1.10). But then it follows from (1.4) that

$$\frac{d}{dh} V_n(Y_h)_{|h=0} = \int_Y d^{n-1} A. \tag{1.11}$$

A moment's thought shows that the left hand side of (1.11) is exactly what we want to mean by "area": it is the "volume of an infinitesimally thickened region". This justifies taking (1.10) as a definition. Furthermore, although the definition (1.10) is only valid in a coordinate neighborhood, and seems to depend on the choice of local coordinates, equation (1.11) shows that it is independent of the local description by coordinates, and hence is a well defined object on Y. The functions H_j have been defined independent of any choice of local coordinates. Hence (1.4) works globally: To compute the right hand side of (1.4) we may have to break Y up into patches, and do the integration in each patch, summing the pieces. But we know in advance that the final answer is independent of how we break Y up or which local coordinates we use.

1.5 Gauss's theorema egregium.

Suppose we consider the two sided region about the surface, that is

$$V_n(Y_h^+) + V_n(Y_h^-)$$

corresponding to the two different choices of normals. When we replace $\nu(x)$ by $-\nu(x)$ at each point, the Gauss map ν is replaced by $-\nu$, and hence the Weingarten maps W_x are also replaced by their negatives. The principal curvatures change sign. Hence, in the above sum the coefficients of the even powers of h cancel, since they are given in terms of products of the principal curvatures with an odd number of factors. For $n=3$ we are left with a sum of two terms, the coefficient of h which is the area, and the coefficient of h^3 which is the integral of the Gaussian curvature. It was the remarkable discovery of Gauss that this curvature depends only on the intrinsic geometry of the surface, and not on how the surface is embedded into three space. Thus, for both the cylinder and the plane the cubic terms vanish, because (locally) the cylinder is isometric to the plane. We can wrap the plane around the cylinder without stretching or tearing.

It was this fundamental observation of Gauss that led Riemann to investigate the intrinsic metric geometry of higher dimensional space, eventually leading to Einstein's general relativity which derives the gravitational force from the curvature of space time. A first objective will be to understand this major theorem of Gauss.

An important generalization of Gauss's result was proved by Hermann Weyl in 1939. He showed: if Y is any k dimensional submanifold of n dimensional space (so for $k=1$, $n=3$ Y is a curve in three space), let $Y(h)$ denote the "tube" around Y of radius h, the set of all points at distance h from Y. Then, for small h, $V_n(Y(h))$ is a polynomial in h whose coefficients are integrals over Y of intrinsic expressions, depending only on the notion of distance within Y. This theorem of Weyl had a major impact on topology and analysis over the next seventy years (Chern classes etc.) An excellent presentation of Weyl's theorem and its impact on the subsequent development of mathematics is in the book *Tubes* by Alfred Gray.

Let us multiply both sides of (1.6) on the left by the matrix $(X_1, \ldots, X_{n-1})^T$ to obtain

$$L = QW$$

where $L_{ij} = (X_i, N_j)$ as before, and

$$Q = (Q_{ij}) := (X_i, X_j)$$

is called the matrix of the **first fundamental form** relative to our choice of local coordinates. Since the X_i are linearly independent, Q is non-singular, in fact it is symmetric and positive definite. All three matrices in this equality are of size $(n-1) \times (n-1)$. If we take the determinant of the equation $L = QW$ we obtain

$$\det W = \frac{\det L}{\det Q}, \qquad (1.12)$$

an expression for the determinant of the Weingarten map (a geometrical property of the embedded surface) as the quotient of two local expressions. We thus obtain a local expression for the Gaussian curvature, $K = \det W$.

The first fundamental form encodes the intrinsic geometry of the hypersurface in terms of local coordinates: it gives the Euclidean geometry of the tangent space in terms of the basis $X_1, \ldots, X_{n-1}$. If we describe a curve $t \mapsto \gamma(t)$ on the surface in terms of the coordinates $y^1, \ldots, y^{n-1}$ by giving the functions $t \mapsto y^j(t)$, $j = 1, \ldots, n-1$ then the chain rule says that

$$\gamma'(t) = \sum_{j=1}^{n-1} X_j(y(t)) \frac{dy^j}{dt}(t)$$

where

$$y(t) = (y^1(t), \ldots, y^{n-1}(t)).$$

Therefore the (Euclidean) square length of the tangent vector $\gamma'(t)$ is

$$\|\gamma'(t)\|^2 = \sum_{i,j=1}^{n-1} Q_{ij}(y(t)) \frac{dy^i}{dt}(t) \frac{dy^j}{dt}(t).$$

Thus the length of the curve γ given by

$$\int \|\gamma'(t)\| dt$$

can be computed in terms of $y(t)$ as

$$\int \sqrt{\sum_{i,j=1}^{n-1} Q_{ij}(y(t)) \frac{dy^i}{dt}(t) \frac{dy^j}{dt}(t)} \; dt$$

(so long as the curve lies within the coordinate system).

1.5. GAUSS'S THEOREMA EGREGIUM.

So two hypersurfaces have the same local intrinsic geometry if they have the same Q in any local coordinate system.

In order to conform with a (somewhat variable) classical literature, we shall make some slight changes in our notation for the case of surfaces in three dimensional space. We will denote our local coordinates by u,v instead of y_1, y_2 and so X_u will replace X_1 and X_v will replace X_2, and we will denote the scalar product of two vectors in three dimensional space by a $\cdot$ instead of (,). We write

$$Q = \begin{pmatrix} E & F \\ F & G \end{pmatrix} \tag{1.13}$$

where

$$E := X_u \cdot X_u \tag{1.14}$$
$$F := X_u \cdot X_v \tag{1.15}$$
$$G := X_v \cdot X_v \tag{1.16}$$

so

$$\det Q = EG - F^2. \tag{1.17}$$

We can write the equations (1.14)-(1.16) as

$$Q = (X_u, X_v)^\dagger (X_u, X_v). \tag{1.18}$$

Similarly, let us set

$$e := N \cdot X_{uu} \tag{1.19}$$
$$f := N \cdot X_{uv} \tag{1.20}$$
$$g := N \cdot X_{vv} \tag{1.21}$$

so

$$L = -\begin{pmatrix} e & f \\ f & g \end{pmatrix} \tag{1.22}$$

and

$$\det L = eg - f^2.$$

Hence (1.12) specializes to

$$K = \frac{eg - f^2}{EG - F^2}, \tag{1.23}$$

an expression for the Gaussian curvature in local coordinates. We can make this expression even more explicit, using the notion of vector product. Notice that the unit normal vector, N is given by

$$N = \frac{1}{\|X_u \times X_v\|} X_u \times X_v$$

and

$$\|X_u \times X_v\| = \sqrt{\|X_u\|^2 \|X_v\|^2 - (X_u \cdot X_v)^2} = \sqrt{EG - F^2}.$$

Therefore

$$e = N \cdot X_{uu}$$
$$= \frac{1}{\sqrt{EG-F^2}} X_{uu} \cdot (X_u \times X_v)$$
$$= \frac{1}{\sqrt{EG-F^2}} \det(X_{uu}, X_u, X_v),$$

This last determinant, is the the determinant of the three by three matrix whose columns are the vectors X_{uu}, X_u and X_v. Replacing the first column by X_{uv} gives a corresponding expression for f, and replacing the first column by X_{vv} gives the expression for g. Substituting into (1.23) gives

$$K = \frac{\det(X_{uu}, X_u, X_v)\det(X_{vv}, X_u, X_v) - \det(X_{uv}, X_u, X_v)^2}{[(X_u \cdot X_u)(X_v \cdot X_v) - (X_u \cdot X_v)^2]^2}. \quad (1.24)$$

This expression is rather complicated for computation by hand, since it involves all those determinants. However a symbolic manipulation program such as maple or mathematica can handle it with ease. Here is the instruction for mathematica, taken from a recent book by Gray (1993), in terms of a function X[u,v] defined in mathematica:

gcurvature[X_][u_,v_]:=Simplify[
(Det[D[X[uu,vv],uu,uu],D[X[uu,vv],uu],D[X[uu,vv],vv]]*
Det[D[X[uu,vv],vv,vv],D[X[uu,vv],uu],D[X[uu,vv],vv]]-
Det[D[X[uu,vv],uu,vv],D[X[uu,vv],uu],D[X[uu,vv],vv]]^2)/
(D[X[uu,vv],uu].D[X[uu,vv],uu]*
D[X[uu,vv],vv].D[X[uu,vv],vv]-
D[X[uu,vv],uu].D[X[uu,vv],vv]^2)^2] /. uu->u,vv->v

We are now in a position to give two proofs, both correct but both somewhat unsatisfactory of Gauss's *theorema egregium* which asserts that the Gaussian curvature is an intrinsic property of the metrical character of the surface. However each proof does have its merits.

1.5.1 First proof, using inertial coordinates.

For the first proof, we analyze how the first fundamental form changes when we change coordinates. Suppose we pass from local coordinates u, v to local coordinates u', v' where $u = u(u', v')$, $v = v(u', v')$. Expressing X as a function

1.5. GAUSS'S THEOREMA EGREGIUM.

of u', v' and using the chain rule gives,

$$X_{u'} = \frac{\partial u}{\partial u'} X_u + \frac{\partial v}{\partial u'} X_v$$

$$X_{v'} = \frac{\partial u}{\partial v'} X_u + \frac{\partial u}{\partial v'} X_v \text{ or}$$

$$(X_{u'}, X_{v'}) = (X_u, X_v) J \text{ where}$$

$$J := \begin{pmatrix} \frac{\partial u}{\partial u'} & \frac{\partial u}{\partial v'} \\ \frac{\partial v}{\partial u'} & \frac{\partial v}{\partial v'} \end{pmatrix}$$

so

$$Q' = (X_{u'}, X_{v'})^\dagger (X_{u'}, X_{v'})$$
$$= J^\dagger Q J.$$

This gives the rule for change of variables of the first fundamental form from the unprimed to the primed coordinate system, and is valid throughout the range where the coordinates are defined. Here J is a matrix valued function of u', v'.

Let us now concentrate attention on a single point, P. The first fundamental form is a symmetric positive definite matrix. By linear algebra, we can always find a matrix R such that $R^\dagger Q(u_P, v_P) R = I$, the two dimensional identity matrix. Here (u_P, v_P) are the coordinates describing P. With no loss of generality we may assume that these coordinates are $(0,0)$. We can then make the linear change of variables whose $J(0,0)$ is R, and so find coordinates such that $Q(0,0) = I$ in this coordinate system. But we can do better. We claim that we can choose coordinates so that

$$Q(0) = I, \quad \frac{\partial Q}{\partial u}(0,0) = \frac{\partial Q}{\partial v}(0,0) = 0. \tag{1.25}$$

Indeed, suppose we start with a coordinate system with $Q(0) = I$, and look for a change of coordinates with $J(0) = I$, hoping to determine the second derivatives so that (1.25) holds. Writing $Q' = J^\dagger Q J$ and using Leibniz's formula for the derivative of a product, the equations become

$$\frac{\partial (J + J^\dagger)}{\partial u'}(0) = -\frac{\partial Q}{\partial u}(0) \quad \frac{\partial (J + J^\dagger)}{\partial v'}(0) = -\frac{\partial Q}{\partial v}(0),$$

when we make use of $J(0) = I$. Writing out these equations gives

$$\begin{pmatrix} 2\frac{\partial^2 u}{(\partial u')^2} & \frac{\partial^2 u}{\partial u' \partial v'} + \frac{\partial^2 v}{(\partial u')^2} \\ \frac{\partial^2 u}{\partial u' \partial v'} + \frac{\partial^2 v}{(\partial u')^2} & 2\frac{\partial^2 v}{\partial u' \partial v'} \end{pmatrix}(0) = -\frac{\partial Q}{\partial u}(0)$$

$$\begin{pmatrix} 2\frac{\partial^2 u}{\partial u' \partial v'} & \frac{\partial^2 u}{(\partial v')^2} + \frac{\partial^2 v}{\partial u' \partial v'} \\ \frac{\partial^2 u}{(\partial v')^2} + \frac{\partial^2 v}{\partial u' \partial v'} & 2\frac{\partial^2 v}{(\partial v')^2} \end{pmatrix}(0) = -\frac{\partial Q}{\partial v}(0).$$

The lower right hand corner of the first equation and the upper left hand corner of the second equation determine

$$\frac{\partial^2 v}{\partial u' \partial v'}(0) \quad \text{and} \quad \frac{\partial^2 u}{\partial u' \partial v'}(0).$$

All of the remaining second derivatives are then determined (consistently since Q is a symmetric matrix). We may now choose u and v as functions of u', v'. which vanish at $(0,0)$ together with all their first partial derivatives, and with the second derivatives as above. For example, we can choose the u and v as homogeneous polynomials in u' and v' with the above partial derivatives. A coordinate system in which (1.25) holds (at a point P having coordinates $(0,0)$) is called an **inertial coordinate system** based at P. Obviously the collection of all inertial coordinate systems based at P is intrinsically associated to the metric, since the definition depends only on properties of Q in the coordinate system. We now claim the following

Proposition 1. *If u, v is an inertial coordinate system of an embedded surface based at P then then the Gaussian curvature is given by*

$$K(P) = F_{uv} - \frac{1}{2}G_{uu} - \frac{1}{2}E_{vv} \qquad (1.26)$$

the expression on the right being evaluated at $(0,0)$.

As the collection of inertial systems is intrinsic, and as (1.26) expresses the curvature in terms of a local expression for the metric in an inertial coordinate system, the proposition implies the *Theorema egregium*.

To prove the proposition, let us first make a rotation and translation in three dimensional space (if necessary) so that $X(P)$ is at the origin and the tangent plane to the surface at P is the x, y plane. The fact that $Q(0) = I$ implies that the vectors $X_u(0), X_v(0)$ form an orthonormal basis of the x, y plane, so by a further rotation, if necessary, we may assume that X_u is the unit vector in the positive x direction and by replacing v by $-v$ if necessary, that X_v is the unit vector in the positive y direction. These Euclidean motions we used do not change the value of the determinant of the Weingarten map and so have no effect on the curvature. If we replace v by $-v$, E and G are unchanged and G_{uu} and E_{vv} are also unchanged. Under the change $v \mapsto -v$ F goes to $-F$, but the cross derivative F_{uv} picks up an additional minus sign. So F_{uv} is unchanged.

We have arranged that we need prove (1.26) under the assumptions that

$$X(u,v) = \begin{pmatrix} u + r(u,v) \\ v + s(u,v) \\ q(u,v) \end{pmatrix},$$

where r, s, and q are functions which vanish together with their first derivatives at the origin in u, v space. So far we have only used the property $Q(0) = I$, not the full strength of the definition of an inertial coordinate system. We claim that if the coordinate system is inertial, all the second partials of r and s also

1.5. GAUSS'S THEOREMA EGREGIUM.

vanish at the origin. To see this, observe that

$$\begin{aligned}
E &= (1+r_u)^2 + s_u^2 + q_u^2 \\
F &= r_v + r_u r_v + s_u + s_u s_v + q_u q_v \\
G &= r_v^2 + (1+s_v)^2 + q_v^2 \quad \text{so} \\
E_u(0) &= 2r_{uu}(0) \\
E_v(0) &= 2r_{uv}(0) \\
F_u(0) &= r_{uv}(0) + s_{uu}(0) \\
F_v(0) &= r_{vv}(0) + s_{uv}(0) \\
G_u(0) &= 2s_{uv}(0) \\
G_v(0) &= 2s_{vv}(0).
\end{aligned}$$

The vanishing of all the first partials of $E, F,$ and G at 0 thus implies the vanishing of second partial derivatives of r and s.

Now $N(0)$ is just the unit vector in the positive $z-$ direction and so

$$\begin{aligned}
e &= q_{uu} \\
f &= q_{uv} \\
g &= q_{vv}
\end{aligned}$$

so
$$K = q_{uu} q_{vv} - q_{uv}^2$$

(all the above meant as values at the origin) since $EG - F^2 = 1$ at the origin. On the other hand, taking the partial derivatives of the above expressions for E, F and G and evaluating at the origin (in particular discarding terms which vanish at the origin) gives

$$\begin{aligned}
F_{uv} &= r_{uvv} + s_{uuv} + q_{uu} q_{vv} + q_{uv}^2 \\
E_{vv} &= 2\left[r_{uvv} + q_{uv}^2\right] \\
G_{uu} &= 2\left[s_{uuv} + q_{uv}^2\right]
\end{aligned}$$

when evaluated at $(0,0)$. So (1.26) holds by direct computation.

1.5.2 Second proof. The Brioschi formula.

Since the Gaussian curvature depends only on the metric, we should be able to find a general formula expressing the Gaussian curvature in terms of a metric, valid in any coordinate system, not just an inertial system. This we shall do by massaging (1.24). The numerator in (1.24) is the difference of products of two determinants. Now $\det B = \det B^\dagger$ so $\det A \det B = \det AB^\dagger$ and we can write the numerator of (1.24) as

$$\det \begin{pmatrix} X_{uu} \cdot X_{vv} & X_{uu} \cdot X_u & X_{uu} \cdot X_v \\ X_u \cdot X_{vv} & X_u \cdot X_u & X_u \cdot X_v \\ X_v \cdot X_{vv} & X_v \cdot X_u & X_v \cdot X_v \end{pmatrix} - \det \begin{pmatrix} X_{uv} \cdot X_{uv} & X_{uv} \cdot X_u & X_{uv} \cdot X_v \\ X_u \cdot X_{uv} & X_u \cdot X_u & X_u \cdot X_v \\ X_v \cdot X_{uv} & X_v \cdot X_u & X_v \cdot X_v \end{pmatrix}.$$

All the terms in these matrices except for the entries in the upper left hand corner of each is either a term of the form E, F, or G or expressible as in terms of derivatives of E, F and G. For example, $X_{uu} \cdot X_u = \frac{1}{2} E_u$ and $F_u = X_{uu} \cdot X_v + X_u \cdot X_{uv}$ so $X_{uu} \cdot X_v = F_u - \frac{1}{2} E_v$ and so on. So if not for the terms in the upper left hand corners, we would already have expressed the Gaussian curvature in terms of E, F and G. So our problem is how to deal with the two terms in the upper left hand corner. Notice that the lower right hand two by two block in these two matrices are the same. So (expanding both matrices along the top row, for example) the difference of the two determinants would be unchanged if we replace the upper left hand term, $X_{uu} \cdot X_{vv}$ in the first matrix by $X_{uu} \cdot X_{vv} - X_{uv} \cdot X_{uv}$ and the upper left hand term in the second matrix by 0. We now show how to express $X_{uu} X_{vv} - X_{uv} \cdot X_{uv}$ in terms of E, F and G and this will then give a proof of the *Theorema egregium*. We have

$$\begin{aligned}
X_{uu} \cdot X_{vv} - X_{uv} \cdot X_{uv} &= (X_u \cdot X_{vv})_u - X_u \cdot X_{vvu} \\
&\quad - (X_u \cdot X_{uv})_v + X_u \cdot X_{uvv} \\
&= (X_u \cdot X_{vv})_u - (X_u \cdot X_{uv})_v \\
&= ((X_u \cdot X_v)_v - X_{uv} \cdot X_v)_u - \frac{1}{2}(X_u \cdot X_u)_{vv} \\
&= (X_u \cdot X_v)_{vu} - \frac{1}{2}(X_v \cdot X_v)_{uu} - \frac{1}{2}(X_u \cdot X_u)_{vv} \\
&= -\frac{1}{2} E_{vv} + F_{uv} - \frac{1}{2} G_{uu}.
\end{aligned}$$

We thus obtain **Brioschi's formula**

$$K = \frac{\det A - \det B}{(EG - F)^2} \quad \text{where} \tag{1.27}$$

$$A = \begin{pmatrix} \frac{1}{2} E_{vv} + F_{uv} - \frac{1}{2} G_{uu} & \frac{1}{2} E_u & F_u - \frac{1}{2} E_v \\ F_v - \frac{1}{2} G_u & E & F \\ \frac{1}{2} G_v & F & G \end{pmatrix}$$

$$B = \begin{pmatrix} 0 & \frac{1}{2} E_v & \frac{1}{2} G_u \\ \frac{1}{2} E_v & E & F \\ \frac{1}{2} G_u & F & G \end{pmatrix}.$$

Brioschi's formula is not fit for human use but can be fed to a machine if necessary. It does give a proof of Gauss' theorem. Notice that if we have coordinates which are inertial at some point, P, then Brioschi's formula reduces to (1.26) since $E = G = 1, F = 0$ and all first partials vanish at P.

In case we have **orthogonal coordinates**, a coordinate system in which $F \equiv 0$, Brioschi's formula simplifies and becomes useful: If we set $F = F_u = F_v = 0$ in Brioschi's formula and expand the determinants we get

$$\frac{1}{(EG)^2} \left[\left(-\frac{1}{2} E_{vv} - \frac{1}{2} G_{uu} \right) EG + \frac{1}{4} E_u G_u G + \frac{1}{4} E_v G_v E + \frac{1}{4} E_v^2 G + \frac{1}{4} G_u^2 E \right]$$

$$= \left[-\frac{1}{2} \frac{E_{vv}}{EG} + \frac{1}{4} \frac{E_v^2}{E^2 G} + \frac{1}{4} \frac{E_v G_v}{EG^2} \right] + \left[-\frac{1}{2} \frac{G_{uu}}{EG} + \frac{1}{4} \frac{G_u^2}{EG^2} + \frac{1}{4} \frac{E_u G_u}{E^2 G} \right].$$

1.6. BACK TO THE AREA FORMULA.

We claim that the first bracketed expression can be written as

$$-\frac{1}{\sqrt{EG}}\frac{\partial}{\partial v}\left(\frac{1}{\sqrt{G}}\frac{\partial\sqrt{E}}{\partial v}\right).$$

Indeed,

$$\begin{aligned}
\frac{1}{\sqrt{EG}}\frac{\partial}{\partial v}\left(\frac{1}{\sqrt{G}}\frac{\partial\sqrt{E}}{\partial v}\right) &= \frac{1}{\sqrt{EG}}\left(-\frac{G_v}{2G^{\frac{3}{2}}}\frac{\partial\sqrt{E}}{\partial v} + \frac{1}{\sqrt{G}}\frac{\partial^2\sqrt{E}}{(\partial v)^2}\right) \\
&= \frac{1}{\sqrt{EG}}\left(-\frac{E_v G_v}{4G^{\frac{3}{2}}\sqrt{E}} + \frac{1}{2\sqrt{G}}\frac{\partial}{\partial v}(E^{-\frac{1}{2}}E_v)\right) \\
&= \frac{1}{\sqrt{EG}}\left(-\frac{E_v G_v}{4G^{\frac{3}{2}}\sqrt{E}} + \frac{1}{2\sqrt{G}}\left[-\frac{E_v^2}{2E^{\frac{3}{2}}} + \frac{E_{vv}}{\sqrt{E}}\right]\right) \\
&= -\frac{G_v E_v}{4G^2 E} - \frac{E_v^2}{4E^2 G} + \frac{E_{vv}}{2EG}.
\end{aligned}$$

Doing a similar computation for the second bracketed term gives

$$K = \frac{-1}{\sqrt{EG}}\left[\frac{\partial}{\partial u}\left(\frac{1}{\sqrt{E}}\frac{\partial\sqrt{G}}{\partial u}\right) + \frac{\partial}{\partial v}\left(\frac{1}{\sqrt{G}}\frac{\partial\sqrt{E}}{\partial v}\right)\right] \quad (1.28)$$

as the expression for the Gaussian curvature in orthogonal coordinates. We shall give a more direct proof of this formula and of Gauss' theorema egregium once we develop the Cartan calculus.

1.6 Back to the area formula.

For the rest of this chapter, I will discuss the area formula, and the meaning of the mean curvature, and briefly describe the concept of a "minimal surface", a concept that goes back Meusnier in the 18th century.

1.6.1 An alternative expression for the surface "area".

Let us go back to our formula

$$|\det(X_1,\ldots,X_{n-1},N)|\,dy_1\cdots dy_{n-1} \quad (1.10)$$

for the integrand of the surface "area". The determinant in (1.10) is the determinant of the $n \times n$ matrix $C := (X_1,\ldots,X_{n-1},N)$. For any matrix C we have

$$|\det C| = ((\det C)^2)^{\frac{1}{2}} = ((\det C^*)(\det C))^{\frac{1}{2}} = (\det(C^*C))^{\frac{1}{2}}$$

For the case of the C in (1.10) the unit vector N is orthogonal to all the X_i so all the entries in the bottom row and right hand column of C^*C vanish except the entry in the lower right hand corner which is one. So $|\det C| = \det Q$ where Q is the $(n-1) \times (n-1)$ matrix giving the first fundamental form:

$$Q := (X_1,\ldots,X_{n-1})^* \cdot (X_1,\ldots X_{n-1})$$

1.6.2 The mean curvature and minimal surfaces.

Normal variations of a hypersurface.

Let h be a smooth function on the hypersurface, so h is given as a function of $(y_1,\ldots,y_{n-1})$. For small $|t|$ we can consider the hypersurface given by the function

$$y \mapsto X(y,t,h) := X(y) + thN(y).$$

In other words, we are displacing our original hypersurface by the amount th in the normal direction. We can then compute the area $\text{Area}(t,h)$ of this displaced surface which is a function of t and h. We can differentiate this function with respect to t and set $t = 0$. I claim that we have

$$\frac{d}{dt}\text{Area}(t,h)_{|t=0} = (n-1)\int h(y)H_1(y)dA. \tag{1.29}$$

Proof. By our alternative formula for the area, the area element for the normally displaced hypersurface is $\det(G^*G)^{\frac{1}{2}}dy_1,\ldots dy_{n-1}$ where G is the matrix

$$G := (X_1 + th_1N + thN_1, \ldots, X_{n-1} + th_{n-1}N + thN_{n-1})$$

where we are using our notation of a subscript to denote a partial derivative, so $h_i = \frac{\partial h}{\partial y_i}$ and $N_i = \frac{\partial N}{\partial y_i}$.

We want to compute the derivative of $\det(G^*G)^{\frac{1}{2}}$ with respect to t and set $t = 0$. In doing this computation, we can ignore all terms involving t^2 or higher powers of t. Since N is orthogonal to all the X_i this means that we can ignore the terms th_iN when computing the product G^*G and so consider the expresssion M^*M where M is the matrix $M := (X_1+thN_1,\ldots,X_{n-1}+thN_{n-1})$. According to

$$(N_1,\ldots,N_{n-1}) = (X_1,\ldots,X_{n-1})W, \quad (1.6)$$

$M = (X_1,\ldots,X_{n-1})(I_{n-1} + thW)$. So

$$M^*M = (I_{n-1} + thW)^*Q(I_{n-1} + thW),$$

where $Q = (X_1,\ldots,X_{n-1})^* \cdot (X_1,\ldots,X_{n-1})$

as before, and Q does not depend on t. So our computation reduces to computing the derivative of $|\det(I_{n-1} + thW)|$ with respect to t and setting $t = 0$. For small values of t the determinant will be positive, so we can ignore the absolute value sign. We need to compute this derivative for each fixed y and set $t = 0$. For each fixed y we have

$$\det(I_{n-1} + thW) = (1 + thk_1)\cdots(1 + thk_{n-1})$$

where the curvatures k_i are evaluated at y. Differentiating with respect to t and setting $t = 0$ gives

$$h(k_1 + \cdots k_{n-1}) = (n-1)hH_1$$

as desired. $\square$

1.6. BACK TO THE AREA FORMULA.

1.6.3 Minimal hypersurfaces.

It follows from the variational formula that our original surface has the property that this derivative vanishes for all possible h (of compact support, say) if and only if

$$H_1 \equiv 0,$$

a result due to Meusnier (1776) (for n = 3). With a slight abuse of language a hypersurface with identically vanishing mean curvature is called a minimal hypersurface.

We will not study minimal surfaces in this course. But here are some pictures:

The helicoid.

44   CHAPTER 1. GAUSS'S THEOREMA EGREGIUM.

The catenoid.

Enneper's surface.

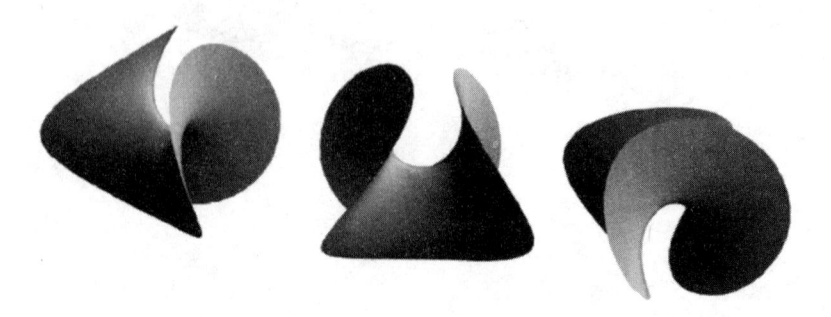

I have been unable to find a portrait of Meusnier on the web, but here is a drawing for which he is most famous, his proposed dirigible:

1.6. BACK TO THE AREA FORMULA.

Here is a biography of Meusnier (in French):

Jean Baptiste Marie Charles **Meusnier** de La Place (1754-1793)

Officier, géomètre et mathématicien français, né le 19 juin 1754 à Tours et mort à Cassel près de Mayence, le 13 juin 1793.

Élève de Monge à l'École Royale du Génie de Mézières, il présente en 1776, une étude géométrique de la courbure des surfaces qui contient un théorème auquel on donnera son nom.

Officier du Génie, il participe aux grands travaux de la rade de Cherbourg (1778-1789).

Il collabore aux expériences de Lavoisier, concernant la recherche d'un meilleur procédé d'élaboration de l'hydrogène.
En 1783, il participe à une expérience de décomposition de l'eau par le fer rougi, mettant clairement en évidence la nature et la composition de l'eau.

Il entre à l'Académie de Sciences en 1784 pour sa contribution dans l'étude locale des surfaces.

En 1784, il conçoit un ballon elliptique d'environ 80 mètres et d'une capacité de 1700 mètres cubes dont la nacelle et prévue pour flotter sur l'eau. Le ballon est mû par trois propulseurs qui exigent la force de 80 hommes. Meusnier a présenté ce projet à l'Académie, mais le manque de fonds et son décès empêchèrent sa réalisation. Il est considéré comme l'inventeur du dirigeable.

Lieutenant-colonel du génie sous la Révolution, il organise les armées de la République.

En 1791, il est chargé par l'Assemblée Constituante de participer avec Monge à la détermination des bases qui serviront de repères à la mesure du méridien terrestre alors que la triangulation et la détermination des latitudes sont confiées à Cassini, Legendre et Méchain.

Général de division en 1792, il combat face aux armées prussiennes.

Maréchal de camp auprès de l'Armée du Rhin, il est tué à Cassel près de Mayence, le 13 juin 1793, touché par un boulet de canon.

Anecdote :
On dit que le roi de Prusse aurait demandé un cessez-le-feu jusqu'à ce qu'il soit inhumé.

1.7 Problem set - Surfaces of revolution.

The simplest (non-trivial) case is when $n = 2$ - the study of a curve in the plane. For the case of a curve $X(t) = (x(t), y(t))$ in the plane, we have

$$X'(t) = (x'(t), y'(t)), \quad N(t) = \pm \frac{1}{(x'(t)^2 + y'(t)^2)^{1/2}}(-y'(t), x'(t)),$$

where the $\pm$ reflects the two possible choices of normals. Equation (1.5) says that the one by one matrix L is given by

$$L_{11} = -(N, X'') = \mp \frac{1}{\sqrt{x'^2 + y'^2}}(-y'x'' + x'y'').$$

The first fundamental form is the one by one matrix given by

$$Q_{11} = \|X'\|^2.$$

So the curvature is

$$\pm \frac{1}{(x'^2 + y'^2)^{\frac{3}{2}}}(x''y' - y''x').$$

Verify that a straight line has curvature zero and that the curvature of a circle of radius r is $\pm 1/r$ with the plus sign when the normal points outward.

1. What does this formula reduce to in the case that x is used as a parameter, i.e. $x(t) = t, y = f(x)$?

1.7. PROBLEM SET - SURFACES OF REVOLUTION.

We want to study a surface in three space obtained by rotating a curve, γ, in the x, z plane about the z-axis. Such a surface is called a **surface of revolution**. Surfaces of revolution form one of simplest yet very important classes of surfaces. The sphere, torus, paraboloid, ellipsoid with two equal axes are all surfaces of revolution. Because of modes of production going back to the potter's wheel, the surfaces of many objects of daily life are surfaces of revolution. We will find that the geometry of famous Schwartzschild black hole can be considered as a particular analogue of a surface of revolution in four dimensional space-time. In fact, as we shall see, the so called Friedmann Robertson Walker metrics in general relativity have a form analogous to that of a surface of revolution.

Let us temporarily assume that the curve γ is given by a function $x = f(z) > 0$ so that we can use z, θ as coordinates, where the surface is given by

$$X(z, \theta) = \begin{pmatrix} f(z) \cos \theta \\ f(z) \sin \theta \\ z \end{pmatrix},$$

and we choose the normal to point away from the z-axis.

2. Find $\nu(z, \theta)$ and show that the Weingarten map is diagonal in the X_z, X_θ basis, in fact

$$N_z = \kappa X_z, \quad N_\theta = \frac{d}{f} X_\theta$$

where κ is the curvature of the curve γ and where d is the distance of the normal vector, ν, from the z-axis. Therefore the Gaussian curvature is given by

$$K = \frac{d\kappa}{f}. \tag{1.30}$$

Check that the Gaussian curvature of a cylinder vanishes and that of a sphere of radius R is $1/R^2$.

Notice that (1.30) makes sense even if we can't use z as a parameter everywhere on γ. Indeed, suppose that γ is a curve in the x, z plane that does not intersect the z-axis, and we construct the corresponding surface of revolution. At points where the tangent to γ is horizontal (parallel to the x-axis) the normal vector to the surface of revolution is vertical, so $d = 0$. Also the Gaussian curvature vanishes, since the Gauss map takes the entire circle of revolution into the north or south pole. So (1.30) is correct at these points. At all other points we can use z as a parameter. But we must watch the sign of κ. Remember that the Gaussian curvature of a surface does not depend on the choice of normal vector, but the curvature of a curve in the plane does. In using (1.30) we must be sure that the sign of κ is the one determined by the normal pointing away from the z-axis.

3. For example, take γ to be a circle of radius r centered at a point at distance $D > r$ from the z-axis, say
$$x = D + r\cos\phi, \quad z = r\sin\phi$$
in terms of an angular parameter, ϕ. The corresponding surface of revolution is a torus. Notice that in using (1.30) we have to take κ as negative on the semicircle closer to the z-axis. So the Gaussian curvature is negative on the "inner" half of the torus and positive on the outer half. Using (1.30) and ϕ, θ as coordinates on the torus, express K as a function on ϕ, θ. Also, express the area element dA in terms of $d\phi d\theta$. Without any computation, show that the total integral of the curvature vanishes, i.e. $\int_T K dA = 0$.

Recall our definitions of E, F, and G given in equations (1.14)-(1.16). In the classical literature, one write the first fundamental form as
$$ds^2 = E du^2 + 2F du dv + G dv^2.$$
the meaning of this expression is as follows: let $t \mapsto (u(t), v(t))$ describe the curve
$$C : t \mapsto X(u(t), v(t))$$
on the surface. Then ds gives the element of arc length of this curve if we substitute $u = u(t), v = v(t)$ into the expression for the first fundamental form. So the first fundamental form describes the intrinsic metrical properties of the surface in terms of the local coordinates. Recall equation (1.28) which says that if u, v is an orthogonal coordinate system then the expression for the Gaussian curvature is
$$K = \frac{-1}{\sqrt{EG}} \left[\frac{\partial}{\partial u}\left(\frac{1}{\sqrt{E}} \frac{\partial \sqrt{G}}{\partial u} \right) + \frac{\partial}{\partial v}\left(\frac{1}{\sqrt{G}} \frac{\partial \sqrt{E}}{\partial v} \right) \right].$$

4. Show that the z, θ coordinates introduced in problem 2 for a surface of revolution is an orthogonal coordinate system, find E and G and verify (1.28) for this case.

A curve $s \mapsto C(s)$ on a surface is called a **geodesic** if its acceleration, C'', is everywhere orthogonal to the surface. Notice that
$$\frac{d}{ds}(C'(s), C'(s)) = 2(C''(s), C'(s))$$
and this $= 0$ if C is a geodesic. The term geodesic refers to a parametrized curve and the above equation shows that the condition to be a geodesic implies that $\|C'(s)\|$ is a constant; i.e that the curve is parametrized by a constant multiple of arc length. If we use a different parameterization, say $s = s(t)$ with dot denoting derivative with respect to t, then the chain rule implies that
$$\dot{C} = C'\dot{s}, \quad \ddot{C} = C''\dot{s}^2 + C'\ddot{s}.$$

1.7. PROBLEM SET - SURFACES OF REVOLUTION.

So if use a parameter other than arc length, the projection of the acceleration onto the surface is proportional to the tangent vector if C is a geodesic. In other words, the acceleration is in the plane spanned by the tangent vector to the curve and the normal vector to the surface. Conversely, suppose we start with a curve C which has the property that its acceleration lies in the plane spanned by the tangent vector to the curve and the normal vector to the surface at all points. Let us reparametrize this curve by arc length. Then $(C'(s), C'(s)) \equiv 1$ and hence $(C'', C') \equiv 0$. As we are assuming that $\ddot{C}$ lies in the plane spanned by $\dot{C}$ and the normal vector to the surface at each point of the curve, and that $\dot{s}$ is nowhere 0 we conclude that C, in its arc length parametrization is a geodesic. Standard usage calls a curve which is a geodesic "up to reparametrization" a *pregeodesic*. I don't like this terminology but will live with it.

5. Show that the curves $\theta = $ const in the terminology of problem 2 are all pregeodsics. Show that the curves $z = $ const. are pregeodesics if and only if z is a critical point of f, (i.e. $f'(s) = 0$).

The general setting for the concept of surfaces of revolution is that of a Riemannian submersion, which will be the subject of Chapter 19.

Chapter 2

Rules of calculus.

It would be useful for the reader to to review, at this stage, the concept of a differentiable manifold and, in particular, the idea of the tangent space. My favorite reference is to my "Advanced Calculus" with Loomis, available on my Harvard web site.

2.1 Superalgebras.

A (commutative associative) *superalgebra* is a vector space

$$A = A_{even} \oplus A_{odd}$$

with a given direct sum decomposition into even and odd pieces, and a map

$$A \times A \to A$$

which is bilinear, satisfies the associative law for multiplication, and

$$\begin{aligned}
A_{even} \times A_{even} &\to A_{even} \\
A_{even} \times A_{odd} &\to A_{odd} \\
A_{odd} \times A_{even} &\to A_{odd} \\
A_{odd} \times A_{odd} &\to A_{even} \\
\omega \cdot \sigma &= \sigma \cdot \omega \text{ if either } \omega \text{ or } \sigma \text{ are even,} \\
\omega \cdot \sigma &= -\sigma \cdot \omega \text{ if both } \omega \text{ and } \sigma \text{ are odd.}
\end{aligned}$$

We write these last two conditions as

$$\omega \cdot \sigma = (-1)^{\deg \sigma \deg \omega} \sigma \cdot \omega.$$

Here $\deg \tau = 0$ if τ is even, and $\deg \tau = 1 \pmod 2$ if τ is odd.

2.2 Differential forms.

2.2.1 Linear differential forms.

A **linear differential form** on a manifold, M, is a rule which assigns to each $p \in M$ a linear function on TM_p. So a linear differential form, ω, assigns to each p an element of TM_p^*. We will, as usual, only consider linear differential forms which are *smooth* meaning the following:

In a local coordinate chart with coordinates $x_1, \ldots, x_n$ the vector fields $\frac{\partial}{\partial x_1}, \ldots, \frac{\partial}{\partial x_n}$ span the tangent space at each point, so the dual basis $dx_1, \ldots, dx_n$ span the dual space at each point. Hence, in this chart, the most general linear differential form can be written as

$$a_1 dx_1 + \cdots + a_n dx_n$$

where the a_i are **functions**. We demand that these functions be smooth. It is easy to check that this condition is consistent relative to changes of charts. See my book with Loomis for more details.

2.2.2 $\Omega(M)$, the algebra of exterior differential forms.

The superalgebra, $\Omega(M)$ is the superalgebra generated by smooth functions on M (taken as even) and by the linear differential forms, taken as odd.

Multiplication of differentials is usually denoted by $\wedge$. The number of differential factors is called the **degree** of the form. So functions have degree zero, linear differential forms have degree one.

As we have seen, in terms of local coordinates, the most general *linear* differential form has an expression as $a_1 dx_1 + \cdots + a_n dx_n$ (where the a_i are functions). Expressions of the form

$$a_{12} dx_1 \wedge dx_2 + a_{13} dx_1 \wedge dx_3 + \cdots + a_{n-1,n} dx_{n-1} \wedge dx_n$$

have degree two (and are even). Notice that the multiplication rules require

$$dx_i \wedge dx_j = -dx_j \wedge dx_i$$

and, in particular, $dx_i \wedge dx_i = 0$. So the most general sum of products of two linear differential forms is a differential form of degree two, and can be brought to the above form, locally, after collections of coefficients. Similarly, the most general differential form of degree $k \leq n$ in n dimensional manifold is a sum, locally, with function coefficients, of expressions of the form

$$dx_{i_1} \wedge \cdots \wedge dx_{i_k}, \quad i_1 < \cdots < i_k.$$

There are $\binom{n}{k}$ such expressions, and they are all even, if k is even, and odd if k is odd.

2.3 The d operator.

There is a linear operator d acting on differential forms called **exterior differentiation**, which is completely determined by the following rules: It satisfies Leibnitz' rule in the "super" form

$$d(\omega \wedge \sigma) = (d\omega) \wedge \sigma + (-1)^{\deg \omega} \omega \wedge (d\sigma).$$

On functions it is given by

$$df = \frac{\partial f}{\partial x_1} dx_1 + \cdots + \frac{\partial f}{\partial x_n} dx_n$$

and, finally,

$$d(dx_i) = 0.$$

Since functions and the dx_i generate, this determines d completely. For example, on linear differential forms

$$\omega = a_1 dx_1 + \cdots a_n dx_n$$

we have

$$\begin{aligned} d\omega &= da_1 \wedge dx_1 + \cdots + da_n \wedge dx_n \\ &= \left(\frac{\partial a_1}{\partial x_1} dx_1 + \cdots + \frac{\partial a_1}{\partial x_n} dx_n\right) \wedge dx_1 + \cdots \\ &\quad \left(\frac{\partial a_n}{\partial x_1} dx_1 + \cdots + \frac{\partial a_n}{x_n} dx_n\right) \wedge dx_n \\ &= \left(\frac{\partial a_2}{\partial x_1} - \frac{\partial a_1}{\partial x_2}\right) dx_1 \wedge dx_2 + \cdots + \left(\frac{\partial a_n}{\partial x_{n-1}} - \frac{\partial a_{n-1}}{\partial x_n}\right) dx_{n-1} \wedge dx_n. \end{aligned}$$

In particular, equality of mixed derivatives shows that $d^2 f = 0$, and hence that $d^2 \omega = 0$ for any differential form.

Hence the rules to remember about d are:

$$\begin{aligned} d(\omega \wedge \sigma) &= (d\omega) \wedge \sigma + (-1)^{\deg \omega} \omega \wedge (d\sigma) \\ d^2 &= 0 \\ df &= \frac{\partial f}{\partial x_1} dx_1 + \cdots + \frac{\partial f}{\partial x_n} dx_n. \end{aligned}$$

2.4 Even and odd derivations of a superalgebra.

A linear operator $\ell : A \to A$ on a superalgebra A is called an **odd derivation** if, like d, it satisfies

$$\ell : A_{even} \to A_{odd}, \quad \ell : A_{odd} \to A_{even}$$

and
$$\ell(\omega \cdot \sigma) = (\ell\omega) \cdot \sigma + (-1)^{\deg \omega} \omega \cdot \ell\sigma.$$

A linear map $\ell : A \to A$,
$$\ell : A_{even} \to A_{even}, \quad \ell : A_{odd} \to A_{odd}$$

satisfying
$$\ell(\omega \cdot \sigma) = (\ell\omega) \cdot \sigma + \omega \cdot (\ell\sigma)$$

is called an **even derivation**. So the Leibnitz rule for derivations, even or odd is
$$\ell(\omega \cdot \sigma) = (\ell\omega) \cdot \sigma + (-1)^{\deg \ell \deg \omega} \omega \cdot \ell\sigma,$$

where $\deg \ell = 0 \pmod 2$ if ℓ is even and $\deg \ell = 1 \pmod 2$ if ℓ is odd.

Knowing the action of a derivation on a set of generators of a superalgebra determines it completely. For example, the equations
$$d(x_i) = dx_i, \quad d(dx_i) = 0 \quad \forall i$$

implies that
$$dp = \frac{\partial p}{\partial x_1} dx_1 + \cdots + \frac{\partial p}{\partial x_n} dx_n$$

for any polynomial, and hence determines the value of d on any differential form with polynomial coefficients. The local formula we gave for df where f is any differentiable function, was just the natural extension (by continuity, if you like) of the above formula for polynomials.

The sum of two even derivations is an even derivation, and the sum of two odd derivations is an odd derivation.

The composition of two derivations will not, in general, be a derivation, but an instructive computation from the definitions shows that the **commutator**
$$[\ell_1, \ell_2] := \ell_1 \circ \ell_2 - (-1)^{\deg \ell_1 \deg \ell_2} \ell_2 \circ \ell_1$$

is again a derivation which is even if both are even or both are odd, and odd if one is even and the other odd.

A derivation followed by a multiplication is again a derivation: specifically, let ℓ be a derivation (even or odd) and let τ be an even or odd element of A. Consider the map
$$\omega \mapsto \tau \ell \omega.$$

We have
$$\begin{aligned}\tau\ell(\omega\sigma) &= (\tau\ell\omega) \cdot \sigma + (-1)^{\deg \ell \deg \omega} \tau\omega \cdot \ell\sigma \\ &= (\tau\ell\omega) \cdot \sigma + (-1)^{(\deg \ell + \deg \tau)\deg \omega} \omega \cdot (\tau\ell\sigma)\end{aligned}$$

so $\omega \mapsto \tau\ell\omega$ is a derivation whose degree is
$$\deg \tau + \deg \ell.$$

2.5 Pullback.

Let $\phi : M \to N$ be a smooth map. Then the pullback map ϕ^* is a linear map that sends differential forms on N to differential forms on M and satisfies

$$\phi^*(\omega \wedge \sigma) = \phi^*\omega \wedge \phi^*\sigma$$
$$\phi^* d\omega = d\phi^*\omega$$
$$(\phi^* f) = f \circ \phi.$$

The first two equations imply that ϕ^* is completely determined by what it does on functions. The last equation says that on functions, ϕ^* is given by "substitution": In terms of local coordinates on M and on N ϕ is given by

$$\phi(x^1, \ldots, x^m) = (y^1, \ldots, y^n)$$
$$y^i = \phi^i(x^1, \ldots, x^m) \quad i = 1, \ldots, n$$

where the ϕ_i are smooth functions. The local expression for the pullback of a function $f(y^1, \ldots, y^n)$ is to substitute ϕ^i for the y^i's as into the expression for f so as to obtain a function of the x's.

It is important to observe that the pull back on differential forms is defined for any smooth map, not merely for diffeomorphisms.

2.6 Chain rule.

Suppose that $\psi : N \to P$ is a smooth map so that the composition

$$\phi \circ \psi : M \to P$$

is again smooth. Then the **chain rule** says

$$(\phi \circ \psi)^* = \psi^* \circ \phi^*.$$

On functions this is essentially a tautology - it is the associativity of composition: $f \circ (\psi \circ \phi) = (f \circ \psi) \circ \phi$. But since pull-back is completely determined by what it does on functions, the chain rule applies to differential forms of any degree.

2.7 Lie derivative.

Let ϕ_t be a one parameter group of transformations of M. If ω is a differential form, we get a family of differential forms, $\phi_t^* \omega$ depending differentiably on t, and so we can take the derivative at $t = 0$:

$$\frac{d}{dt}(\phi_t^*\omega)_{|t=0} = \lim_{t \to 0} \frac{1}{t}[\phi_t^*\omega - \omega].$$

Since $\phi_t^*(\omega \wedge \sigma) = \phi_t^*\omega \wedge \phi_t^*\sigma$ it follows from the Leibnitz argument that

$$\ell_\phi : \omega \mapsto \frac{d}{dt}(\phi_t^*\omega)_{|t=0}$$

is an even derivation. We want a formula for this derivation.

Notice that since $\phi_t^* d = d\phi_t^*$ for all t, it follows by differentiation that

$$\ell_\phi d = d\ell_\phi$$

and hence the formula for ℓ_ϕ is completely determined by how it acts on functions.

Let X be the vector field generating ϕ_t. Recall that the geometrical significance of this vector field is as follows: If we fix a point x, then

$$t \mapsto \phi_t(x)$$

is a curve which passes through the point x at $t = 0$. The tangent to this curve at $t = 0$ is the vector $X(x)$. In terms of local coordinates, X has coordinates $X = (X^1, \ldots, X^n)$ where $X^i(x)$ is the derivative of $\phi^i(t, x^1, \ldots, x^n)$ with respect to t at $t = 0$. The chain rule then gives, for any function f,

$$\begin{aligned}\ell_\phi f &= \frac{d}{dt} f(\phi^1(t, x^1, \ldots, x^n), \ldots, \phi_n(t, x^1, \ldots, x^n))_{|t=0} \\ &= X^1 \frac{\partial f}{\partial x_1} + \cdots + X^n \frac{\partial f}{\partial x_n}.\end{aligned}$$

For this reason we use the notation

$$X = X^1 \frac{\partial}{\partial x_1} + \cdots + X^n \frac{\partial}{\partial x_n}$$

so that the differential operator

$$f \mapsto Xf$$

gives the action of ℓ_ϕ on functions.

As we mentioned, this action of ℓ_ϕ on functions determines it completely. In particular, ℓ_ϕ depends only on the vector field X, so we may write

$$\ell_\phi = L_X$$

where L_X is the even derivation determined by

$$L_X f = Xf, \quad L_X d = dL_X.$$

2.8 Weil's formula.

But we want a more explicit formula for L_X. For this it is useful to introduce an odd derivation associated to X called the *interior product* and denoted by $i(X)$. It is defined as follows: First consider the case where

$$X = \frac{\partial}{\partial x_j}$$

2.8. WEIL'S FORMULA.

and define its interior product by

$$i\left(\frac{\partial}{\partial x_j}\right) f = 0$$

for all functions while

$$i\left(\frac{\partial}{\partial x_j}\right) dx_k = 0, \quad k \neq j$$

and

$$i\left(\frac{\partial}{\partial x_j}\right) dx_j = 1.$$

The fact that it is a derivation then gives an easy rule for calculating $i(\partial/\partial x_j)$ when applied to any differential form: Write the differential form as

$$\omega + dx_j \wedge \sigma$$

where the expressions for ω and σ do not involve dx_j. Then

$$i\left(\frac{\partial}{\partial x_j}\right)[\omega + dx_j \wedge \sigma] = \sigma.$$

The operator

$$X^j i\left(\frac{\partial}{\partial x_j}\right)$$

which means first apply $i(\partial/\partial x_j)$ and then multiply by the function X^j is again an odd derivation, and so we can make the definition

$$i(X) := X^1 i\left(\frac{\partial}{\partial x_1}\right) + \cdots + X^n i\left(\frac{\partial}{\partial x_n}\right). \tag{2.1}$$

It is easy to check that this does not depend on the local coordinate system used.

Notice that we can write

$$Xf = i(X)df.$$

In particular we have

$$\begin{aligned} L_X dx_j &= dL_X x_j \\ &= dX_j \\ &= di(X)dx_j. \end{aligned}$$

We can combine these two formulas as follows: Since $i(X)f = 0$ for any function f we have

$$L_X f = di(X)f + i(X)df.$$

Since $ddx_j = 0$ we have

$$L_X dx_j = di(X)dx_j + i(X)ddx_j.$$

Hence
$$L_X = di(X) + i(X)d = [d, i(X)] \tag{2.2}$$
when applied to functions or to the forms dx_j. But the right hand side of the preceding equation is an even derivation, being the commutator of two odd derivations. So if the left and right hand side agree on functions and on the differential forms dx_j they agree everywhere. This equation, (2.2), known as **Weil's formula** or *Cartan's formula*, is a basic formula in differential calculus.

2.9 Integration.

Let
$$\omega = f dx_1 \wedge \cdots \wedge dx_n$$
be a form of degree n on $\mathbb{R}^n$. (Recall that the most general differential form of degree n is an expression of this type.) Then its integral is defined by
$$\int_M \omega := \int_M f dx_1 \cdots dx_n$$
where M is any (measurable) subset. This of course, is subject to the condition that the right hand side converges if M is unbounded. There is a lot of hidden subtlety built into this definition having to do with the notion of orientation. But for the moment this is a good working definition.

The *change of variables formula* says that if $\phi : M \to \mathbb{R}^n$ is a smooth differentiable map which is one to one with everywhere positive Jacobian determinant then
$$\int_M \phi^* \omega = \int_{\phi(M)} \omega.$$

The object that can be integrated on a general manifold is a *density* of compact support (or which vanishes "sufficiently rapidly at infinity"). See the beginning of Chapter 13 for a review of this concept.

2.10 Stokes theorem.

Let U be a region in $\mathbb{R}^n$ with a chosen orientation and smooth boundary. We then orient the boundary according to the rule that an outward pointing normal vector, together with the a positive frame on the boundary give a positive frame in $\mathbb{R}^n$. If σ is an $(n-1)$-form, then
$$\int_{\partial U} \sigma = \int_U d\sigma.$$

A manifold is called **orientable** if we can choose an atlas consisting of charts such that the Jacobian of the transition maps $\phi_\alpha \circ \phi_\beta^{-1}$ is always positive. Such a choice of an atlas is called an orientation. (Not all manifolds are orientable.) If we have chosen an orientation, then relative to the charts of our orientation, the

2.11 Lie derivatives of vector fields.

Let Y be a vector field and ϕ_t a one parameter group of transformations whose "infinitesimal generator" is some other vector field X. We can consider the "pulled back" vector field $\phi_t^* Y$ defined by

$$\phi_t^* Y(x) = d\phi_{-t}\{Y(\phi_t x)\}.$$

In words, we evaluate the vector field Y at the point $\phi_t(x)$, obtaining a tangent vector at $\phi_t(x)$, and then apply the differential of the (inverse) map ϕ_{-t} to obtain a tangent vector at x.

If ω is a linear differential form, then we may compute $i(Y)\omega$ which is a function whose value at any point is obtained by evaluating the linear function $\omega(x)$ on the tangent vector $Y(x)$. Thus

$$i(\phi_t^* Y)\phi_t^* \omega(x) = \langle \phi_t^* \omega(\phi_t x), d\phi_{-t} Y(\phi_t x) \rangle = \{i(Y)\omega\}(\phi_t x).$$

In other words,

$$\phi_t^*\{i(Y)\omega\} = i(\phi_t^* Y)\phi_t^* \omega.$$

We have verified this when ω is a differential form of degree one. It is trivially true when ω is a differential form of degree zero, i.e. a function, since then both sides are zero. But then, by the derivation property, we conclude that it is true for forms of all degrees. We may rewrite the result in shorthand form as

$$\phi_t^* \circ i(Y) = i(\phi_t^* Y) \circ \phi_t^*.$$

Since $\phi_t^* d = d\phi_t^*$ we conclude from Weil's formula that

$$\phi_t^* \circ L_Y = L_{\phi_t^* Y} \circ \phi_t^*.$$

Until now the subscript t was superfluous, the formulas being true for any fixed diffeomorphism. Now we differentiate the preceding equations with respect to t and set $t = 0$. We obtain, using Leibnitz's rule,

$$L_X \circ i(Y) = i(L_X Y) + i(Y) \circ L_X$$

and

$$L_X \circ L_Y = L_{L_X Y} + L_Y \circ L_X.$$

This last equation says that Lie derivative (on forms) with respect to the vector field $L_X Y$ is just the commutator of L_x with L_Y:

$$L_{L_X Y} = [L_X, L_Y].$$

For this reason we write
$$[X,Y] := L_X Y$$
and call it the Lie bracket (or commutator) of the two vector fields X and Y. The equation for interior product can then be written as
$$i([X,Y]) = [L_X, i(Y)].$$
The Lie bracket is antisymmetric in X and Y. We may multiply Y by a function g to obtain a new vector field gY. Form the definitions we have
$$\phi_t^*(gY) = (\phi_t^* g)\phi_t^* Y.$$
Differentiating at $t = 0$ and using Leibniz's rule we get
$$[X, gY] = (Xg)Y + g[X,Y] \tag{2.3}$$
where we use the alternative notation Xg for $L_X g$. The antisymmetry then implies that for any differentiable function f we have
$$[fX, Y] = -(Yf)X + f[X,Y]. \tag{2.4}$$
From both this equation and from Weil's formula (applied to differential forms of degree greater than zero) we see that the Lie derivative with respect to X at a point x depends on more than the value of the vector field X at x.

2.12 Jacobi's identity.

From the fact that $[X,Y]$ acts as the commutator of X and Y it follows that for any three vector fields X, Y and Z we have
$$[X, [Y, Z]] + [Z, [X, Y]] + [Y, [Z, X]] = 0.$$
This is known as **Jacobi's identity**. We can also derive it from the fact that $[Y, Z]$ is a natural operation and hence for any one parameter group ϕ_t of diffeomorphisms we have
$$\phi_t^*([Y, Z]) = [\phi_t^* Y, \phi_t^* Z].$$
If X is the infinitesimal generator of ϕ_t then differentiating the preceding equation with respect to t at $t = 0$ gives
$$[X, [Y, Z]] = [[X, Y], Z] + [Y, [X, Z]].$$
In other words, X acts as a derivation of the "multiplication" given by Lie bracket. This is just Jacobi's identity when we use the antisymmetry of the bracket. In the future we we will have occasion to take cyclic sums such as those which arise on the left of Jacobi's identity. So if F is a function of three vector fields (or of three elements of any set) with values in some vector space (for example in the space of vector fields) we will define the cyclic sum $\mathcal{C}yc\ F$ by
$$\mathcal{C}yc\ F(X, Y, Z) := F(X, Y, Z) + F(Y, Z, X) + F(Z, X, Y).$$
With this definition Jacobi's identity becomes
$$\mathcal{C}yc\ [X, [Y, Z]] = 0. \tag{2.5}$$

2.13 Forms as multilinear functions on vector fields.

We can use the interior product to consider differential forms of degree k as k−multilinear functions on the tangent space at each point. To illustrate, let ν be a differential form of degree two. Then for any vector field, X, $i(X)\nu$ is a linear differential form, and hence can be evaluated on any vector field, Y to produce a function. So we define

$$\nu(X,Y) := [i(X)\nu](Y). \tag{2.6}$$

For example, if $\nu = \sigma \wedge \tau$ then

$$i(X)\nu = (i(X)\sigma)\tau - (i(X)\tau)\sigma$$

by the derivation property and $i(X)\sigma = \sigma(X)$, $i(X)\tau = \tau(X)$ (evaluation of forms on vector fields). So

$$i(X)\nu = \sigma(X)\tau - \tau(X)\sigma.$$

Taking interior product with respect to Y then gives

$$\nu(X,Y) = \sigma(X)\tau(Y) - \tau(X)\sigma(Y)$$

We can use (2.6) to express exterior derivative in terms of ordinary derivative and Lie bracket: If θ is a linear differential form, we have

$$\begin{aligned} d\theta(X,Y) &= [i(X)d\theta](Y) \\ i(X)d\theta &= L_X\theta - d(i(X)\theta) \\ d(i(X)\theta)(Y) &= Y[\theta(X)] \\ [L_X\theta](Y) &= L_X[\theta(Y)] - \theta(L_X(Y)) \\ &= X[\theta(Y)] - \theta(X,Y) \end{aligned}$$

where we have introduced the notation $L_X Y =: [X,Y]$.

The notation $L_X Y =: [X,Y]$ is legitimate since on functions we have

$$(L_X Y)f = L_X(Yf) - YL_X f = X(Yf) - Y(Xf)$$

so $L_X Y$ as an operator on functions is exactly the commutator of X and Y. Putting the previous pieces together gives

$$d\theta(X,Y) = X\theta(Y) - Y\theta(X) - \theta([X,Y]), \tag{2.7}$$

with similar expressions for differential forms of higher degree. We will make much use of this formula also. As we have seen, it is a consequence of Weil's identity.

2.14 Problems.

2.14.1 Matrix valued differential forms.

We list the rules of the calculus for matrices whose entries are differential forms. We will also call such objects matrix valued differential forms for the following reason. Suppose that $R = (R_{ij})$ is a matrix whose entries are *linear* differential forms on some manifold M. Then if $v \in T_pM$ we can evaluate each of the R_{ij} at v and so obtain the matrix $R(v) := (R_{ij}(v))$. In other words, we can think of R as a linear differential form whose value is a matrix.

But we are going to allow our matrices to have forms of arbitrary degree. So a matrix valued form of degree zero is just a matrix R whose entries R_{ij} are functions on M etc.

For example, we might take M to be Mat(n), the n^2 dimensional space consisting of all $n \times n$ matrices (a_{ij}) and let A be the (tautological) matrix valued function $A = (a_{ij})$ assigning to each matrix itself. A slightly less trivial example is to let $Gl(n)$ be the set of all *invertible* $n \times n$ matrices and consider the function A^{-1} which assigns to each matrix its inverse. The entries of A^{-1} are functions of A (given by Cramer's rule from linear algebra).

For any two matrix valued differential forms $R = (R_{ij})$ and $S = (S_{ij})$ define their matrix exterior product $R \wedge S$ by the usual formula for matrix product, but with exterior multiplication of the entries instead of ordinary multiplication, so

$$(R \wedge S)_{ik} := \sum_j R_{ij} \wedge S_{jk}.$$

Also, if $R = (R_{ij})$ is a matrix valued differential form, define dR by applying d to each of the entries. So

$$(dR)_{ij} := (dR_{ij}).$$

Finally, if $\psi : X \to Y$ is a smooth map and $R = (R_{ij})$ is a matrix valued form on Y then we define its pullback by pulling back each of the entries:

$$(\psi^* R)_{ij} := (\psi^* R_{ij}).$$

So, for example, getting back to our trivial example, dA is the matrix of differential forms where $dA = (da_{ij})$. The i, j entry of dA is just the differential of the coordinate a_{ij}.

Recall from Math 1a that we have $d\left(\frac{1}{x}\right) = -\frac{1}{x^2} dx$.

1. What is $d\left(A^{-1}\right)$ where, as above, A^{-1} is the matrix valued function on $Gl(n)$ assigning to each invertible matrix its inverse? [Hint: Apply d to the equation $AA^{-1} \equiv I$ where I is the identity matrix.]

Amazingly, this elementary result will be a key ingredient in many of the computations we do later in this book.

2.14. PROBLEMS.

2.14.2 Actions of Lie groups on themselves.

Let G be a group and M be a set. A **(left) action** of G on M consists of a map

$$\phi : G \times M \to M$$

satisfying the conditions

$$\phi(a, \phi(b, m))) = \phi(ab, m)$$

(an associativity law) and

$$\phi(e, m) = m, \quad \forall m \in M$$

where e is the identity element of the group. When there is no risk of confusion we will write am for $\phi(a, m)$. (But in much of the beginning of the following exercises there *will* be a risk of confusion since there will be several different actions of the same group G on the set M). We think of an action as assigning to each element $a \in G$ a transformation, ϕ_a, of M:

$$\phi_a : M \to M, \quad \phi_a : m \mapsto \phi(a, m).$$

So we also use the notation

$$\phi_a m = \phi(a, m).$$

For example, we may take M to be the group G itself and let the action be left multiplication, L, so

$$L(a, m) = am.$$

We will write

$$L_a : G \to G, \quad L_a m = am.$$

We may may also consider the (left) action of right multiplication:

$$R : G \times G \to G, \quad R(a, m) = ma^{-1}.$$

(The inverse is needed to get the order right in $R(a, R(b, m)) = R(ab, m)$.) So we will write

$$R_a : G \to G, \quad R_a m = ma^{-1}.$$

We will be interested in the case that G is a Lie group, which means that G is a smooth manifold and the multiplication map $G \times G \to G$ and the inverse map $G \to G$, $a \mapsto a^{-1}$ are both smooth maps. Then the differential, $(dL_a)_m$ maps the tangent space to G at m, to the tangent space to G at am:

$$dL_a : TG_m \to TG_{am}$$

and similarly

$$dR_a : TG_m \to TG_{ma^{-1}}.$$

In particular,

$$dL_{a^{-1}} : TG_a \to TG_e.$$

2.14.3 The Maurer-Cartan form.

Let $G = Gl(n)$ be the group of all invertible $n \times n$ matrices. It is an open subset (hence a submanifold) of the n^2 dimensional space $\text{Mat}(n)$ of all $n \times n$ matrices. As above, we can think of the tautological map which sends every $A \in G$ into itself thought of as an element of $\text{Mat}(n)$ as a matrix valued function on G. Put another way, A is a matrix of functions on G, each of the matrix entries a_{ij} of A is a function on G. Hence $dA = (da_{ij})$ is a matrix of differential forms. So we may consider
$$A^{-1}dA$$
which is also a matrix valued differential form on G. Let B be a fixed element of G.

2. Show that
$$L_B^*(A^{-1}dA) = A^{-1}dA. \tag{2.8}$$
So each of the entries of $A^{-1}dA$ is *left invariant*.

It is easy to check that the entries of $A^{-1}dA$ are linearly independent, so we have found n^2 linearly independent (scalar valued) linear differential differential forms on $Gl(n)$. Therefor any wedge product of k distinct entries of $A^{-1}dA$ will be a non-vanishing left invariant (scalar valued) k form (where $k \leq n^2$). In particular, taking $k = n^2$, i.e. wedging all of these forms together gives a non-vanishing left invariant scalar valued form of degree n^2, a left invariant "volume form" on $Gl(n)$.

3. Show that
$$R_B^*(A^{-1}dA) = B(A^{-1}dA)B^{-1}. \tag{2.9}$$
So the entries of $A^{-1}dA$ are not right invariant (in general), but (2.9) shows how they are transformed into one another by right multiplication.

2.14.4 The Maurer-Cartan equation(s).

4. Show that if we set $\omega = A^{-1}dA$ then
$$d\omega + \omega \wedge \omega = 0. \tag{2.10}$$

[Hint: Use Problem 1.]

It will turn out that this equation (known as the **Maurer-Cartan equation**) in its various guises and generalizations will encapsulate most of this course!

Here is another way of thinking about $A^{-1}dA$: Since $G = Gl(n)$ is an open subset of the vector space $\text{Mat}(n)$, we may identify the tangent space TG_A with the vector space $\text{Mat}(n)$. That is we have an isomorphism between TG_A and

Mat(n). If you think about it for a minute, it is the form dA which effects this isomorphism at every point. On the other hand, left multiplication by A^{-1} is a linear map. Under this identification, the differential of a linear map L looks just like L. So in terms of this identification, $A^{-1}dA$, when evaluated at the tangent space TG_A is just the isomorphism $dL_A^{-1} : TG_A \to TG_I$ where I is the identity matrix.

2.14.5 Restriction to a subgroup of $Gl(n)$.

Let H be a Lie subgroup of $G = Gl(n)$. This means that H is a subgroup of G and it is also a submanifold. In other words we have an embedding

$$\iota : H \to G$$

which is a(n injective) group homomorphism. Let

$$\mathfrak{h} = TH_I$$

denote the tangent space to H at the identity element.

5. Conclude from the preceding discussion that if we now set

$$\omega = \iota^*(A^{-1}dA)$$

then ω takes values in $\mathfrak{h}$. In other words, when we evaluate ω on any tangent vector to H at any point of H we get a matrix belonging to the subspace $\mathfrak{h}$.

For example, let $H = O(n)$. So O consists of all matrices which satisfy

$$AA^\dagger = I$$

where $A^\dagger$ denotes the transpose of A. If $A(t)$ is a curve in $O(n)$ with $A(0) = I$, then differentiating the above equation and setting $t = 0$ gives

$$A'(0) + (A^\dagger)'(0) = 0.$$

In other words, $A'(0)$ is an antisymmetric matrix. Conversely, if ξ is an antisymmetric matrix, then $\exp t\xi$ is a curve (actually a one parameter subgroup) in $O(n)$ whose derivative at 0 is ξ. So if we take $H = O(n)$ then $\mathfrak{h}$ consists of the anti-symmetric matrices. So if Ω denotes the restriction of ω from $Gl(n)$ to $O(n)$ then Ω is an anti-symmetric matrix:

$$\Omega = (\Omega_{ij}) \qquad \Omega_{ji} = -\Omega_{ij}.$$

2.14.6 Interlude, the Haar integral.

We know that Ω is left invariant:

$$L_A^*\Omega = \Omega \quad \forall\ A \in O(n).$$

But it will not be right invariant (for $n > 2$) in general. Indeed, by Problem **3**, we know that
$$R_B^* \Omega = B\Omega B^{-1}, \quad \text{for } B \in O(n).$$
In particular, the entries Ω_{ij} will not be right invariant in general.

For the sake of simple notation let me concentrate on the case $n = 3$. The arguments I will present work for general n. So
$$\Omega = \begin{pmatrix} 0 & \Omega_{12} & \Omega_{13} \\ -\Omega_{12} & 0 & \Omega_{23} \\ -\Omega_{13} & -\Omega_{23} & 0 \end{pmatrix}.$$

There are three independent entries in Ω, for example $\Omega_{12}, \Omega_{13}, \Omega_{23}$. I claim that the product of all three *is* right invariant. In other words, I claim that
$$R_B^* \left(\Omega_{12} \wedge \Omega_{13} \wedge \Omega_{23} \right) = \Omega_{12} \wedge \Omega_{13} \wedge \Omega_{23} \quad (2.11)$$
for all $B \in O(n)$. To prove this, I will compute $B\Omega B^{-1}$ for various $B \in O(n)$.

Let me first take B to be rotation though angle ϕ in the x_1, x_2 plane. So
$$B = \begin{pmatrix} c & s & 0 \\ -s & c & 0 \\ 0 & 0 & 1 \end{pmatrix}, \quad c = \cos\phi, \; s = \sin\phi,$$
with
$$B^{-1} = \begin{pmatrix} c & -s & 0 \\ s & c & 0 \\ 0 & 0 & 1 \end{pmatrix}.$$

A straightforward computation (check this out) shows that
$$B\Omega B^{-1} = \begin{pmatrix} 0 & \Omega_{12} & c\Omega_{13} + s\Omega_{23} \\ -\Omega_{12} & 0 & -s\Omega_{13} + c\Omega_{23} \\ -(c\Omega_{13} + s\Omega_{23}) & -(-s\Omega_{13} + c\Omega_{23}) & 0 \end{pmatrix}.$$

In other words, by Problem **3**,
$$R_B^* \Omega_{12} = \Omega_{12}, \quad R_B^* \Omega_{13} = c\Omega_{13} + s\Omega_{23}, \quad R_B^* \Omega_{23} = -s\Omega_{13} + c\Omega_{23}.$$

Now
$$(c\Omega_{13} + s\Omega_{23}) \wedge (-s\Omega_{13} + c\Omega_{23}) = \Omega_{13} \wedge \Omega_{23}.$$

So (2.11) holds for this choice of B. But there is nothing special about the x_1, x_2 plane, so (2.11) must hold for rotations in the x_1, x_3 plane and the x_2, x_3 plane (or you can check this directly). Since these generate the entire rotation group (those $A \in O(n)$ with determinant one), we see that (2.11) holds for the entire rotation group. If we take
$$B = \begin{pmatrix} -1 & 0 & 0 \\ 0 & 1 & 0 \\ 0 & 0 & 1 \end{pmatrix}$$

2.14. PROBLEMS.

Then replacing Ω by $B\Omega B^{-1}$ has the effect of replacing Ω_{12} and Ω_{13} by $-\Omega_{12}, -\Omega_{13}$ and leaving Ω_{23} unchanged, as you can check. So again (2.11) holds. But since this B together with all the rotations give all of $O(n)$, we see that (2.11) holds for all $B \in O(n)$ as claimed.

Now $\Omega_{12} \wedge \Omega_{13} \wedge \Omega_{23}$ can not vanish at any point of G, since if it did, it would have to vanish all all point by invariance. So $\Omega_{12} \wedge \Omega_{13} \wedge \Omega_{23}$ determines an orientation on G. So for any continuous function f on G, we can define

$$I(f) := \int_G f\Omega_{12} \wedge \Omega_{13} \wedge \Omega_{23},$$

the integral relative to this preferred orientation. From the invariance of $\Omega_{12} \wedge \Omega_{13} \wedge \Omega_{23}$ it follows that $I(f)$ is right and left invariant in the sense that

$$I(f) = I(L_A^* f) = I(R_A^* f)$$

for all $A \in O(n)$. A theorem of Haar guarantees the existence of such an I, which is determined up to a multiplicative constant. It is convenient to choose the constant so that $I(\mathbf{1}) = 1$, where $\mathbf{1}$ is the function which is identically one.

My point is that we can construct the Haar integral out of the Maurer-Cartan form.

6. Show that on a group, the only transformations which commute with all the right multiplications, R_b, $b \in G$, are the left multiplications, L_a.

For any vector $\xi \in TH_I$, define the vector field X by

$$X(A) = dR_{A^{-1}}\xi.$$

(Recall that $R_{A^{-1}}$ is right multiplication by A and so sends I into A.) For example, if we take H to be the full group $G = Gl(n)$ and identify the tangent space at every point with $\text{Mat}(n)$ then the above definition becomes

$$X(A) = \xi A.$$

By construction, the vector field X is right invariant, i.e. is invariant under all the diffeomorphisms R_B.

7. Conclude that the flow generated by X is left multiplication by a one parameter subgroup. Also conclude that in the case $H = Gl(n)$ the flow generated by X is left multiplication by the one parameter group

$$\exp t\xi.$$

Finally conclude that for a general subgroup H, if $\xi \in \mathfrak{h}$ then all the $\exp t\xi$ lie in H.

8. What is the space $\mathfrak{h}$ in the case that H is the group of Euclidean motions in three dimensional space, thought of as the set of all four by four matrices of the form
$$\begin{pmatrix} A & v \\ 0 & 1 \end{pmatrix}, \quad AA^\dagger = I, \ v \in \mathbb{R}^3?$$

2.14.7 Frames.

Let V be an n dimensional vector space over $\mathbb{R}$. A **frame** on V is, by definition, an isomorphism $\mathbf{f} : \mathbb{R}^n \to V$. Giving $\mathbf{f}$ is the same as giving each of the vectors $f_i = \mathbf{f}(\delta_i)$ where the δ_i range over the standard basis of $\mathbb{R}^n$. So giving a frame is the same as giving an ordered basis of V and we will sometimes write
$$\mathbf{f} = (f_1, \ldots, f_n).$$
If $A \in Gl(n)$ then A is an isomorphism of $\mathbb{R}^n$ with itself, so $\mathbf{f} \circ A^{-1}$ is another frame. So we get an action, $R : Gl(n) \times \mathbf{F} \to \mathbf{F}$ where $\mathbf{F} = \mathbf{F}(V)$ denotes the space of all frames:
$$R(A, \mathbf{f}) = \mathbf{f} \circ A^{-1}. \tag{2.12}$$
If $\mathbf{f}$ and $\mathbf{g}$ are two frames, then $\mathbf{g}^{-1} \circ \mathbf{f} = M$ is an isomorphism of $\mathbb{R}^n$ with itself, i.e. a matrix. So given any two frames, $\mathbf{f}$ and $\mathbf{g}$, there is a unique $M \in Gl(n)$ so that $\mathbf{g} = \mathbf{f} \circ M^{-1}$.

Once we fix an $\mathbf{f}$, we can use this fact to identify $\mathbf{F}$ with $Gl(n)$, but the identification depends on the choice of $\mathbf{f}$. In any event the (non-unique) identification shows that $\mathbf{F}$ is a manifold and that (2.12) defines an action of $Gl(n)$ on $\mathbf{F}$. Each of the f_i (the i−th basis vector in the frame) can be thought of as a V valued function on $\mathbf{F}$. So we may write
$$df_j = \sum \omega_{ij} f_i \tag{2.13}$$
where the ω_{ij} are ordinary (number valued) linear differential forms on $\mathbf{F}$. We think of this equation as giving the expansion of an infinitesimal change in f_j in terms of the basis $\mathbf{f} = (f_1, \ldots, f_n)$. If we use the "row" representation of $\mathbf{f}$ as above, we can write these equations as
$$d\mathbf{f} = \mathbf{f}\omega \tag{2.14}$$
where $\omega = (\omega_{ij})$.

9. Show that the ω defined by (2.14) satisfies
$$R_B^* \omega = B\omega B^{-1}. \tag{2.15}$$

To see the relation with what went on before, notice that we *could* take $V = \mathbb{R}^n$ itself. Then $\mathbf{f}$ is just an invertible matrix, A and (2.14) becomes our old equation $\omega = A^{-1}dA$. So (2.15) reduces to (2.9).

2.14. PROBLEMS.

If we take the exterior derivative of (2.14) we get

$$0 = d(d\mathbf{f}) = d\mathbf{f} \wedge \omega + \mathbf{f}d\omega = \mathbf{f}(\omega \wedge \omega + d\omega)$$

from which we conclude

$$d\omega + \omega \wedge \omega = 0. \tag{2.16}$$

2.14.8 Euclidean frames.

We specialize to the case where $V = \mathbb{R}^n, n = d+1$ so that the set of frames becomes identified with the group $Gl(n)$ and restrict to the subgroup, H, of Euclidean motions which consist of all $n \times n$ matrices of the form

$$\begin{pmatrix} A & v \\ 0 & 1 \end{pmatrix}, \quad A \in O(d), \ v \in \mathbb{R}^d.$$

Such a matrix, when applied to a vector

$$\begin{pmatrix} w \\ 1 \end{pmatrix}$$

sends it into the vector

$$\begin{pmatrix} Aw + v \\ 1 \end{pmatrix}$$

and $Aw + v$ is the orthogonal transformation A applied to w followed by the translation by v. The corresponding **Euclidean frames** (consisting of the columns of the elements of H) are thus defined to be the frames of the form

$$f_i = \begin{pmatrix} e_i \\ 0 \end{pmatrix}, \quad i = 1, \ldots d,$$

where the e_i form an orthonormal basis of $\mathbb{R}^d$ and

$$f_n = \begin{pmatrix} v \\ 1 \end{pmatrix},$$

where $v \in \mathbb{R}^d$ is an arbitrary vector. The idea is that v represents a choice of origin in the d dimensional $\mathbb{R}^d$ space and $\mathbf{e} = (e_1, \ldots, e_d)$ is an orthonormal basis. We can write this in shorthand notation as

$$\mathbf{f} = \begin{pmatrix} \mathbf{e} & v \\ 0 & 1 \end{pmatrix}.$$

If ι denotes the embedding of H into G, we know from exercise **5** that

$$\iota^* \omega = \begin{pmatrix} \Omega & \theta \\ 0 & 0 \end{pmatrix},$$

where

$$\Omega_{ij} = -\Omega_{ji}.$$

So the pull back of (2.14) becomes

$$d\begin{pmatrix} \mathbf{e} & v \\ 0 & 1 \end{pmatrix} = \begin{pmatrix} \mathbf{e}\Omega & \mathbf{e}\theta \\ 0 & 0 \end{pmatrix} \quad (2.17)$$

or, in more expanded notation,

$$de_j = \sum_i \Omega_{ij} e_i, \quad dv = \sum_i \theta_i e_i.$$

Let (,) denote the Euclidean scalar product. Then we can write

$$\theta_i = (dv, e_i) \quad (2.18)$$

and

$$(de_j, e_i) = \Omega_{ij}.$$

If we set

$$\Theta = -\Omega$$

this becomes

$$(de_i, e_j) = \Theta_{ij}. \quad (2.19)$$

Then (2.16) becomes

$$d\theta = \Theta \wedge \theta, \quad d\Theta = \Theta \wedge \Theta. \quad (2.20)$$

Or, in more expanded notation,

$$d\theta_i = \sum_j \Theta_{ij} \wedge \theta_j, \quad d\Theta_{ik} = \sum_j \Theta_{ij} \wedge \Theta_{jk}. \quad (2.21)$$

Equations (2.18)-(2.20) or (2.21) are known as the **structure equations of Euclidean geometry**.

2.14.9 Frames adapted to a submanifold.

Let M be a k dimensional submanifold of $\mathbb{R}^d$. This determines a submanifold of the manifold, H, of all Euclidean frames by the following requirements:
 i) $v \in M$ and
 ii) $e_i \in TM_v$ for $i \leq k$.
We will usually write m instead of v to emphasize the first requirement - that the frames be based at points of M. The second requirement says that the first k vectors in the frame based at m be tangent to M (and hence that the last $n - k$ vectors in the frame are normal to M). We will denote this manifold by $\mathcal{O}(M)$. It has dimension

$$k + \frac{k(k-1)}{2} + \frac{(d-k-1)(d-k)}{2}.$$

The first term comes from the point m varying on M, the second is the dimension of the orthogonal group $O(k)$ corresponding to the choices of the first k vectors

2.14. PROBLEMS.

in the frame, and the third term is dim $O(d-k)$ correspond to the last $(n-k)$ vectors. We have an embedding of $\mathcal{O}(M)$ into H, and hence the forms θ and Θ pull back to $\mathcal{O}(M)$. As we are running out of letters, we will continue to denote these pull backs by the same letters. So the pulled back forms satisfy the same structure equations (2.18)-(2.20) or (2.21) as above, but they are supplemented by

$$\theta_i = 0, \quad \forall i > k. \tag{2.22}$$

2.14.10 Curves and surfaces - their structure equations.

We will be particularly interested in curves and surfaces in three dimensional Euclidean space. In particular, we will show how the classical Frenet equations and more recent Bishop framing are related to the Cartan structure equations.

For a curve, C, the manifold of (adapted) frames is two dimensional, and we have

$$dC = \theta_1 e_1 \tag{2.23}$$
$$de_1 = \Theta_{12} e_2 + \Theta_{13} e_3 \tag{2.24}$$
$$de_2 = \Theta_{21} e_1 + \Theta_{23} e_3 \tag{2.25}$$
$$de_3 = \Theta_{31} e_1 + \Theta_{32} e_2. \tag{2.26}$$

One can visualize the manifold of frames as a sort of tube: about each point of the curve there is a circle in the plane normal to the tangent line corresponding the possible choices of e_2.

For the case of a surface the manifold of frames is three dimensional: we can think of it as a union of circles each centered at a point of S and in the plane tangent to S at that point. Then equation (2.23) is replaced by

$$dX = \theta_1 e_1 + \theta_2 e_2 \tag{2.27}$$

but otherwise the equations are as above, including the structure equations (2.21). These become

$$d\theta_1 = \Theta_{12} \wedge \theta_2 \tag{2.28}$$
$$d\theta_2 = -\Theta_{12} \wedge \theta_1 \tag{2.29}$$
$$0 = \Theta_{31} \wedge \theta_1 + \Theta_{32} \wedge \theta_2 \tag{2.30}$$
$$d\Theta_{12} = \Theta_{13} \wedge \Theta_{32} \tag{2.31}$$
$$d\Theta_{13} = \Theta_{12} \wedge \Theta_{23} \tag{2.32}$$
$$d\Theta_{23} = \Theta_{21} \wedge \Theta_{13} \tag{2.33}$$

Equation (2.31) is known as Gauss' equation, and equations (2.32) and (2.33) are known as the Codazzi-Mainardi equations.

2.14.11 The sphere as an example.

In computations with local coordinates, we may find it convenient to use a "cross-section" of the manifold of frames, that is a map which assigns to each point of neighborhood on the surface a preferred frame. If we are given a parametrization $m = m(u,v)$ of the surface, one way of choosing such a cross-section is to apply the Gram-Schmidt orthogonalization procedure to the tangent vector fields m_u and m_v, and take into account the chosen orientation.

For example, consider the sphere of radius R. We can parameterize the sphere with the north and south poles (and one longitudinal semi-circle) removed by the $(u,v) \in (0, 2\pi) \times (0, \pi)$ by $X = X(u,v)$ where

$$X(u,v) = \begin{pmatrix} R \cos u \sin v \\ R \sin u \sin v \\ R \cos v \end{pmatrix}.$$

Here v denotes the angular distance from the north pole, so the excluded value $v = 0$ corresponds to the north pole and the excluded value $v = \pi$ corresponds to the south pole. Each constant value of v between 0 and π is a circle of latitude with the equator given by $v = \frac{\pi}{2}$. The parameter u describes the longitude from the excluded semi-circle.

In any frame adapted to a surface in $\mathbb{R}^3$, the third vector e_3 is normal to the surface at the base point of the frame. There are two such choices at each base point. In our sphere example let us choose the outward pointing normal, which at the point $m(u,v)$ is

$$e_3(m(u,v)) = \begin{pmatrix} \cos u \sin v \\ \sin u \sin v \\ \cos v \end{pmatrix}.$$

We will write the left hand side of this equation as $e_3(u,v)$. The coordinates u, v are orthogonal, i.e. X_u and X_v are orthogonal at every point, so the orthonormalization procedure amounts only to normalization: Replace each of these vectors by the unit vectors pointing in the same direction at each point. So we get

$$e_1(u,v) = \begin{pmatrix} -\sin u \\ \cos u \\ 0 \end{pmatrix}, \quad e_2(u,v) = \begin{pmatrix} \cos u \cos v \\ \sin u \cos v \\ -\sin v \end{pmatrix}.$$

We thus obtain a map ψ from $(0, 2\pi) \times (0, \pi)$ to the manifold of frames,

$$\psi(u,v) = (X(u,v), e_1(u,v), e_2, (u,v), e_3(u,v)).$$

Since $X_u \cdot e_1 = R \sin v$ and $X_v \cdot e_2 = R$ we have

$$dX(u,v) = (R \sin v du) e_1(u,v) + (R dv) e_2(u,v).$$

Thus we see from (2.27) that

$$\psi^* \theta_1 = R \sin v du, \quad \psi^* \theta_2 = R dv$$

and hence that

$$\psi^*(\theta_1 \wedge \theta_2) = R^2 \sin v du \wedge dv.$$

2.14. PROBLEMS.

Now $R^2 \sin v \, du \, dv$ is just the area element of the sphere expressed in u, v coordinates. The choice of e_1, e_2 determines an orientation of the tangent space to the sphere at the point $X(u,v)$ and so $\psi^*(\theta_1 \wedge \theta_2)$ is the pull-back of the corresponding oriented area form.

10. Compute $\psi^*\Theta_{12}$, $\psi^*\Theta_{13}$, and $\psi^*\Theta_{23}$ and verify that

$$\psi^*(d\Theta_{12}) = -\psi^*(K)\psi^*(\theta_1 \wedge \theta_2)$$

where $K = 1/R^2$ is the curvature of the sphere.

2.14.12 Ribbons

The idea here is to study a curve on a surface, or rather a curve with an "infinitesimal" neighborhood of a surface along it. So let C be a curve and $\mathcal{O}(C)$ its associated two dimensional manifold of frames. We have a projection $\pi : \mathcal{O}(C) \to C$ sending every frame into its origin. By a **ribbon** based on C we mean a section $n : C \to \mathcal{O}(C)$, so n assigns a unique frame to each point of the curve in a smooth way. We will only be considering curves with non-vanishing tangent vector everywhere. With no loss of generality we may assume that we have parametrized the curve by arc length, and the choice of e_1 determines an orientation of the curve, so $\theta = ds$. The choice of e_2 at every point then determines e_3 up to a $\pm$ sign. So a good way to visualize s is to think of a rigid metal ribbon determined by the curve and the vectors e_2 perpendicular to the curve (determined by n) at each point. The forms Θ_{ij} all pull back under n to function multiples of ds:

$$n^*\Theta_{12} = k\,ds, \quad n^*\Theta_{23} = -\tau\,ds, \quad n^*\Theta_{13} = w\,ds \qquad (2.34)$$

where k, τ and w are functions of s. We can write equations (2.23)- (2.26) above as

$$\frac{dC}{ds} = e_1,$$

and

$$\frac{de_1}{ds} = ke_2 + we_3, \quad \frac{de_2}{ds} = -ke_1 - \tau e_3, \quad \frac{de_3}{ds} = -we_1 + \tau e_3. \qquad (2.35)$$

For later applications we will sometimes be sloppy and write Θ_{ij} instead of $n^*\Theta_{ij}$ for the pull back to the curve, so along the ribbon we have $\Theta_{12} = k\,ds$ etc. Also it will sometimes be convenient in computations (as opposed to proving theorems) to use parameters other than arc length.

11. Show that two ribbons (defined over the same interval of s values) are congruent (that is there is a Euclidean motion carrying one into the other) if and only if the functions k, τ, and w are the same.

I now want to draw a consequence of this Problem: Recall that two surfaces which differ by a Euclidean motion have the same induced metric and the same Weingarten map. Here is a converse:

2.14.13 The induced metric and the Weingarten map determine a surface up to a Euclidean motion.

Let M_1 and M_2 be two connected two dimensional manifolds, and let

$$\mathcal{F}_1 : M_1 \to \mathbb{R}^3, \quad \mathcal{F}_2 : M_2 \to \mathbb{R}^3$$

be embeddings of of these manifolds into Euclidean three space. Suppose that there is a diffeomorphism

$$\phi : M_1 \to M_2$$

and a Euclidean motion E such that

$$E \circ \mathcal{F}_1 = \mathcal{F}_2 \circ \phi.$$

Then if $\mathbf{g}_1$ denotes the induced metric on M_1 and $\mathbf{g}_2$ denotes the induced metric on M_2 we know that

$$\phi^* \mathbf{g}_2 = \mathbf{g}_1$$

in the sense that for any pair of tangent vectors v, w at a point $m \in M_1$ we have

$$(v, w)_{\mathbf{g}_1} = (d\phi_m v, d\phi_m w)_{\mathbf{g}_2}.$$

Also, if W_1 denotes the Weingarten map on M_1 corresponding to the embedding $\mathcal{F}_1$ and W_2 denotes the Weingarten map on M_2 corresponding to the embedding $\mathcal{F}_1$ then

$$\phi^* W_2 = W_1$$

in the sense that for any tangent vector v at a point $m \in M_1$ we have

$$d\phi_m(W_{1m} v) = W_{2\phi(m)} d\phi_m(v).$$

We can assert the converse:

Theorem 2. *Let M_1 and M_2 be two connected two dimensional manifolds, and let*

$$\mathcal{F}_1 : M_1 \to \mathbb{R}^3, \quad \mathcal{F}_2 : M_2 \to \mathbb{R}^3$$

be embeddings of of these manifolds into Euclidean three space. Suppose that there is a diffeomorphism $\phi : M_1 \to M_2$ such that $\phi^* \mathbf{g}_2 = \mathbf{g}_1$ and $\phi^* W_2 = W_1$. Then there is a Euclidean motion E such that

$$E \circ \mathcal{F}_1 = \mathcal{F}_2 \circ \phi.$$

Proof. Choose a point $m \in M_1$ and an orthonormal basis v_1, v_2 of TM_1 relative to the induced metric $\mathbf{g}_1$. This picks out:

- A point $p = \mathcal{F}_1(m) \in \mathbb{R}^3$,

- Vectors e_1, e_2 tangent to $\mathcal{F}_1(M_1)$ at $\mathcal{F}_1(m)$ and also

2.14. PROBLEMS.

- the unit normal e_3 which gives the Weingarten map at m (since $-e_3$ would give $-W_1$ at m). In addition we have

- the orthonormal basis $w_1 = d\phi_m(v_1), w_2$ of $TM_{\phi(m)}$ and consequently

- the corresponding frame f_1, f_2, f_3 at $q = \mathcal{F}_2(\phi(m))$.

- Also ϕ induces a diffeomorphism (which we will denote by Φ) from $\mathcal{O}(M_1)$ to $\mathcal{O}(M_2)$.

We have a unique Euclidean motion E which carries (p, e_1, e_2, e_3) into (q, f_1, f_2, f_3). Suppose we have made this preliminary Euclidean motion, so that we can add to the hypotheses of the theorem the hypothesis that $(p, e_1, e_2, e_3) = (q, f_1, f_2, f_3)$.

Let n be any point of M_1 and join m to n by a curve γ. This is possible since we are assuming that our manifolds are connected. We wish to show that

$$\mathcal{F}_1 \circ \gamma = \mathcal{F}_2 \circ (\phi(\gamma))$$

as this will imply that

$$\mathcal{F}_1(n) = \mathcal{F}_2(\phi(n))$$

(and this holds for all $n \in M_1$).

The form Θ_{12}^1 on $\mathcal{O}(M_1)$ is determined by the induced metric and the forms Θ_{13}^1 and Θ_{23}^1 by the Weingarten map. Hence if Θ_{ij}^2 denote the corresponding forms on $\mathcal{O}(M_2)$ we have

$$\Phi^*(\Theta_{ij}^2) = \Theta_{ij}^1.$$

But then the ribbons determined by the curves $C_1 = \mathcal{F}_1 \circ \gamma$ on the surface $\mathcal{F}_1(M_1)$ and $C_1 = \mathcal{F}_2 \circ \phi(\gamma)$ on $\mathcal{F}_2(M_2)$ must be identical by Problem **1**. In particular, the curves are the same. □

Later, I will prove the n-dimensional version of this theorem (that we just proved in three dimensions) as a consequence of a theorem of Frobenius.

2.14.14 Back to ribbons.

A ribbon is really just a curve in the space, H, of all Euclidean frames, having the property that the base point, that is the v of the frame (v, e_1, e_2, e_3) has non-vanishing derivative. Problem **11** says that two curves, $i : I \to H$ and $j : I \to H$ in H differ by an overall left translation (that is satisfy $j = L_h \circ i$) if and only if the forms $\theta, \Theta_{12}, \Theta_{13}, \Theta_{23}$ pull back to the same forms on I. The form $i^*\theta$ is just the arc length form ds as we mentioned above. It is absolutely crucial for the rest of this course to understand the meaning of the form $i^*\Theta_{12}$.

12. Consider a circle of latitude on a sphere of radius r. To fix the notation, suppose that the circle is at angular distance ϕ from the north pole and that we use ψ as angular coordinates along the circle. Take the ribbon adapted to the sphere, so e_1 is the unit tangent vector to the circle of latitude and e_2 is the unit tangent vector to the circle of longitude chosen so that e_1, e_2, e_3 with e_3 pointing out of the sphere gives the statndard orientation on $\mathbf{R}^3$. So $i^*\Theta_{12} = Ld\psi$ for some function L. What is L?

13. Let C be a straight line (say a piece of the z-axis) parametrized according to arc length and let e_2 be rotating at a rate $f(s)$ about C (so, for example, $e_2 = \cos f(s)\mathbf{i} + \sin f(s)\mathbf{j}$ where $\mathbf{i}$ and $\mathbf{j}$ are the unit vectors in the x and y directions). What is $i^*\Theta_{12}$?

To continue our understanding of Θ_{12}, let us consider what it means for two ribbons, $i : I \to H$ and $j : I \to H$ to have the same value of the pullback of Θ_{12} at some point $s_0 \in I$ (where I is some interval on the real line). So

$$(i^*\Theta_{12})|_{s=s_0} = (j^*\Theta_{12})|_{s=s_0}.$$

There is a (unique) left multiplication, that is a unique Euclidean motion, which carries $i(s_0)$ to $j(s_0)$. Let assume that we have applied this motion so we assume that $i(s_0) = j(s_0)$. Let us write

$$i(s) = (C(s), e_1(s), e_2(s), e_3(s)), \quad j(s) = (D(s), f_1(s), f_2(s).f_3(s))$$

and we are assuming that $C(s_0) = D(s_0)$, $C'(s_0) = e_1(s_0) = f_1(s_0) = D'(s_0)$ so the curves C and D are tangent at s_0, and that $e_2(s_0) = f_2(s_0)$ so that the planes of the ribbon (spanned by the first two orthonormal vectors) coincide. Then our condition about the equality of the pullbacks of Θ_{12} asserts that

$$((e_2' - f_2')(s_0), e_1(s_0)) = 0$$

and of course $((e_2' - f_2')(s_0), e_2(s_0)) = 0$ automatically since $e_2(s)$ and $f_2(s)$ are unit vectors. So the condition is that the relative change of e_2 and f_2 (and similarly e_1 and f_1) at s_0 be normal to the common tangent plane fo the ribbon.

2.14.15 Developing a ribbon.

We will now drop one dimension, and consider ribbons in the plane (or, if you like, ribbons lying in a fixed plane in three dimensional space). So all we have is θ and Θ_{12}. Also, the orientation of the curve and of the plane completely determines e_2 as the unit vector in the plane perpendicular to the curve and such that e_1, e_2 give the correct orientation. So a ribbon in the plane is the same as an oriented curve.

14. Let $k = k(s)$ be any continuous function of s. Show that there is a ribbon in the plane whose base curve is parametrized by arc length and for which $i^*\Theta_{12} = kds$. Furthermore, show that this planar ribbon (curve) is uniquely determined up to a planar Euclidean motion.

It follows from the preceding exercise, that we have a way of associating a curve in the plane (determined up to a planar Euclidean motion) to any

2.14. PROBLEMS.

ribbon in space. It consists of rocking and rolling the ribbon along the plane in such a way that infinitesimal change in the e_1 and e_2 are always normal to the plane. Mathematically, it consists in solving problem **14** for the $k = k(s)$ where $i^*\Theta_{12} = kds$ for the ribbon. We call this operation **developing** the ribbon onto a plane. In particular, if we have a curve on a surface, we can consider the ribbon along the curve induced by the surface. In this way, we may talk of developing the surface on a plane along the given curve. Intuitively, if the surface were convex, this amounts to rolling the surface on a plane along the curve.

15. What are results of developing the ribbons of problems **12** and **13**?

2.14.16 The general Maurer-Cartan equations.

Until now, we have been studying the Maurer-Cartan equations for subgroups of $Gl(n)$. If $\mathfrak{h}$ is the Lie algebra of such a subgroup, we can think of an element of $\mathfrak{h}$ as a matrix, and we know how to multiply two matrices, and so the expression $\omega \wedge \omega$ made sense.

But for a general Lie group G, with Lie algebra $\mathfrak{g} = TG_e$, we do not have any associative multiplication defined on $\mathfrak{g}$. But we *do* have a Lie bracket: if we identify elements of $\mathfrak{g}$ with left invariant vector fields, then the Lie bracket of two such vector fields defines a Lie bracket on $\mathfrak{g}$.

Now in general, if we have a bilinear map $M : U \times V \to W$ from two vector spaces into a third, and if σ and τ are differential forms on a smooth manifold with values in U and V respectively, then we can form $M(\tau, \sigma)$ which will be a W valued form.(In terms of bases of the three spaces, M would be given by $M(x, y) = \sum_{ij}^{k} M_{ij}^k x^i y^j b_k$, and now simply replace the real variables x and y by differential forms and use exterior product.)

Since we have the bracket, which is a bilinear map $\mathfrak{g} \times \mathfrak{g} \to \mathfrak{g}$, we can do the corresponding "bracket" on forms. For want of better notation, if σ and τ forms with values in $\mathfrak{g}$, I will denote the corresponding "product" by $[\sigma \wedge, \tau]$.

Now let us consider the case where our manifold is the Lie group G, and ω is the $\mathfrak{g}$ valued form which assigns to any vector $v \in T_g G$ the element $\omega(v) \in T_e G$, $\omega(v) = dL_g^{-1}(v)$. If X is a left invariant vector field with $X(e) = \xi$, then $\omega(X(g)) = \xi$ for all $g \in G$.

Let X and Y be vector fields on G. For any 2-form Ω, recall that

$$\Omega(X, Y) = (\iota(X)\Omega)(Y) = \iota(Y)\iota(X)\Omega.$$

Let us apply this to the case that X and Y are left invariant vector fields with $X(e) = \xi$, $Y(e) = \eta$, and we take $\Omega = [\omega \wedge, \omega]$.

We have

$$\iota(X)[\omega \wedge, \omega] = [\iota(X)\omega, \omega] - [\omega, \iota(X)\omega] = [\xi, \omega] - [\omega, \xi].$$

Applying $\iota(Y)$ to the above equation shows that

$$[\omega \wedge, \omega](X,Y) = [\xi, \eta] - [\eta, \xi] = 2[\xi, \eta] = 2\omega([X,Y]).$$

In other words,

$$\frac{1}{2}[\omega \wedge, \omega](X,Y) = \omega([X,Y]).$$

Now we recall a formula, a consequence of Weil's formula. It said that for any one form θ (including a vector valued one form), and any vector fields X and Y,

$$d\theta(X,Y) = X(\theta(Y)) - Y(\theta(X)) - \theta([X,Y]).$$

Let us apply this formula to our case where $\theta = \omega$ and X and Y are left invariant vector fields. Then the first two terms on the right above vanish, because $\omega(Y) = \eta$ is a constant (element of $\mathfrak{g}$) and $\omega(X) = \xi$ is also constant. So the equation becomes,

$$d\omega(X,Y) = -\omega([X,Y]).$$

If we compare this with the preceding, we get

$$d\omega + \frac{1}{2}[\omega \wedge, \omega] = 0. \qquad (2.36)$$

This is the general version of the Maurer-Cartan equation. With the correct definitions, it is an immediate consequence of the Weil formula.

16. Explain why for (any Lie subgroup of) $Gl(n)$

$$\frac{1}{2}[\omega \wedge, \omega] = \omega \wedge \omega.$$

In any event, *the basic thing to remember is Weil's formula.*

2.14.17 Back to curves in $\mathbb{R}^3$.

In the preceding, we studied properties of ribbons. This raises the question: Starting with just the curve in $\mathbb{R}^3$, is there a natural way of associating a ribbon to it? Recall that we are only considering "regular" curves, i.e. curves C such that C' never vanishes. Having chosen an orientation (= direction) on this curve, we can parametrize the curve by arc length. Then the tangent vector T has length one everywhere. So differentiating the equation $(T,T) \equiv 1$ gives

$$(T, T') = 0.$$

Suppose that we make the hypothesis

$$T' \neq 0 \qquad (2.37)$$

2.14. PROBLEMS.

anywhere. Then we can choose

$$N := \frac{1}{\|T'\|} T'$$

as defining a ribbon with

$$e_1 = T, \quad e_2 = N$$

and e_3 the unit vector (field) perpendicular to e_1 and e_2 (and determined, say, by the orientation of $\mathbb{R}^3$). In classical language $-e_3$ is usually denoted by B, the function $\|T'\|$ is denoted by κ. The frame field corresponding to the choice is called the **Frenet** frame field. For the Frenet frame field the function w in (2.35) vanishes and (2.35) becomes the classical **Frenet equations**

$$T' = \kappa N, \quad N' = -\kappa T + \tau B, \quad B' = -\tau N. \qquad (2.38)$$

These were discovered by Frenet in the 19th century. (In 19th century language, differing from our usage, κ was called the curvature and τ was called the torsion of the curve.) In the Frenet equations, the meaning of the condition $\tau \equiv 0$ is clear - it is that the curve lie in a plane. The problem with the Frenet frame is condition (2.37). It requires that the curvature vanish nowhere. Many interesting curves (for example straight lines) do not have this property.

In 1975 Richard Bishop introduced a different family of frames which works for any regular curve in $\mathbb{R}^3$. Here is how it goes: Let us call a normal vector field M (i.e. a vector field alnog C perpendicular to C') "relatively parallel" if if M' is tangential, i.e. if $M' = f \cdot T$ for some function f, where T is the unit tangent vector field.

17. Show that if M_1 and M_2 are relatively parallel normal vector fields then their scalar product (M_1, M_2) is constant. In particular, the length of a relatively parallel normal vector field is constant, and if $M_1(t_0) = M_2(t_0)$ then $M_1(t) = M_2(t)$ for all t. In other words, a relatively parallel normal vector field is determined by its value at a single point.

We now turn to the issue of existence of a relatively parallel normal vector field M with a prescribed value of $M(t_0)$. It is enough to prove the existence locally, since the uniqueness proved in Problem **17** shows that local solutions fit together to give a global solution. Now choose some local frame field e_1, e_2, e_3. (Such local frame fields always exist - just choose any three independent vector fields f_1, f_2, f_3 with $f_1 = e_1$ and apply Gram-Schmidt.)

Any normal vector field M of constant length L can be written as

$$M = L(\cos\theta e_2 + \sin\theta e_3).$$

where θ is a differentiable function of t.

18. Use (2.35) to obtain a differential equation for θ so as to prove the local existence of a relatively parallel normal vector field with a given initial value.

A ribbon whose (unit) normal is relatively parallel will be called a **Bishop** ribbon and its associated frame field will be called a **Bishop** frame field. The special properties of such frame field are that in (2.35) the function τ vanishes identically. Following Bishop's notation, we will write a Bishop frame field as (T, M_1, M_2). From the above discussion it follows that Bishop frame fields always exist, and unique up to a transformation of the form

$$N_1 = aM_1 + bM_2, \quad N_2 = cM_1 + dM_2$$

where $\begin{pmatrix} a & b \\ c & d \end{pmatrix}$ is a constant orthogonal matrix. The equations (2.35) for a Bishop frame field take the form

$$T' = k_1 M_1 + k_2 M_2, \quad M_1' = -k_1 T, \quad M_2' = -k_2 T. \qquad (2.39)$$

So we get a curve $t \mapsto (k_1(t), k_2(t))$ in the plane, determined up to a constant orthogonal transformation of the plane (or up to a constant rotation of the plane in the oriented case).

Some properties of the curve C have nice expressions in terms of the curve (k_1, k_2). For example,

Theorem 3. *The curve C lies on a sphere if and only if the curve (k_1, k_2) lies on a line not through the origin (for some and hence any) Bishop frame. The distance of this line from the origin is the reciprocal of the radius of the sphere.*

Proof. Suppose that C lies on a sphere with center P and radius r so

$$((C-P),(C-P)) \equiv r^2.$$

Differentiating with respect to arc length gives

$$(T,(C-P)) \equiv 0$$

so

$$C - P = fM_1 + gM_2$$

for functions f and g where

$$f = ((C-P), M_1), \quad g = ((C-P), M_2).$$

So

$$f' = ((C-P), M_1)' = (T \cdot M_1) + ((C-P), (-k_1 T)).$$

The first term vanishes since M_1 is normal to the curve. The second term vanishes since T is tangent to the sphere and hence perpendicular to the radius. Hence f is constant, and similarly so is g. So $C - P = fM_1 + gM_2$ with f and g constant. Differentiating $((C-P), T) \equiv 0$ gives

$$((k_1 M_1 + k_2 M_2), (C-P)) + (T,T) = fk_1 + gk_2 + 1 \equiv 0.$$

2.14. PROBLEMS.

In other words, (k_1, k_2) lies on the line $fx + gy + 1 = 0$. Furthermore
$$\|C - P\|^2 = f^2 + g^2 = 1/d^2$$
where d is the distance from this line to the origin.

Conversely, suppose that $fk_1 + gk_2 + 1 = 0$ with f, g constant. Let $P := C - fM_1 - gM_2$. Then
$$P' = T + (fk_1 + gk_2)T = (1 + fk_1 + gk_2)T = 0$$
so P is constant. Differentiating $((C - P), (C - P))$ gives
$$(2T, (fM_1 + gM_2)) = 0$$
so C lies on a sphere of radius $r = \sqrt{f^2 + g^2}$ about P

□

Suppose that C satisfies the Frenet condition (2.37). We examine the relation between the Frenet frame (T, N, B) and a Bishop frame (T, M_1, M_2). The principal curvature is defined by $\kappa = \|T'\|$. So
$$\kappa = \|k_1 M_1 + k_2 M_2\| = (k_1^2 + k_2^2)^{\frac{1}{2}}.$$
Thus κ is the (radial) distance from the origin to (k_1, k_2). Write
$$N = \cos\theta M_1 + \sin\theta M_2,$$
$$B = -\sin\theta M_1 + \cos\theta M_2,$$
where θ is the angle from the x-axis to (k_1, k_2). The torsion τ is defined by $N' = -\kappa T + \tau B$. So differentiating $N = \cos\theta M_1 + \sin\theta M_2$ shows that
$$\tau = \theta'.$$
Hence κ and an indefinite integral $\int \tau ds$ are polar coordinates for (k_1, k_2). The freedom of choice of constant in this indefinite integral illustrates the fact that (k_1, k_2) is determined only up to a (constant) rotation in the plane.

Notice also that $\tau \equiv 0$ means that θ is a constant. So that the condition that C lie in a plane is equivalent to the condition that (k_1, k_2) lie on line through the origin (which is the "limiting version" of the condition that C lie on a sphere obtained above).

2.14.18 Summary in more logical order, starting from Weil's formula.

I now summarize some of what we have done in this problem set in decreasing order of abstraction or implication, starting at the top of the food chain with Weil's formula:

- **Weil's formula:**
$$L_X\Omega = i(X)d\Omega + di(X)\Omega$$
for any (possibly vector valued) differential form Ω and any vector field X on any manifold.

⇓

- As a consequence (obtained by computing $L_X(\sigma(Y))$) for any (possibly vector valued) linear differential form σ
$$(d\sigma)(X,Y) = X(\sigma(Y)) - Y(\sigma(X)) - \sigma([X,Y])$$
for any pair of vector fields X and Y on any manifold.

⇓

- On any Lie group G the (left) Maurer-Cartan form ω is the linear differential form with values in the Lie algebra $\mathfrak{g}$ which assigns to any $v \in TG_a$ the element $\omega(v) = dL_{a^{-1}}v \in \mathfrak{g} = TG_e$. The preceding item applied to ω yields the **Maurer-Cartan equation**
$$d\omega + \frac{1}{2}[\omega \wedge, \omega] = 0.$$

⇓

- For linear Lie groups, i.e. for Lie subgroups of $Gl(n)$ we have
$$\frac{1}{2}[\omega \wedge, \omega] = \omega \wedge \omega$$
so the Maurer-Cartan equation can be written as
$$d\omega + \omega \wedge \omega = 0.$$

- If we take G to be the group of Euclidean motions, we obtain the equations of structure of Euclidean space (2.18)-(2.20) on the space of all Euclidean frames.

⇓

- We may consider the frames adapted to a submanifold and so obtain the equations of structure of a submanifold of Euclidean space.

⇓

- Applied to a ribbon we get the concept of developing a ribbon in $\mathbb{R}^3$ onto a plane.

2.14. PROBLEMS.

- All regular curves in $\mathbb{R}^3$ have a preferred family of attached ribbons - the Bishop ribbons. The equations of structure of Euclidean three space restricted to such a ribbon give the Bishop equations (2.39). A curve lies on a sphere if and only if the corresponding curve (k_1, k_2) lies on a line not passing through the origin.

$$\Downarrow$$

- If the curvature κ of C never vanishes, the curve has a Frenet ribbon and the structure equations become the Frenet equations (2.38).

$$\Downarrow$$

- The relation between the (k_1, k_2) of the Bishop ribbon and the (κ, τ) of the Frenet ribbon is that κ and an indefinite integral $\int \tau ds$ are polar coordinates for (k_1, k_2).

Chapter 3

Connections on the tangent bundle.

3.1 Definition of a linear connection on the tangent bundle.

A **linear connection** ∇ on a manifold M is a rule which assigns a vector field $\nabla_X Y$ to each pair of vector fields X and Y which is bilinear (over $\mathbb{R}$) subject to the rules

$$\nabla_{fX} Y = f \nabla_X Y \tag{3.1}$$

and

$$\nabla_X (gY) = (Xg)Y + g(\nabla_X Y). \tag{3.2}$$

While condition (3.2) is the same as the corresponding condition

$$L_X(gY) = [X, gY] = (Xg)Y + g L_X Y$$

for Lie derivatives, condition (3.1) is quite different from the corresponding formula

$$L_{fX} Y = [fX, Y] = -(Yf)X + f L_X Y$$

for Lie derivatives. In contrast to the Lie derivative, condition (3.1) implies that the value of $\nabla_X Y$ at $x \in M$ depends only on the value $X(x)$.

If $\xi \in TM_x$ is a tangent vector at $x \in M$, and Y is a vector field defined in some neighborhood of x we use the notation

$$\nabla_\xi Y := (\nabla_X Y)(x), \quad \text{where } X(x) = \xi. \tag{3.3}$$

By the preceding comments, this does not depend on how we choose to extend ξ to X so long as $X(x) = \xi$.

Notational difference. Notice that I am using ∇ (which is relatively standard notation for a connection) while O'Neill uses D for a connection.

While the Lie derivative is an intrinsic notion depending only on the differentiable structure, a connection is an additional piece of geometric structure.

3.2 Christoffel symbols.

These give the expression of a connection in local coordinates: Let $x^1, \ldots, x^n$ be a coordinate system, and let us write

$$\partial_i := \frac{\partial}{\partial x^i}$$

for the corresponding vector fields. Then

$$\nabla_{\partial_i} \partial_j = \sum_k \Gamma_{ij}^k \partial_k$$

where the functions Γ_{ij}^k are called the **Christoffel symbols**. We will frequently use the shortened notation

$$\nabla_i := \nabla_{\partial_i}.$$

So the definition of the Christoffel symbols is written as

$$\nabla_i \partial_j = \sum_k \Gamma_{ij}^k \partial_k. \tag{3.4}$$

If

$$Y = \sum_j Y^j \partial_j$$

is the local expression of a general vector field Y then (3.2) implies that

$$\nabla_i Y = \sum_k \left\{ \frac{\partial Y^k}{\partial x^i} + \sum_j \Gamma_{ij}^k Y^j \right\} \partial_k. \tag{3.5}$$

3.3 Parallel transport.

Let $C : I \to M$ be a smooth map of an interval I into M. We refer to C as a parameterized curve. We will say that this curve is non-singular if $C'(t) \neq 0$ for any t where $C'(t)$ denotes the tangent vector at $t \in I$. By a **vector field** Z **along** C we mean a rule which smoothly attaches to each $t \in I$ a tangent vector $Z(t)$ to M at $C(t)$. We will let $\mathcal{V}(C)$ denote the set of all smooth vector fields along C. For example, if V is a vector field on M, then the restriction of V to C, i.e. the rule

$$V_C(t) := V(C(t))$$

is a vector field along C. Since the curve C might cross itself, or be closed, it is clear that not every vector field along C is the restriction of a vector field.

3.3. PARALLEL TRANSPORT.

On the other hand, if C is non-singular, then the implicit function theorem says that for any $t_0 \in I$ we can find an interval J containing t_0 and a system of coordinates about $C(t_0)$ in M such that in terms of these coordinates the curve is given by
$$x^1(t) = t, \ x^i(t) = 0, \ i > 1$$
for $t \in J$. If Z is a smooth vector field along C then for $t \in J$ we may write
$$Z(t) = \sum_j Z^j(t) \partial_j(t, 0, \ldots, 0).$$
We may then define the vector field Y on this coordinate neighborhood by
$$Y(x^1, \ldots, x^n) = \sum_j Z^j(x^1) \partial_j$$
and it is clear that Z is the restriction of Y to C on J. In other words, *locally*, every vector field along a non-singular curve *is* the restriction of a vector field of M. If $Z = Y_C$ is the restriction of a vector field Y to C we can define its "derivative" Z', also a vector field along C by
$$Y'_C(t) := \nabla_{C'(t)} Y. \tag{3.6}$$
If g is a smooth function defined in a neighborhood of the image of C, and h is the pull back of g to I via C, so
$$h(t) = g(C(t))$$
then the chain rule says that
$$h'(t) = \frac{d}{dt} g(C(t)) = C'(t)g,$$
the derivative of g with respect to the tangent vector $C'(t)$. Then if
$$Z = Y_C$$
for some vector field Y on M (and $h = g(C(t))$) equation (3.2) implies that
$$(hZ)' = h'Z + hZ'. \tag{3.7}$$

Proposition 2. *There is a unique linear map $Z \mapsto Z'$ defined on all of $\mathcal{V}(C)$ such that (3.7) and (3.6) hold.*

Proof. To prove uniqueness, it is enough to prove uniqueness in a coordinate neighborhood, where
$$Z(t) = \sum_j Z^j(t)(\partial_j)_C(t).$$
Equations (3.7) and (3.6) then imply that
$$Z'(t) = \sum_j \left(Z^{j'}(t)(\partial_j)_C + Z^j(t) \nabla_{C'(t)} \partial_j \right). \tag{3.8}$$

In other words, any notion of "derivative along C" satisfying (3.7) and (3.6) must be given by (3.8) in any coordinate system. This proves the uniqueness. On the other hand, it is immediate to check that (3.8) satisfies (3.7) and (3.6) if the curve lies entirely in a coordinate neighborhood. But the uniqueness implies that on the overlap of two neighborhoods the two formulas corresponding to (3.8) must coincide, proving the global existence. □

This method of proof is a manifestation of a general principle - "uniqueness + local existence" implies "global existence".

We can make formula (3.8) even more explicit in local coordinates using the Christoffel symbols which tell us that

$$\nabla_{C'(t)} \partial_j = \sum_k \Gamma_{ij}^k \frac{dx^i \circ C}{dt} (\partial_k)_C.$$

Substituting into (3.8) gives

$$Z' = \sum_k \left(\frac{dZ^k}{dt} + \sum_{ij} \Gamma_{ij}^k \frac{dx^i \circ C}{dt} Z^j \right) (\partial_k)_C. \tag{3.9}$$

3.4 Parallel vector fields along a curve.

A vector field Z along C is said to be **parallel** if

$$Z'(t) \equiv 0.$$

Locally this amounts to the Z^i satisfying the system of linear differential equations

$$\frac{dZ^k}{dt} + \sum_{ij} \Gamma_{ij}^k \frac{dx^i \circ C}{dt} Z^j = 0. \tag{3.10}$$

Hence the existence and uniqueness theorem for linear homogeneous differential equations (in particular existence over the entire interval of definition) implies that

Proposition 3. *For any $\zeta \in TM_{C(0)}$ there is a unique parallel vector field Z along C with $Z(0) = \zeta$.*

On a general manifold the concept of parallelism for two tangent vectors at different points makes no sense. When we are given a connection, this gives us the notion of a vector field being parallel along a curve.

The rule $t \mapsto C'(t)$ is a vector field along C and hence we can compute its derivative, which we denote by C'' and call the **acceleration** of C. Whereas the notion of tangent vector, C', makes sense on any manifold, the acceleration only makes sense when we are given a connection.

So to formulate some variant of Newton's equation $\mathbf{F} = m\mathbf{a}$ on a manifold we need the concept of a connection.

3.5 Geodesics.

A curve with acceleration zero is called a **geodesic**. In local coordinates we substitute
$$Z^k = \frac{dx^k}{dt}$$
into (3.10) to obtain the equation for geodesics in local coordinates:
$$\frac{d^2 x^k}{dt^2} + \sum_{ij} \Gamma^k_{ij} \frac{dx^i}{dt} \frac{dx^j}{dt} = 0, \qquad (3.11)$$

where we have written x^k instead of $x^k \circ C$ in (3.11) to unburden the notation. The existence and uniqueness theorem for ordinary differential equations implies that

Proposition 4. *For any tangent vector ξ at any point $x \in M$ there is an interval I about 0 and a unique geodesic C such that $C(0) = x$ and $C'(0) = \xi$.*

By the usual arguments we can then extend the domain of definition of the geodesic through ξ to be maximal.

This is the first of many definitions (or characterizations, if take this to be the basic definition) that we shall have of geodesics - the notion of being self-parallel. (In the case that all the $\Gamma^k_{ij} = 0$ we get the equations for straight lines.)

How much freedom do we have in the parametrization of a geodesic?

Suppose that $C : I \to M$ is a (non-constant) geodesic, and we consider a "reparametrization" of C, i.e. consider the curve $B = C \circ h : J \to M$ where $h : J \to I$ is a diffeomorphism of the interval J onto the interval I. We write $t = h(s)$ so that
$$\frac{dB}{ds} = \frac{dC}{dt} \frac{dh}{ds}$$
and hence
$$\frac{d^2 B}{ds^2} = \frac{d^2 C}{dt^2} \left(\frac{dh}{ds}\right)^2 + \frac{dC}{dt} \frac{d^2 h}{ds^2} = \frac{dC}{dt} \frac{d^2 h}{ds^2}$$
since $C'' = 0$ as C is a geodesic. The fact that C is not constant (and the uniqueness theorem for differential equations) says that C' is never zero. Hence B is a geodesic if and only if
$$\frac{d^2 h}{ds^2} \equiv 0$$
or
$$h(s) = as + b$$
where a and b are constants with $a \neq 0$. In short, the fact of being a non-constant geodesic determines the parameterization up to an affine change of parameter.

3.5.1 Making a linear change of parameter.

Suppose we take $h(s) = as$. Then B is a geodesic with $B(0) = C(0)$ and $B'(0) = aC'(0)$. So the uniqueness theorem of differential equations tells us that B is the unique geodesic with $B(0) = C(0)$ and $B'(0) = aC'(0)$. So we can improve on our proposition as follows:

Proposition 5. *For any tangent vector ξ at any point $x \in M$ there is an interval I about 0 and a unique geodesic C such that $C(0) = x$ and $C'(0) = \xi$. The geodesic B with $B(0) = x$ and $B'(0) = a\xi$ is given by*

$$B(t) = C(at).$$

3.6 Torsion.

Recall the equations

$$\nabla_{fX} Y = f \nabla_X Y, \tag{3.1}$$

and

$$\nabla_X (gY) = (Xg)Y + g(\nabla_X Y). \tag{3.2}$$

These give

$$\nabla_{fX}(gY) - \nabla_{gY}(fX) = fg(\nabla_X Y - \nabla_Y X) + (fXg)Y - g(Yf)X. \quad (*)$$

The equations

$$L_X(gY) = [X, gY] = (Xg)Y + gL_X Y,$$

and

$$L_{fX} Y = [fX, Y] = -(Yf)X + fL_X Y$$

for Lie derivatives give

$$[fX, gY] = fg[X,Y] + f(Xg)Y - g(Yf)X.$$

Subtracting this last equation from $(*)$ shows that

$$\nabla_{fX}(gY) - \nabla_{gY}(fX) - [fX, gY] = fg(\nabla_X Y - \nabla_Y X - [X,Y]).$$

In other words, if we define the map $\tau : \mathcal{V}(M) \times \mathcal{V}(M) \to \mathcal{V}(M)$ by

$$\tau(X,Y) := \nabla_X Y - \nabla_Y X - [X,Y]$$

then

$$\tau(fX, gY) = fg\tau(X,Y).$$

Thus the value of $\tau(X,Y)$ at any point x depends only on the values $X(x), Y(x)$ of the vector fields at x, and is a bilinear function over $\mathbb{R}$ of these values. So τ defines a tensor field of type (1,2) in the sense that it assigns to any pair of

3.7. CURVATURE.

tangent vectors at a point, a third tangent vector at that point. This tensor field is called the **torsion tensor** of the connection. See Section 3.8 below for a more detailed discussion of tensors and tensor fields

So a connection has *zero torsion* if and only if

$$\nabla_X Y - \nabla_Y X = [X, Y] \tag{3.12}$$

for all pairs of vector fields X and Y.

In terms of local coordinates, $[\partial_i, \partial_j] = 0$. So

$$\tau(\partial_i, \partial_j) = \nabla_i \partial_j - \nabla_j \partial_i = \sum_k \left(\Gamma_{ij}^k - \Gamma_{ji}^k \right) \partial_k.$$

Thus a connection has zero torsion if and only if its Christoffel symbols are symmetric in i and j.

In this book the only connections we will deal with in any serious manner are ones with zero torsion.

3.7 Curvature.

The **curvature** $R = R(\nabla)$ of the connection ∇ is defined to be the map $\mathcal{V}(M)^3 \to \mathcal{V}(M)$ assigning to three vector fields X, Y, Z the value

$$R_{XY} Z := [\nabla_X, \nabla_Y] Z - \nabla_{[X,Y]} Z. \tag{3.13}$$

Warning! Notice that this differs by a sign from the definition given in O'Neill on page 74. There are two competing conventions in the literature. Each has its advantages.

The expression $[\nabla_X, \nabla_Y]$ occurring on the right in (3.13) is the commutator of the two operators ∇_X and ∇_Y, that is $[\nabla_X, \nabla_Y] = \nabla_X \circ \nabla_Y - \nabla_Y \circ \nabla_X$.

3.7.1 R is a tensor.

We first observe that R is a tensor, i.e. that the value of $R_{XY}Z$ at a point depends only on the values of X, Y, and Z at that point and is trilinear over $\mathbb{R}$ as a function of these values. By its definition and the properties of ∇, it is clear that $R_{XY}Z$ is trilinear over $\mathbb{R}$ in , Y and Z. To see that the value of $R_{XY}Z$ at a point depends only on the values of X, Y, and Z at that point, we must show that

$$R_{fX gY} hZ = fgh R_{XY} Z$$

for any three smooth functions f, g and h. For this it suffices to check this one at a time, i.e. when two of the three functions are identically equal to one. For example, if $f \equiv 1 \equiv h$ we have

$$\begin{aligned} R_{X, gY} Z &= \nabla_X \nabla_{gY} Z - \nabla_{gY} \nabla_X Z - \nabla_{[X, gY]} Z \\ &= (Xg) \nabla_Y Z + g \nabla_X \nabla_Y Z - g \nabla_Y \nabla_X Z - (Xg) \nabla_Y Z - g \nabla_{[X,Y]} Z \\ &= g R_{XY} Z. \end{aligned}$$

Since R is anti-symmetric in X and Y we conclude that $R_{fXY}Z = fR_{XY}Z$. Finally,

$$\begin{aligned}R_{XY}(hZ) &= \nabla_X((Yh)Z + h\nabla_Y Z) - \nabla_Y((Xh)Z + h\nabla_X Z) - ([X,Y]h)Z - h\nabla_{[X,Y]}Z \\ &= hR_{[X,Y]}Z + (X(Yh) - Y(Xh) - [X,Y]h)Z + (Yh)\nabla_X Z + (Xh)\nabla_Y Z \\ &\quad - (Xh)\nabla_Y Z - (Yh)\nabla_X Z \\ &= hR_{[X,Y]}Z.\end{aligned}$$

Thus we get a curvature **tensor** (of type (1,3)) which assigns to every three tangent vectors ξ, η, ζ at a point x the value

$$R_{\xi\eta}\zeta := (R_{XY}Z)(x)$$

where X, Y, Z are any three vector fields with $X(x) = \xi, Y(x) = \eta, Z(x) = \zeta$. Alternatively, we speak of the curvature **operator** at the point x defined by

$$R_{\xi\eta} : TM_x \to TM_x, \quad R_{\xi\eta} : \zeta \mapsto R_{\xi\eta}\zeta.$$

As we mentioned, the curvature operator is anti-symmetric in ξ and η:

$$R_{\xi\eta} = -R_{\eta\xi}.$$

The classical expression of the curvature tensor in terms of the Christoffel symbols is obtained as follows: Since $[\partial_k, \partial_\ell] = 0$,

$$\begin{aligned}R_{\partial_k \partial_\ell}\partial_j &= \nabla_k(\nabla_\ell \partial_j) - \nabla_\ell(\nabla_k \partial_j) \\ &= -\nabla_\ell \left(\sum_m \Gamma^m_{kj}\partial_m\right) + \nabla_k\left(\sum_m \Gamma^m_{\ell j}\partial_m\right) \\ &= -\sum_m \left(\frac{\partial}{\partial x^\ell}\Gamma^m_{kj}\partial_m + \sum_r \Gamma^m_{kj}\Gamma^r_{\ell m}\partial_r\right) + \sum_m\left(\frac{\partial}{\partial x^k}\Gamma^m_{\ell j}\partial_m + \sum_r \Gamma^m_{\ell j}\Gamma^r_{km}\right)\partial_r \\ &= \sum_i R^i_{jk\ell}\partial_i\end{aligned}$$

where

$$R^i_{jk\ell} = -\frac{\partial}{\partial x^\ell}\Gamma^i_{kj} + \frac{\partial}{\partial x^k}\Gamma^i_{\ell j} - \sum_m \Gamma^i_{\ell m}\Gamma^m_{kj} + \sum_m \Gamma^i_{km}\Gamma^m_{\ell j}. \qquad (3.14)$$

3.7.2 The first Bianchi identity.

If the connection has zero torsion we claim that

$$R_{\xi\eta}\zeta + R_{\eta\zeta}\xi + R_{\zeta\xi}\eta = 0, \qquad (3.15)$$

or, using the cyclic sum notation we introduced with the Jacobi identity, that

$$\mathcal{C}yc\, R_{\xi\eta}\zeta = 0.$$

Proof. We may extend ξ, η, and ζ to vector fields whose brackets all commute (say by using vector fields with constant coefficients in a coordinate neighborhood). Then
$$R_{XY}Z = \nabla_X \nabla_Y Z - \nabla_Y \nabla_X Z.$$
Therefore
$$\begin{aligned}\mathcal{C}yc\, R_{XY}Z &= \mathcal{C}yc\nabla_X\nabla_Y Z - \mathcal{C}yc\nabla_Y\nabla_X Z \\ &= \mathcal{C}yc\nabla_X\nabla_Y Z - \mathcal{C}yc\nabla_X\nabla_Z Y\end{aligned}$$
since making a cyclic permutation in an expression $\mathcal{C}yc\, F(X, Y, Z)$ does not affect its value. But the fact that the connection is torsion free means that we can write the last expression as
$$\mathcal{C}yc\, \nabla_X[Y, Z] = 0$$
by our assumption that all Lie brackets vanish. $\square$

Equation (3.15) is known is the **first Bianchi identity**. It is a consequence of the assumption that the connection is torsion free.

3.8 Tensors and tensor analysis.

In this section (following O'Neill, chapter 2) I discuss the effects of the existence of a connection on the algebra of all tensor fields on a manifold. In the nineteenth century (and today in most of the physics literature) this was (or is) a mainstay of the study of Riemannian geometry and of general relativity.

3.8.1 Multilinear functions and tensors of type (r,s).

Let $\mathfrak{F}(M)$ denote the space of smooth functions on a smooth manifold. It is a commutative associative ring with unit, the unit being the constant function **1**.

Recall that $\mathcal{V}(M)$ denotes the space of smooth vector fields on M. It is a **module** over $\mathfrak{F}(M)$ in the sense that we can multiply a smooth vector field by a smooth function to get another smooth vector field, and this multiplication satisfies the usual expected identities:

- $(fg)X = f(g(X))$, an associative law,
- $(f + g)X = fX + gX$, a distributive law,
- $1X = X$, the unit acts as the identity,
- $f(X + Y) = fX + fY$, another distributive law.

A (smooth) linear differential form θ defines (by evaluation) a $\mathfrak{F}(M)$ linear map $X \mapsto \langle \theta, X \rangle = i(X)\theta$ from $\mathcal{V}(M)$ to $\mathfrak{F}(M)$. To say that this map is $\mathfrak{F}(M)$ linear means that
$$\langle \theta, fX + gY \rangle = f\langle \theta, X \rangle + g\langle \theta, Y \rangle, \quad \forall X, Y \in \mathcal{V}(M),\ f, g \in \mathfrak{F}(M).$$

Conversely, any such $\mathfrak{F}(M)$ linear map is given by a (smooth) linear differential form as can be checked in any coordinate neighborhood. So we may (temporarily - to follow O'Neill) denote the space of smooth linear differential forms by $\mathcal{V}(M)^*$.

We can generalize this as follows: Let K be a commutative ring with unit. We let V be a module over K which means that V is a commutative group (under a binary operation denoted by $+$) and a map $K \times V \to V$, $(f, X) \mapsto fX$ satisfying the above identities. The examples we have in mind are $K = \mathfrak{F}(M)$ and $V = \mathcal{V}(M)$ as above, and the case where $K = \mathbb{R}$ and V is a (finite dimensional) vector space over $\mathbb{R}$, in particular, where M is a differentiable manifold, $p \in M$ and $V = T_pM$ is the tangent space to M at p. Let V^* denote the set of all K-linear functions from V to K. The usual definition of addition of functions and multiplication by elements of K makes V^* into a module over K.

Let $V_1, \ldots, V_q$ be modules over the ring K. We make their direct product $V_1 \times \cdots \times V_q$ into a module over K by component-wise definitions of addition and multiplication by elements of K. Let W be another module over K. A function
$$A: V_1 \times \cdots \times V_q \to W$$
is said to be K-**multilinear** if it is linear in each slot when all the other slots are held fixed.

Example - the torsion. Recall that we defined the torsion of a connection ∇ on a manifold by $\tau: \mathcal{V}(M) \times \mathcal{V}(M) \to \mathcal{V}(M)$ by
$$\tau(X, Y) := \nabla_X Y - \nabla_Y X - [X, Y]$$
and proved that it is linear in X over $\mathfrak{F}(M)$ when Y is fixed, and is linear in Y over $\mathfrak{F}(M)$ in Y when X is fixed.

So the map $A: \mathcal{V}(M)^* \times \mathcal{V}(M) \times \mathcal{V}(M) \to \mathfrak{F}(M)$ sending (θ, X, Y) to
$$A(\theta, X, Y) := \langle \theta, \tau(X, Y) \rangle$$
is $\mathfrak{F}(M)$ trilinear.

Let V be a module over a ring K.

Definition 1. *For integers (r, s) not both zero a K-multilinear functon*
$$A: (V^*)^r \times V^s \to K$$
is called a **tensor of type** (r, s) **over** V. *A* **tensor of type** $(0, 0)$ *is defined to be an element of K*

The set $\mathfrak{T}^r_s(V)$ of all tensors of type (r, s) over V is again a module over K with the usual definitions of addition of functions and multiplication of a function by an element of K.

3.8. TENSORS AND TENSOR ANALYSIS.

So we can think of the torsion as a tensor of type $(1,2)$ over $\mathcal{V}(M)$. Similarly, we can think of the curvature as a tensor of type $(1,3)$ over $\mathcal{V}(M)$.

A tensor A of type (r,s) over $\mathcal{V}(M)$ is usually called a **tensor field** of type (r,s) on M. We will soon see why. It is a multilinear machine which, when fed in r one forms $\theta^1,\ldots,\theta^r$ and s vector fields $X_1,\ldots,X_s$, produces a function

$$f = A(\theta^1,\ldots,\theta^r, X_1,\ldots,X_s).$$

A tensor field of type $(0,0)$ is simply a smooth function on M.

When we are given a map

$$A : (\mathcal{V}(M)^*)^r \times (\mathcal{V}(M))^s \to \mathfrak{F}(M).$$

the additivity in each slot when the others are held fixed is usually obvious. What is crucial is that functions can be factored out of each slot separately, e.g. that, for example

$$A(\theta^1,\ldots,\theta^r, X_1,\ldots,fX_i,\ldots X_s) = fA(\theta^1,\ldots,\theta^r, X_1,\ldots,X_i,\ldots X_s).$$

We have encountered this in the case of the torsion and of the curvature of a connection.

3.8.2 Tensor multiplication.

If A is a tensor of type (r,s) over V and B is a tensor of type (r',s') we define their tensor product $A \otimes B$, tensor of type $(r+r', s+s')$, by

$$(A \otimes B)(\theta^1,\ldots,\theta^{r+r'}, X_1,\ldots X_{s+s'})$$
$$:= A(\theta^1,\ldots,\theta^r, X_1,\ldots X_s)B(\theta^{r+1},\ldots,\theta^{r+r'}, X_{s+1},\ldots X_{s+s'}).$$

Notice that the order matters!

If A or B is of type $(0,0)$ then tensor multiplication reduces to ordinary multiplication by an element of K.

If $F : V^s \to V$ is K-multilinear, we produce a tensor A of type $(1,s)$ by setting

$$A(\theta, X_1,\ldots X_s) := \theta(F(X_1,\ldots X_s)).$$

This is how we made the torsion into a tensor field of type $(1,2)$ and the curvature into a tensor field of type $(1,3)$.

3.8.3 Why a tensor over $\mathcal{V}(M)$ is a "field" of tensors over M.

Let A be a tensor of type (r,s) over $\mathcal{V}(M)$ and let $\theta_1,\ldots\theta_r$ be one forms and $X_1,\ldots,X_s$ be vector fields. Let $p \in M$.

Lemma 1. *If any one of the one forms or vector fields vanishes at p then*
$$A(\theta_1, \ldots \theta_r, X_1, \ldots, X_s)(p) = 0.$$

Before proving the lemma, I need to prove a sublemma (which we will use often) about the existence of what O'Neill calls "bump functions". Namely,

Bump functions.

Given any neighborhood U of a point $p \in M$ there is an $f \in \mathfrak{F}(M)$ (called a **bump function at** p) such that

- $0 \leq f \leq 1$,
- $f \equiv 1$ in some neighborhood of p,
- The closure of the set where $f > 0$ is contained in U.

Proof of the existence of bump functions. The function f on $\mathbb{R}$ defined by
$$f(u) := \begin{cases} e^{-1/u} & \text{if } u > 0 \\ 0 & \text{if } u \leq 0 \end{cases}$$
is C^∞. Indeed, for $u \neq 0$ it is clear that f has derivatives of all orders. To check that f is C^∞ at 0, it suffices to show that $f^{(k)}(u) \to 0$ as $u \to 0$ from the right. But for $u > 0$ we have
$$f^{(k)}(u) = P_k(1/u) e^{-1/u}$$
where P_k is a polynomial of degree $2k$ and
$$\lim_{u \searrow 0} f^{(k)}(u) = \lim_{s \to \infty} P_k(s) e^{-s} = 0$$
since e^s goes to infinity faster than any polynomial.

So the function g given by $g(x) := f(x+2) \cdot f(2-x)$ satisfies $g(x) > 0$ for $|x| < 2$ and $g \equiv 0$ for $|x| \geq 2$. Let k be the function $k(x) = g(x+3) + g(x) + g(x-3)$. Then $k \geq 0$, $k(x) = 0$ for $|x| \geq 5$, $k(x) > 0$ for $|x| < 5$, and $k(x) = g(x)$ for $|x| < 1$. So h defined by
$$h(x) = \begin{cases} \frac{g(x)}{k(x)} & \text{for } |x| < 2 \\ 0 & \text{for } |x| \geq 2 \end{cases}$$
is a C^∞ function on $\mathbb{R}$ which vanishes for $|x| \geq 2$, is identically 1 for $|x| < 1$ and $0 \leq h \leq 1$ everywhere.

Now let $p \in M$ and U a neighborhood of p. We may choose coordinates about p such that p has coordinates 0 and that the "ball" B given by $x_1^2 + \cdots + x_n^2 < 3$ is contained in U. Then (recycling the letter f), setting
$$f(q) = h(x_1^2(q) + \cdots + x_n^2(q))$$
if $x_1^2(q) + \cdots + x_n^2(q) < 3$ and extending f to be identically zero outside B, we obtain a bump function, proving the existence of bump functions.

Now back to the proof of the lemma.

3.8. TENSORS AND TENSOR ANALYSIS.

Proof. Suppose, say, that $X_s(p) = 0$. Let $x^1, \ldots, x^n$ be a system of coordinates about p. We can write

$$X_s = \sum_i X^i \partial_i$$

in a neighborhood of p where the X^i are functions defined in this neighborhood. Choose a bump function at p which vanishes outside this neighborhood. Then the functions fX^i extend, by zero, to be defined on all of M, and the vector fields $f\partial_i$ also extend, by zero to be defined as vector fields on all of M. So

$$f^2 A(\theta_1, \ldots \theta_r, X_1, \ldots, X_s) = A\left((\theta_1, \ldots \theta_r, X_1, \ldots, \sum_i (fX^i) f \partial_i \right)$$

$$= \sum_i (fX^i) A(\theta_1, \ldots \theta_r, X_1, \ldots, f\partial_i).$$

Since $X_s(p) = 0$, each of the functions X^i vanishes at p so the last expression above vanishes. On the other hand $f^2(p) = 1$ showing from the first expression above that $A(\theta_1, \ldots \theta_r, X_1, \ldots, X_s)(p) = 0$. □

Proposition 6. *Suppose that* $\theta^1, \ldots, \theta^r, X_1, \ldots, X_s$ *and* $\overline{\theta}^1, \ldots, \overline{\theta}^r, \overline{X}_1, \ldots \overline{X}_s$ *are one forms and vectors fields which agree at* p. *Then*

$$A(\theta^1, \ldots, \theta^r, X_1, \ldots, X_s)(p) = A(\overline{\theta}^1, \ldots, \overline{\theta}^r, \overline{X}_1, \ldots \overline{X}_s)(p).$$

In other words, A assigns a tensor of type (r, s) over T_pM for each $p \in M$ and this assignment varies smoothly with p.

Proof. For clarity and simplicity, we may assume that $r = 1$ and $s = 2$ as, for example, in the case of the torsion. Then

$$A(\overline{\theta}, \overline{X}, \overline{Y}) - A(\theta, X, Y) =$$

$$A(\overline{\theta} - \theta, \overline{X}, \overline{Y}) + A(\theta, \overline{X} - X, \overline{Y}) + A(\theta, X, \overline{Y} - Y)$$

and each of the terms on the right vanishes at p by the lemma. It is clear that this "telescoping" argument works in general. □

3.8.4 Tensor notation.

Suppose, to fix the ideas, that $r = 1, s = 2$, and we have extended the vector fields ∂_k and dx^i defined near p (using a bump function at p) so as to be defined on all of M. Then

$$A(dx^k, \partial_i, \partial_j) =: A_{ij}^k$$

is a smooth function. Thus

$$A = \sum_{ijk} A_{ij}^k \partial_k dx^i dx^j$$

near p with a similar expression for arbitrary r and s.

Conversely, suppose that we are given smooth functions A^k_{ij} on each coordinate neighborhood, and that on the overlap of two coordinate neighborhoods the A^k_{ij} transform as derived from the transition rules for ∂_i and dx^j demand, then we get a tensor A of type $(1,2)$ over $\mathcal{V}(M)$. This is where the "index notation" of a tensor - as an array of indexed quantities transforming according to the above derived rules - so common in the physics and older mathematical literature comes from.

We say that A has covariant degree 1 and contravariant degree 2 with similar definitions for arbitrary r and s.

From the "field" point of view of tensors, it is clear that it makes sense to restrict a tensor to an arbitrary open subset of M.

3.8.5 Contraction.

There is an operation called **contraction** which shrinks an (r,s) tensor to an $(r-1, s-1)$ tensor. To explain it, we first deal with the $(1,1)$ contraction which shrinks tensors of type $(1,1)$ into functions:

Lemma 2. *There is a unique $\mathfrak{F}(M)$ linear function*

$$\mathbf{C} : \mathfrak{T}^1_1(M) \to \mathfrak{F}(M)$$

called $(1,1)$ contraction such that

$$\mathbf{C}(X \otimes \theta) = \langle \theta, X \rangle$$

for all one forms θ and vector fields X.

Proof. On a coordinate neighborhood, we must have

$$\mathbf{C}(dx^i, \partial_j) = \delta^i_j$$

so we must have

$$\mathbf{C}\left(\sum A^i_j \partial_i dx^j\right) = \sum A^i_i.$$

(By the way, the *Einstein convention* is that if there is a repeated upper and lower index then a summation is indicated. So, for example, the above equation would be written as

$$\mathbf{C}(A) = A^i_i.$$

We shall avoid the Einstein convention as much as possible.)

The $\mathbf{C}$ as defined above works on a coordinate neighborhood to fulfill the requirements of the lemma. So to prove the lemma, we must show that the above definition is consistent on overlaps of coordinate neighborhoods.

We have

$$A\left(dy^m, \frac{\partial}{\partial y^\ell}\right) = \sum_{m,\ell} A\left(\sum_i \frac{\partial y^m}{\partial x_i} dx^i, \sum_j \frac{\partial x^j}{\partial y_\ell} \frac{\partial}{\partial x^j}\right).$$

3.8. TENSORS AND TENSOR ANALYSIS.

Apply the contraction formula in the x coordinates to the right hand side and use the fact that the matrices

$$\left(\frac{\partial y^m}{\partial x_i}\right) \text{ and } \left(\frac{\partial x^j}{\partial y_\ell}\right)$$

are inverses of one another to conclude that we obtain

$$A\left(dy^m, \frac{\partial}{\partial y^\ell}\right) = \delta_\ell^m.$$

$\square$

In the definition of **C** in general, we must specify a contravariant position $1 \le i \le r$ and $1 \le j \le s$. If we hold the forms and vector fields that we feed into A in all other positions fixed, then A becomes a tensor of type $(1,1)$ in the i and j positions and we can apply the $(1,1)$ contraction as given by the lemma. The function of $r-1$ forms and $s-1$ vector fields that we so obtain is then a tensor of type $(r-1, s-1)$.

For example, if A is a tensor of type $(2,3)$ then $\mathbf{C}_3^1(A)$ is the $(1,2)$ tensor field given by

$$(\mathbf{C}_3^1(A))(\theta, X, Y) = \mathbf{C}(A(\cdot, \theta, X, Y, \cdot))$$

where the contraction on the right is the contraction given by the lemma relative to the positions indicated by the dots. In terms of local coordinates,

$$(\mathbf{C}_3^1 A)_{ij}^k = \sum_m A_{ijm}^{mk}$$

or, with the Einstein convention (which I will avoid whenever possible)

$$(\mathbf{C}_3^1 A)_{ij}^k = A_{ijm}^{mk}.$$

3.8.6 Tensor derivations.

A **tensor derivation** on M is a collection of $\mathbb{R}$ linear maps

$$\mathcal{D} = T_s^r \to T_s^r$$

which satisfies

- Leibnitz's rule in the form $\mathcal{D}(A \otimes B) = (\mathcal{D}(A)) \otimes B + A \otimes (\mathcal{D}(B))$ and

- Commutes with contraction in the sense that $\mathcal{D}(\mathbf{C}(A)) = \mathbf{C}(\mathcal{D}(A))$ for any cntraction **C**.

The case of Leibnitz's rule applied to tensors of type $(0,0)$, i.e. to functions, is the usual Leibnitz's rule, so this means that there is a vector field V such that $\mathcal{D}(f) = Vf$ for all $f \in \mathfrak{F}(M)$.

In particular, if f is identically one on an open set U, then $\mathcal{D}(fA) = f\mathcal{D}(A)$ on that open set. So by use of bump functions, it follows the $(\mathcal{D}(A))(p)$ depends only on the value of A in a neighborhood of p. But, in general, it will definitely depend on more than the value of A just at p.

Just as in elementary calculus, we can use various forms of the Leibnitz rule (combined with the commuting with contraction rule) to compute various tensor derivations. For example, let A be a tensor of type $(0,2)$. Then for any vector fields X and Y, we can regard $A(X,Y)$ as a "double contraction". Indeed, in local coordinates A is given as $\sum_{ij} A_{ij} dx^i dx^j$ while $X = \sum_k X^k \partial_k$ and $Y = \sum_\ell \partial_\ell$ so

$$A \otimes X \otimes Y = \sum_{ijk\ell} A_{ij} X^k Y^\ell dx^i dx^j \partial_k \partial_\ell$$

and

$$\mathbf{CC}(A \otimes X \otimes Y) = \sum_{ij} A^{ij} X_i Y_j = A(X,Y).$$

Since we can pass $\mathcal{D}$ through the contractions, we obtain, by Leibnitz's rule,

$$\mathcal{D}(A(X,Y)) = \mathbf{CC}\mathcal{D}(A \otimes X \otimes Y) =$$
$$\mathbf{CC}\left[(\mathcal{D}A) \otimes X \otimes Y + A \otimes \mathcal{D}(X) \otimes Y + A \otimes X \otimes \mathcal{D}(Y)\right]$$

yielding

$$[\mathcal{D}(A)](X,Y) = \mathcal{D}(A(X,Y)) - A(\mathcal{D}(X), Y) - A(X, \mathcal{D}(Y)). \quad (3.16)$$

In particular, the condition $\mathcal{D}(A) = 0$ is equivalent to the condition

$$\mathcal{D}[A(X,Y)] = A(\mathcal{D}(X), Y) + A(X, \mathcal{D}(Y)) \quad (3.17)$$

which itself looks like a version of Leibnitz's rule.

The above considerations applied to a general tensor shows that the tensor derivation $\mathcal{D}$ is completely determined by how it acts on functions, one forms, and vector fields. But if θ is a one form and X is a vector field, then Leibnitz tells us that

$$\langle \mathcal{D}\theta, X \rangle = \mathcal{D}(\langle \theta, X \rangle) - \langle \theta, \mathcal{D}(X) \rangle,$$

So

Proposition 7. *A tensor derivation is completely determined by how it acts on functions and vector fields.*

Conversely,

Theorem 4. *Given a vector field V and an $\mathbb{R}$ linear function $\delta : \mathcal{V}(M) \to \mathcal{V}(M)$ which satisfies Leibnitz's rule in the form*

$$\delta(fX) = (Vf) \cdot X + f\delta(X) \quad \text{for all} \quad f \in \mathfrak{F}(M), X \in \mathcal{V}(M) \quad (3.18)$$

then there exists a unique tensor derivation $\mathcal{D}$ which agrees with V on functions and with δ on vector fields.

3.8. TENSORS AND TENSOR ANALYSIS.

Proof. We already know the uniqueness. In fact we know that on one forms we must have
$$\langle \mathcal{D}\theta, X\rangle = V\langle \theta, X\rangle - \langle \theta, \delta(X)\rangle,$$
and it is clear that the right hand side is $\mathfrak{F}(M)$ linear in X and so defines a one form. Then extend $\mathcal{D}$ to arbitrary tensors by Leibniz's rule. It is then clear that we get an operator $\mathcal{D}$ on tensors which is local (i.e. commutes with restriction to open subsets) and satisfies Leibniz's rule.

We must show that it commutes with contractions. On $(1,1)$ tensors of the form $\theta \otimes X$ we defined $\mathcal{D}$ on one forms so that $\mathcal{D}$ commutes with contractions. So $\mathcal{D} \circ \mathbf{C} = \mathbf{C} \circ \mathcal{D}$ on tensors which are sums of terms of the form the form $\theta \otimes X$. Since $\mathcal{D}$ is local and $\mathcal{C}$ is pointwise, it suffices to prove $\mathcal{D} \circ \mathcal{C} = \mathcal{C} \circ \mathcal{D}$ on coordinate neighborhood. But in a coordinate neighborhood, every $(1,1)$ tensor can be written as a sum of terms each of which is a function multiple of $dx^i \otimes \partial_j$ for some i and j. This shows that $\mathcal{D}$ commutes with contraction on all tensors of type $(1,1)$.

But every contraction is really a contraction over one "upper" and one "lower" index, so the general case follows from the above. For example, suppose that A is a tensor of type $(1,2)$ and we perform contraction over the upper and second lower index so that $\mathbf{C}(A)$ is a linear differential form. Then a repeated application of Leibniz's rule (and with X fixed in the following computation) we have

$$\begin{aligned}
[\mathcal{D}(\mathbf{C}(A))](X) &= [\mathcal{D}\mathbf{C}][A](X) - \mathbf{C}(A)[\mathcal{D}(X)] \\
&= \mathcal{D}[\mathbf{C}(A(\cdot, X, \cdot)] - \mathbf{C}[A(\cdot, \mathcal{D}(X), \cdot)] \\
&= \mathbf{C}\left[\mathcal{D}(A(, \cdot, X, \cdot) - A(\cdot, \mathcal{D}(X), \cdot)\right] \\
&= (\mathbf{C} \circ \mathcal{D})(A))(X).
\end{aligned}$$

$\square$

3.8.7 A connection as a tensor derivation, covariant differential.

Suppose that M has a connection ∇. For a vector field V, $\delta = \nabla_V$ satisfies the condition Theorem 4. So ∇_V extends to a tensor derivation which we will continue to denote by ∇_V. So
$$\nabla_V(A \otimes B) = \nabla_V A \otimes B + A \otimes \nabla_V B$$
and
$$\mathbf{C} \circ \nabla_V = \nabla_V \circ \mathbf{C}.$$
We know that
$$\nabla_{fV} = f\nabla_V$$
in its action on functions and vector fields, and hence, by uniqueness, we know that this is true for its action on all tensors.

For any tensor A of type (r,s) the expression

$$\nabla_V A(\theta^1, \ldots, \theta^r, X_1, \ldots, X_s)$$

is thus multilinear in all the variables $\theta^1, \ldots, \theta^r, V, X_1, \ldots, X_s$) and thus we get a tensor of type $(r, s+1)$ which we will denote by ∇A. So ∇A is defined by

$$(\nabla A)(\theta^1, \ldots, \theta^r, V, X_1, \ldots, X_s) := (\nabla_V A)(\theta^1, \ldots, \theta^r, X_1, \ldots, X_s). \qquad (3.19)$$

(An alternative convention is to place the V in the last position). The operator ∇ is sometimes called the covariant differential.

In the classical literature the V is placed in the last position and is denoted by a comma: Thus if T is a tensor of type $(2,3)$ and so is described by

$$t^{ij}_{k\ell m}$$

its covariant differential, which is a tensor of type $(2,4)$ is denoted by

$$t^{ij}_{k\ell m, r}.$$

3.8.8 Covariant differential vs. exterior derivative.

Let σ be a differential form (say of degree k). We know that σ assigns a function to k vector fields (in an anti-symmetric manner). So it is a special kind of tensor of type $(0, k)$.

If ∇ is a linear connection, the covariant differential $\nabla \sigma$ of σ is a tensor of degree $(0, k+1)$ which is anti-symmetric with respect to the last k of its variables. By our definition (3.19),

$$\nabla \sigma(X, Y_1, \ldots, Y_k) = (\nabla_X \sigma)(Y_1, \ldots, Y_k)$$

which is anti-symmetric in the Y's.

So at each $m \in M$, $(\nabla \sigma)_m$ is an element of $T_m^* M \otimes \wedge^k(T_m^* M)$. The fact that ∇_X is a tensor derivation, tells us that

$$\nabla_X(\alpha \wedge \beta) = (\nabla_X \alpha) \wedge \beta + \alpha \wedge \nabla_X \beta.$$

Now we have (at each point) the exterior multiplication map

$$\epsilon : T^* M \otimes \wedge^k(T^* M) \to \wedge^{k+1}(T^* M)$$

and so we can form the operator

$$\epsilon \circ \nabla : \Omega^k(M) \to \Omega^{k+1}(M)$$

sending k-forms into $(k+1)$-forms. The above equation tells us that this is an odd derivation: If α is of degree a then

$$(\epsilon \circ \nabla)(\alpha \wedge \beta) = ((\epsilon \circ \nabla)\alpha) \wedge \beta + (-1)^a \alpha \wedge (\epsilon \circ \nabla)\beta.$$

3.8. TENSORS AND TENSOR ANALYSIS.

Proposition 8. *If the connection ∇ has torsion zero, then $(\epsilon \circ \nabla) = d$, the exterior differential.*

Both $(\epsilon \circ \nabla)$ and d are (odd) derivations of $\Omega(M)$ and both act the same way on functions, sending f into df. So we must check that they act the same way on linear differential forms, or what amounts to the same thing, that they act the same way on df for any f, which amounts to showing that

$$(\epsilon \circ \nabla)(df) = 0.$$

We will prove a more general fact, that

$$(\epsilon \circ \nabla)(df) = -\langle \tau, df \rangle$$

where τ is the torsion of ∇.

In this formula, τ assigns a vector field $\tau(X, Y)$ to every pair of vector fields X and Y and is anti-symmetric with respect to X and Y, and hence $\langle \tau, df \rangle$ assigns a function to every pair of vector fields X and Y, and is anti-symmetric in X and Y, in other words $\langle \tau, df \rangle$ is an exterior two form.

It suffices to prove this formula in local coordinates, where

$$\epsilon \circ \nabla = \sum_i \epsilon(dx^i) \nabla_i$$

and

$$df = \sum_j \partial_j f \, dx^j.$$

Now $\sum_{ij} \epsilon(dx^i) \partial_i \partial_j f \, dx^j = 0$ so

$$(\epsilon \circ \nabla) df = \sum_{ij} \epsilon(dx^i) \partial_j f \nabla_i dx^j$$

To compute $\nabla_i dx^j$ observe that by Leibnitz's rule

$$0 = \nabla_i \langle dx^j, \partial_k \rangle = \langle \nabla_i dx^j, \partial_k \rangle + \langle dx^j, \nabla_i \partial_k \rangle, \quad \text{so}$$

$$\nabla_i dx^j = -\sum_k \langle \nabla_i \partial_k, dx^j \rangle dx^k.$$

Thus

$$(\epsilon \circ \nabla) df = -\sum_{ijk} dx^i \wedge dx^k \Gamma^j_{ik} \partial_j f = -\sum_{i<k} \langle \tau(\partial_i, \partial_k), df \rangle dx^i \wedge dx^k. \quad \square$$

3.9 Variations and the Jacobi equations.

3.9.1 Two parameter maps.

We will let $\mathcal{D} \subset \mathbb{R}^2$ denote an open subset with the property that the intersection of $\mathcal{D}$ with any vertical or horizontal line is (either empty or) an interval. We will let (u, v) be the Cartesian coordinates on the plane. We will let M denote a smooth manifold. A smooth map $\mathbf{x} : \mathcal{D} \to M$ is called a **two parameter map**.

Holding v fixed (say $v = v_0$) we get a curve $u \mapsto \mathbf{x}(u, v_0)$. Such a curve will be called a u-parameter curve. Similarly, holding u fixed we get v-parameter curves.

The **partial velocities** are the vector fields along $\mathbf{x}$ given by

$$\mathbf{x}_u := d\mathbf{x}(\partial_u), \quad \mathbf{x}_v := d\mathbf{x}(\partial_v).$$

If the image of $\mathbf{x}$ lies in a coordinate chart with coordinates $x^1, \ldots, x^n$ then the **coordinate functions** of $\mathbf{x}$ defined as

$$\mathbf{x}^i := x^i \circ \mathbf{x}$$

are smooth real valued functions on $\mathcal{D}$ and

$$\mathbf{x}_u = \sum \frac{\partial \mathbf{x}^i}{\partial u} \partial_i \quad \text{and} \quad \mathbf{x}_v = \sum \frac{\partial \mathbf{x}^i}{\partial v} \partial_i.$$

Now suppose that M is equipped with a torsionless connection ∇ and that Z is a vector field along $\mathbf{x}$. We then have

- $Z_u := \frac{\nabla Z}{\partial u}$, the covariant derivative of Z along the u-parameter curves and
- $Z_v := \frac{\nabla Z}{\partial v}$, the covariant derivative of Z along the v-parameter curves.

In terms of local coordinates where $Z = \sum Z^i \partial_i$ we have

$$Z_u = \sum \left\{ \frac{\partial Z^k}{\partial u} + \sum \Gamma_{ij}^k Z^i \frac{\partial \mathbf{x}^j}{\partial u} \right\} \partial_k,$$

with a similar formula for Z_v.

Let us apply this formula to $Z = \mathbf{x}_u$ where

$$Z^k = \frac{\partial \mathbf{x}^k}{\partial u}.$$

We obtain

$$\mathbf{x}_{uv} = \sum_k \left\{ \frac{\partial^2 \mathbf{x}^k}{\partial v \partial u} + \sum_{i,j} \Gamma_{ij}^k \frac{\partial \mathbf{x}^i}{\partial u} \frac{\partial \mathbf{x}^j}{\partial v} \right\} \partial_k.$$

Notice that $\mathbf{x}_{uv} = \mathbf{x}_{vu}$ since the Christoffel symbols are symmetric in i and j.

3.9. VARIATIONS AND THE JACOBI EQUATIONS.

Let Z be a vector field along $\mathbf{x}$ and let us compute $Z_{uv} := (Z_u)_v$. In local coordinates we have $Z_u = \sum \left\{ \frac{\partial Z^k}{\partial u} + \sum \Gamma_{ij}^k Z^i \frac{\partial \mathbf{x}^j}{\partial u} \right\} \partial_k$. Substituting this into the expression for $(Z_u)_v$ we get, as the coefficient of ∂_k, the expression

$$\frac{\partial^2 Z^k}{\partial u \partial v} + \sum_{ij} \left(\frac{\partial \Gamma_{ij}^k}{\partial v} Z^i \frac{\partial \mathbf{x}^j}{\partial u} + \Gamma_{ij}^k \frac{\partial Z^i}{\partial v} \frac{\partial \mathbf{x}^j}{\partial u} + \Gamma_{ij}^k Z^i \frac{\partial^2 \mathbf{x}^j}{\partial u \partial v} \right)$$

$$+ \sum_{ij} \Gamma_{ij}^k \left(\frac{\partial Z^i}{\partial u} + \sum_{rs} \Gamma_{rs}^i Z^r \frac{\partial \mathbf{x}^s}{\partial u} \right) \frac{\partial \mathbf{x}^j}{\partial v}$$

Let us interchange the roles of u and v and subtract. The terms involving first or second derivatives of Z disappear, as do the terms involving the second derivatives of $\mathbf{x}^j$. Let us eliminate these terms from the above expression so as to obtain:

$$\sum_{ij} \left(\frac{\partial \Gamma_{ij}^k}{\partial v} Z^i \frac{\partial \mathbf{x}^j}{\partial u} + \sum_{rs} \Gamma_{ij}^k \Gamma_{rs}^i Z^r \frac{\partial \mathbf{x}^s}{\partial u} \frac{\partial \mathbf{x}^j}{\partial v} \right)$$

− same expression with u and v interchanged.

We have $\frac{\partial \Gamma_{ij}^k}{\partial v} = \sum_\ell \frac{\partial \Gamma_{ij}^k}{\partial x^\ell} \frac{\partial \mathbf{x}^\ell}{\partial v}$ and the expression for the curvature tensor in local coordinates is

$$R_{jk\ell}^i = -\left(\frac{\partial}{\partial x^\ell} \Gamma_{kj}^i - \frac{\partial}{\partial x^k} \Gamma_{\ell j}^i + \sum_m \Gamma_{\ell m}^i \Gamma_{kj}^m - \sum_m \Gamma_{km}^i \Gamma_{\ell j}^m \right).$$

So from the above expression we obtain the important formula and theorem:

Theorem 5. *Let* $\mathbf{x} : \mathcal{D} \to M$ *be a two parameter family of maps, and let* $Z : \mathcal{D} \to M$ *be a vector field along* $\mathbf{x}$. *Then*

$$Z_{uv} - Z_{vu} = -R(\mathbf{x}_u, \mathbf{x}_v) Z. \tag{3.20}$$

3.9.2 Variations of a curve.

Let $\alpha : [a - \epsilon, b + \epsilon] :\to M$ be a (smooth) curve. Let $\mathcal{D}$ be an open subset of $\mathbb{R}^2$ which contains the rectangle

$$[a, b] \times (-\delta, \delta).$$

A two parameter map $\mathbf{x} : \mathcal{D} \to M$ such that

$$\mathbf{x}(u, 0) = \alpha(u), \quad a \le u \le b$$

is called a (smooth) **variation** of α.

Longitudinal and transverse curves of a variation, the variational vector field.

A two parameter map $\mathbf{x} : \mathcal{D} \to M$ such that
$$\mathbf{x}(u,0) = \alpha(u), \quad a \le u \le b$$
is called a (smooth) **variation** of α.

The u-parameter curves of a variation are called its **longitudinal curves** and the v-parameter curves are called its **transverse** curves.

The vector field V along α given by
$$V(u) = \mathbf{x}_v(u,0)$$
is called the **variational** vector field of $\mathbf{x}$, or the **infinitesimal variation** of α corresponding to $\mathbf{x}$.

3.9.3 Geodesic variations and the Jacobi equations.

If every longitudinal curve of $\mathbf{x}$ is a geodesic, then $\mathbf{x}$ is called a **geodesic variation** or a **one parameter family of geodesics**.

So the condition for $\mathbf{x}$ to be a geodesic variation is
$$\mathbf{x}_{uu} \equiv 0.$$

Let us consider the vector field $Z = \mathbf{x}_u$ for a geodesic variation, so $Z_u = \mathbf{x}_{uu} \equiv 0$. Then
$$\mathbf{x}_{vuu} = \mathbf{x}_{uvu} = \mathbf{x}_{uuv} - R(\mathbf{x}_v, \mathbf{x}_u)\mathbf{x}_u = -R(\mathbf{x}_v, \mathbf{x}_u)\mathbf{x}_u.$$

Since $\mathbf{x}_u(u,0) = \alpha'(u)$ we see that the variational vector field V of a geodesic variation satisfies the **Jacobi equation**
$$V'' + R(V, \alpha')\alpha' = 0$$
i.e.
$$V'' = R(\alpha', V)\alpha'.$$
A solution to this second order linear differential equation is called a **Jacobi vector field**.

It is useful to have a name for the operator
$$R_w : T_pM \to T_pM; \quad v \mapsto R(w,v)w$$
which occurs on the right hand side of the Jacobi equation. O'Neill calls this the "tidal force" because of its meaning in physics. Some call it the Ricci operator. I will use the more neutral name: the **directional curvature** operator in the direction w. So the Jacobi equation can be written as
$$V'' = R_{\alpha'} V.$$

In terms of any frame field along α the Jacobi equations are a system of second order linear differential equations. Such equations have global solutions depending (smoothly) on the initial conditions. So

3.10. THE EXPONENTIAL MAP.

Lemma 3. *Given any $v, w \in T_{\alpha(a)}M$ there is a unique Jacobi vector field V along α such that $V(0) = v$ and $V'(0) = w$. In particular, the set of Jacobi fields along a geodesic form a $2n$ dimensional vector space where $n = \dim(M)$.*

3.9.4 Conjugate points.

Definition 2. *Points $\sigma(a)$ and $\sigma(b)$, $a \neq b$ on a geodesic σ are said to be* **conjugate along** *σ if there is a a non-zero Jacobi field J such that $J(a) = 0$ and $J(b) = 0$.*

Preview of coming attractions.

As I hope to explain, conjugate points (or the non-existence thereof) play a key role in global Riemannian geometry. We will presently state and prove Levi-Civita's theorem which asserts that every Riemannian manifold carries a unique torsion free connection whose parallelism "preserves the metric". A theorem of Jacobi then asserts that in Riemannian geometry geodesics minimize "distance" up to, but not past, the first conjugate point.

A useful technique is then to compare solutions of the Jacobi equation with solutions of the equation
$$\ddot{x} + k^2 x = 0.$$

3.10 The exponential map.

Suppose that M is a manifold with a connection ∇. Let m be a point of M and $\xi \in TM_m$. Then there is a unique (maximal) geodesic γ_ξ with $\gamma_\xi(0) = m$, $\gamma'(0) = \xi$. Recall that it is found by solving a system of second order ordinary differential equations. The existence and uniqueness theorem for solutions of such equations implies that the solutions depend smoothly on ξ. In other words, there exists a neighborhood $\mathcal{N}$ of ξ in the tangent bundle TM and an interval I about 0 in $\mathbb{R}$ such that $(\eta, s) \mapsto \gamma_\eta(s)$ is smooth on $\mathcal{N} \times I$.

If we take $\xi = 0$, the zero tangent vector, the corresponding "geodesic", defined for all t is the constant curve $\gamma_0(t) \equiv m$. The continuity thus implies that for ξ in some neighborhood of the origin in TM_m, the geodesic γ_ξ is defined for $t \in [0, 1]$. Let $\mathcal{D}_0$ be the set of vectors ξ in TM_m such that the maximal geodesic through ξ is defined on $[0, 1]$. By the preceding remarks this contains some neighborhood of the origin.

Define the **exponential map**

$$\exp = \exp_m : \mathcal{D}_0 \to M, \quad \exp(\xi) = \gamma_\xi(1). \tag{3.21}$$

For $\xi \in TM_m$ and fixed $t \in \mathbb{R}$ the curve

$$s \mapsto \gamma_\xi(ts)$$

is a geodesic whose tangent vector at $s = 0$ is $t\xi$. So the exponential map carries straight lines through the origin in TM_m into geodesics through m in M:

$$\exp : t\xi \mapsto \gamma_\xi(t).$$

3.10.1 The differential of the exponential map at 0 is the identity.

We have
$$\exp : t\xi \mapsto \gamma_\xi(t). \tag{3.22}$$

The tangent vector to the line $t \mapsto t\xi$ at $t = 0$ (which is a vector in $T_0(TM_m)$) is just ξ under the standard identification of the tangent space to a vector space with the vector space itself. Also, the tangent vector to the curve $t \mapsto \gamma_\xi(t)$ at $t = 0$ is ξ, by the definition of γ_ξ. So taking the derivatives of both sides of (3.22) shows that the differential of the exponential map is the identity:

$$d\exp_0 : T(TM_m)_0 \to TM_m = \mathrm{id}$$

under the standard identification of the tangent space $T(TM_m)_0$ with TM_m.

So the differential of the exponential map is the identity:

$$d\exp_0 : T(TM_m)_0 \to TM_m = \mathrm{id}$$

under the standard identification of the tangent space $T(TM_m)_0$ with TM_m.

From the inverse function theorem it follows that:

The exponential map is a diffeomorphism in some neighborhood of the origin.

3.10.2 Normal neighborhoods.

A subset S of a vector space is called **starshaped** (about 0) if $v \in S$ implies that $tv \in S$ for all $t \in [0, 1]$. In other words, S is a union of radial line segments.

Let $\tilde{\mathcal{U}}$ be a star shaped neighborhood of the origin in TM_m on which exp is a diffeomorphism, and let $\mathcal{U} := \exp(\tilde{\mathcal{U}})$ be its image in M under the exponential map. Then $\mathcal{U}$ is called a **normal neighborhood** of m. By construction (and the uniqueness theorem for differential equations) for every $m \in \mathcal{U}$ there exists a unique geodesic which joins m to m and lies entirely in $\mathcal{U}$.

3.10.3 Normal coordinates.

Suppose that we choose a basis $e = (e_1, \ldots, e_n)$ of TM_m and let $\ell^1, \ldots, \ell^n$ be the dual basis. We then get a coordinate system on $\mathcal{U}$ defined by

$$\exp^{-1}(m) = \sum x^i(m) e_i$$

3.10. THE EXPONENTIAL MAP.

or, what is the same,
$$x^i = \ell^i \circ \exp^{-1}.$$

These coordinates are known as **normal** coordinates, or sometimes as **inertial** coordinates for the following reason:

Let $\xi = \sum a^i e_i$ be an element of $\tilde{\mathcal{U}} \subset TM_m$. Since $\exp(t\xi) = \gamma_\xi(t)$ the coordinates of $\gamma_\xi(t)$ are given by
$$x^i(\gamma_\xi(t)) = \ell^i(t\xi) = t\ell^i(\xi) = ta^i.$$

Thus the second derivative of $x^i(\gamma_\xi(t))$ with respect to t vanishes and the geodesic equations (satisfied by $\gamma_\xi(t)$) becomes
$$\sum_{ij} \Gamma^k_{ij}(\gamma_\xi(t))a^i a^j = 0, \quad \forall k.$$

In particular, evaluating at $t = 0$ we get
$$\sum_{ij} \Gamma^k_{ij}(0)a^i a^j = 0, \quad \forall k.$$

This must hold for all (sufficiently small) values of the a^i and hence for all values of the a^i.

Now let us suppose that *our connection has zero torsion* so that the Γ^k_{ij} are symmetric in i and j, this implies that
$$\Gamma^k_{ij}(0) = 0. \tag{3.23}$$

We have proved that in a normal coordinate system, the *Christoffel symbols of a torsionless connection vanish at the origin*.

To repeat: In a normal coordinate system, the Christoffel symbols of a torsionless connection vanish at the origin. Hence at this one point, the equations for a geodesic look like the equations of a straight line in terms of these coordinates. This was Einstein's resolution of Mach's problem: How can the laws of physics - particularly mechanics - involve rectilinear motion in absence of forces, as this depends on the coordinate system.

Newton had a theological explanation for this: that "absolute space is God's *sensorium*".

According to Einstein the distribution of matter in the universe determines the metric which then determines the connection which picks out the set of inertial frames.

3.10.4 The exponential map and the Jacobi equation.

Let $m \in M$ and $x, v \in TM_m$. Consider the map
$$\mathbf{y} : \mathbb{R}^2 \to TM_m, \quad (t, s) \mapsto t(x + sv).$$

Let x be sufficiently close to the origin so that (for $|s|$ sufficiently small) the map

$$\mathbf{x} = \exp_m \circ \mathbf{y}$$

is defined. Thus $\mathbf{x}$ is a geodesic variation of the geodesic γ_x where $\gamma_x(0) = m$ and $\gamma'_x(0) = x$. More explicitly,

$$\mathbf{x}(t,s) = \gamma_{x+sv}(t).$$

The corresponding variational vector field V given by

$$V(t) = \mathbf{x}_s(t,0) = d\exp_m(\mathbf{y}_s(t,0))$$

satisfies the Jacobi equation and $V(0) = 0$ since $\mathbf{x}(0,s) \equiv m$.

$$V(t) = \mathbf{x}_s(t,0) = d\exp_m(\mathbf{y}_s(t,0)), \quad V(0) = 0.$$

We also know that $V'(0) = \mathbf{x}_{st}(0,0) = \mathbf{x}_{ts}(0,0)$ and $s \mapsto \mathbf{x}_t(0,s) = x + sv$ so $\mathbf{x}_{ts}(0,s) = v$ and hence $V'(0) = v$.

Let us consider $v \in TM_m$ as an element of $T(TM_m)_x$. Then $V(1) = d(\exp_m)_x(v)$. So we have proved

Theorem 6. *Let $m \in M$, $x \in TM_m$ and $v \in TM_m$ considered as an element of $T(TM_m)_x$. Then*

$$d(\exp_m)_x(v) = V(1)$$

where V is the unique Jacobi field along the geodesic γ_x such that $V(0) = 0$ and $V'(0) = v$.

M. Berger calls this theorem, and its amazing consequences (we shall see some of them) the Élie Cartan philosophy in Riemannian geometry. It appears in the first (1928) edition of Cartan's book on Riemannian geometry and is illustrated by the following diagram:

3.10. THE EXPONENTIAL MAP.

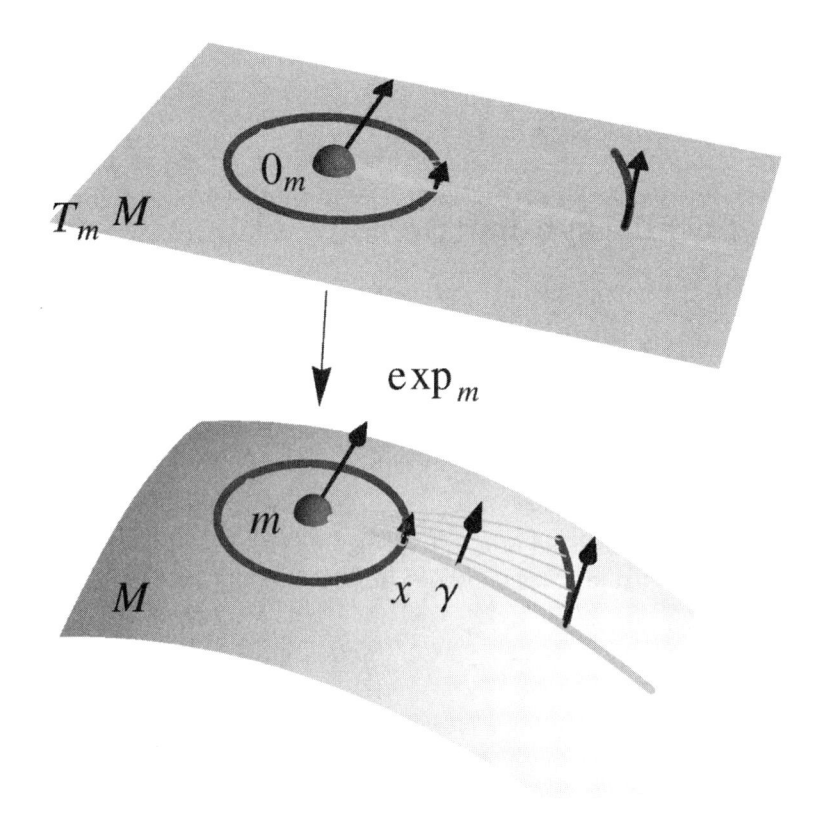

3.10.5 Polar maps.

We can use the concept of a normal neighborhood and exponential map to extend linear transformations: Suppose that M_1 and M_2 are manifolds with (torsion free) connections. Let $p \in M_1$ and $q \in M_2$. Suppose we are given a linear transformation

$$L : T_p M_1 \to T_q M_2.$$

Let $\mathcal{U}_p$ be a normal neighborhood of p and $\mathcal{U}_q$ a normal neighborhood of q with $\exp_p$ and $\exp_q$ denoting the corresponding exponential maps. We define the **polar map**

$$\phi_L : \mathcal{U}_p \to \mathcal{U}_q$$

by

$$\phi_L := \exp_q \circ L \circ (exp_p)^{-1}. \tag{3.24}$$

The following properties of ϕ_L are immediate from its definition:

- ϕ_L carries radial geodesics to radial geodesics; explicitly, if $v \in T_p M_1$ then $\phi_L \gamma_v = \gamma_{Lv}$.

- The differential of ϕ_L at p is L

- If L is an isomorphism then ϕ_L is a diffeomorphism in a neighborhood of p.

3.11 Locally symmetric connections.

Proposition 9. *The following conditions on a connection on a manifold M are equivalent:*

1. $\nabla R = 0$.

2. *If X, Y, Z are parallel vector fields on a non-singular curve α then $R(X,Y)Z$ is also parallel on α.*

Proof. **1 $\Rightarrow$ 2.** Fix a point, say p on α and extend the vector fields X, Y, Z to be defined in a neighborhood of p and also extend the tangent vector $\alpha'(p)$ to a vector field V defined in a neighborhood of p. By hypothesis 1) $\nabla_V R = 0$. But $(\nabla_V R)(X,Y)Z =$

$$= \nabla_V(R(X,Y)Z) - R(\nabla_V X, Y)Z - R(X, \nabla_V Y)Z - R(X,Y)\nabla_V Z,$$

and the last three terms vanish at p by assumption.

2 $\Rightarrow$ 1. Let x, y, z, v be tangent vectors at p with $v \neq 0$, and let α be a curve with $\alpha(0) = p$ and $\alpha'(0) = v$. Extend x, y, z to be parallel vector fields along α then the same computation as above shows that

$$(\nabla_v R)(x,y)z = (R(X,Y)Z)'(0) = 0.$$

□

3.12 Normal neighborhoods and convex open sets.

Recall that a subset S of a vector space is called **starshaped** about 0 if

$$v \in S \Rightarrow tv \in S \quad \forall 0 \leq t \leq 1.$$

In other words, S is a union of radial line segments.

Suppose that $\tilde{\mathcal{U}}$ is a star shaped neighborhood of 0 in $T_o M$ such that $\exp_o$ is a diffeomorphism of $\tilde{\mathcal{U}}$ onto a neighborood $\mathcal{U}$ of o. Then we called $\mathcal{U}$ a **normal neighborhood** of o.

Proposition 10. *If $\mathcal{U}$ is a normal neighborhood of $o \in M$ then for every $p \in \mathcal{U}$ there is a unique geodesic $\sigma : [0,1] \to \mathcal{U}$ such that $\sigma(0) = o$ and $\sigma(1) = p$. Furthermore,*

$$\sigma'(0) = \exp_o^{-1}(p) \in \tilde{\mathcal{U}}.$$

3.12. NORMAL NEIGHBORHOODS AND CONVEX OPEN SETS.

Proof. For the proof of this proposition we will start at the end and define

$$v = \exp_o^{-1}(p).$$

Since $v \in \tilde{\mathcal{U}}$ which is starshaped, the entire ray

$$\rho(t) := tv \quad \in \quad \tilde{\mathcal{U}}.$$

Thus the entire geodesic segment $\sigma := \exp_o \circ \rho$ lies in $\mathcal{U}$ and joins 0 to p.

Now $\rho(t) = tv$ so $\rho'(0) = v$ and we know that the differential of the exponential map is the identity under the identification of $T_0(T_oM)$ with T_oM. Since $\sigma = \exp_o \circ \rho$ we conclude that

$$\sigma'(0) = v = \exp_o^{-1}(p).$$

We must prove that σ is the only geodesic segment joining o to p and lying entirely in $\mathcal{U}$.

Let $\tau : [0,1] \to \mathcal{U}$ be a geodesic with $\tau(0) = o$ and $\tau(1) = p$. Let $w := \tau'(0)$. Then the geodesics τ and $t \mapsto \exp_o(tw)$ have the same initial positions and velocities, and so must coincide.

The radial segment $t \mapsto tw$, $0 \le t \le 1$ does not leave $\tilde{\mathcal{U}}$ because its image under $\exp_o$ lies in $\mathcal{U}$ and we are assuming that $\exp_o$ is a diffeomorphism from $\tilde{\mathcal{U}}$ to $\mathcal{U}$. So $w \in \tilde{\mathcal{U}}$. But

$$\exp_o(w) = \tau(1) = p = \exp_o(v)$$

and $\exp_o$ is one to one on $\tilde{\mathcal{U}}$. So $w = v$ and hence $\tau = \sigma$. $\square$

Convex open sets in a manifold with torsionless connection.

An open subset of a manifold with torsionless connection is called **convex** if it is a normal neighborhood of each of its points.

From the definition of a normal neighborhood and the proposition we just proved, it follows that for any two points p and q in a convex open set $\mathcal{U}$ there is a unique geodesic segment joining them which lies in $\mathcal{U}$.

Proposition 11. *Every point in a manifold M with torsionless connection has a convex neighborhood.*

For the proof of this proposition we need some more refined facts about the exponential map:

The maps $\exp_p$, $p \in M$ fit together to give a map of $TM \to M$ defined in some neighborhood of the zero section of TM. If M is geodesically complete, meaning that geodesics exist for all time, this will be defined on all of TM. Otherwise, let $\mathcal{D}$ be the set of all $v \in TM$ such the geodesic γ_v is defined on $[0,1]$. Define

$$E : \mathcal{D} \to M \times M$$

by
$$E(v) = (p, \exp_p(v)), \quad \text{if} \quad v \in T_pM.$$

We let $\mathcal{D}_p := \mathcal{D} \cap T_pM$.

Lemma 4. *If* $\exp_p : \mathcal{D}_p \to M$ *is non-singular at* x *then* E *is non-singular at* x.

Proof. Let $\text{pr}_1 : M \times M \to M$ denote projection onto the first factor so that
$$\text{pr}_1 \circ E = \pi$$
where $\pi : TM \to M$ is the standard projection of the tangent bundle onto is base. If $dE_x(v) = 0$ then $d(\text{pr}_1 \circ E)_x(v) = d\pi_x(v) = 0$ so v is vertical. But then $dE_x(v) = 0$ means that $d(\exp_p)_x(v) = 0$ implying that $v = 0$ since we are assuming that $\exp_p$ is non-singular at x. □

Now $\exp_o$ is always non-singular at 0. (In fact, as we know, $d\exp_o(0)$ is the identification of $T_0(T_oM)$ with T_oM.) So by the inverse function theorem, we get the following corollary of the preceding lemma:

Corollary 1. E *maps some neighborhood in* TM *of* $0 \in T_oM$ *diffeomorphically onto a neighborhood of* (o,o) *in* $M \times M$.

Let $(x^1, \ldots, x^n)$ be a normal coordinate system on a neighborhood $\mathcal{V}$ of $o \in M$. Let
$$N := (x^1)^2 + \cdots + (x^n)^2$$
and
$$\mathcal{V}(\delta) := \{p \in \mathcal{V} | N(p) < \delta\}.$$
So for δ sufficiently small $\mathcal{V}(\delta)$ is defined and is a neighborhood of o diffeomorphic to an open ball in $\mathbb{R}^n$. So by the corollary there is a neighborhood $\mathcal{W} = \mathcal{W}(\delta)$ in TM of $0 \in T_oM$ which is diffeomorphic under E to $\mathcal{V}(\delta) \times \mathcal{V}(\delta)$. Let B be the symmetric $(0,2)$ tensor defined on $\mathcal{V}$ by
$$B_{ij} = \delta_{ij} - \sum_k \Gamma_{ij}^k x^k$$
in terms of the normal coordinates.

$B_{ij} = \delta_{ij}$ at o, so B is positive definite on some neighborhood of o which we can choose as $\mathcal{V}(\delta)$ if δ is sufficiently small.

Also, by choosing δ sufficiently small, we can arrange that all geodesics of the form $\exp_p(tv)$, $(p,v) \in \mathcal{W}$ lie in the original normal coordinate neighborhood for $0 \leq t \leq 1$ so that the function $t \mapsto N(\exp tv)$ is defined.

We can now give a more precise formulation of the proposition we want to prove:

3.12. NORMAL NEIGHBORHOODS AND CONVEX OPEN SETS.

Proposition 12. *With such a choice of δ, $\mathcal{U} := \mathcal{V}(\delta)$ is a normal neighborhood of every $p \in \mathcal{U}$. In other words, $\mathcal{U}$ is a convex neighborhood of o.*

Proof. For $p \in \mathcal{U}$ let $\mathcal{W}_p := \mathcal{W} \cap T_pM$. By the corollary, $\exp_p$ restricted to $\mathcal{W}_p$ is a diffeomorphism onto $\mathcal{U}$. In order to show that $\mathcal{U}$ is a normal neighborhood of p, we must show that $\mathcal{W}_p$ is starshaped.

If $q \in \mathcal{U}$ let $v = E^{-1}(p,q)$, so $\sigma : [0,1] \to M$ defined by $\sigma(t) = \exp_p tv$ is a geodesic joining p to q. If we know that σ lies entirely in $\mathcal{U}$ then its pre-image under the diffeomorphism $\exp_p$ lies entirely in $\mathcal{W}_p$. But this pre-image is exactly the curve $t \mapsto tv$ proving that $\mathcal{W}_p$ is starshaped.

So the issue is to prove that $N(\sigma(t)) < \delta$ for all $t \in [0,1]$.

Since $N(\sigma(0)) = N(p)$ and $N(\sigma(1)) = N(q)$ are both less than δ, it is enough to show that $N(\sigma(t))$ can not have a maximum at any point in $(0,1)$, and for this it is enough to show that its second derivative is always positive.

Write $x^i(t)$ for $x^i(\sigma(t))$ so that $N(t) := N(\sigma(t)) = \sum (x^i(t))^2$ and hence

$$N'' = 2\sum \left(((x^i)')^2 + x^i(x^i)''\right).$$

Since σ is a geodesic,

$$(x^i)'' + \sum_{jk} \Gamma^i_{jk}(x^j)'(x^k)' = 0.$$

So

$$N''(t) = 2B(\sigma'(t), \sigma'(t)) > 0.$$

$\square$

Chapter 4

Levi-Civita's theorem.

This basic theorem says that on a semi-Riemannian manifold there is a unique "isometric connection" with zero torsion. See below for the definition of these terms. According to Einstein's theory of general relativity, the distribution of energy-matter in the universe determines the Lorentzian geometry of space time. This in turn determines a connection via Levi-Civita's theorem, and small particles move along geodesics. (The word "small" means that we can ignore the effect of the particle on the geometry of the universe.)

4.1 Isometric connections.

Suppose that M is a **semi-Riemannian manifold**, meaning that we are given a smoothly varying non-degenerate scalar product $\langle\,,\,\rangle_x$ on each tangent space TM_x. Given two vector fields X and Y, we let $\langle X, Y \rangle$ denote the function

$$\langle X, Y \rangle(x) := \langle X(x), Y(x) \rangle_x.$$

We say that a connection ∇ is **isometric** for $\langle\,,\,\rangle$ if

$$X\langle Y, Z \rangle = \langle \nabla_X Y, Z \rangle + \langle Y, \nabla_X Z \rangle \tag{4.1}$$

for any three vector fields X, Y, Z. It is a sort of Leibniz's rule for scalar products. If we go back to the definition of the derivative of a vector field along a curve arising from the connection ∇, we see that (4.1) implies that

$$\frac{d}{dt}\langle Y, Z \rangle = \langle Y', Z \rangle + \langle Y, Z' \rangle$$

for any pair of vector fields along a curve C. In particular, if Y and Z are parallel along the curve, so that $Y' = Z' = 0$, we see that $\langle Y, Z \rangle$ is constant. This is the key meaning of the condition that a connection be isometric: parallel translation along any curve is an isometry of the tangent spaces.

4.2 Levi-Civita's theorem.

This asserts that on any semi-Riemannian manifold there exists a unique connection which is isometric and is torsion free. We will give two proofs. The first proof gives an explicit formula for the Levi-Civita connection, namely it is determined by the **Koszul formula**

$$2\langle \nabla_V W, X \rangle =$$
$$V\langle W, X \rangle + W\langle X, V \rangle - X\langle V, W \rangle - \langle V, [W, X] \rangle + \langle W, [X, V] \rangle + \langle X, [V, W] \rangle \quad (4.2)$$

for any three vector fields X, V, W.

Our second proof due to Cartan will be given later.

To prove Koszul's formula, we apply the isometric condition to each of the first three terms occurring on the right hand side of (4.2). For example the first term becomes $\langle \nabla_V W, X \rangle + \langle W, \nabla_V X \rangle$. We apply the torsion free condition to each of the last three terms. For example the last term becomes $\langle X, \nabla_V W - \nabla_W V \rangle$. There will be a lot of cancellation leaving the left hand side. Since the vector field $\nabla_V W$ is determined by knowing its scalar product $\langle \nabla_V W, X \rangle$ for all vector fields X, the Koszul formula proves the uniqueness part of Levi-Civita's theorem. Here are the details of the cancellations: The first three terms on the right hand side of (4.2) expand to

$$\langle \nabla_V W, X \rangle \;\; +\langle W, \nabla_V X \rangle \;\; +\langle \nabla_W X, V \rangle \;\; +\langle X, \nabla_W V \rangle \;\; -\langle \nabla_X V, W \rangle \;\; -\langle V, \nabla_X W \rangle$$
$$** \qquad * \qquad *** \qquad **** \qquad *****$$

while the last three terms on the right expand out as

$$-\langle V, \nabla_W X \rangle \;\; +\langle V, \nabla_X W \rangle \;\; +\langle W, \nabla_X V \rangle \;\; -\langle W, \nabla_V X \rangle \;\; +\langle X, \nabla_V W \rangle \;\; -\langle X, \nabla_W V \rangle$$
$$* \qquad ***** \qquad **** \qquad ** \qquad ***.$$

The cancellations are as indicated. This shows the uniqueness of the Levi-Civita connection.

On the other hand, the right hand side of the Koszul formula is linear in X. We must show that when we replace X by fX on the right hand side of (4.2), no term involving Vf occurs. Indeed upon replacing X by fX in

$$V\langle W, X \rangle + W\langle X, V \rangle - X\langle V, W \rangle - \langle V, [W, X] \rangle + \langle W, [X, V] \rangle + \langle X, [V, W] \rangle$$

only the first and fifth term can involve Vf. The first term contributes $\langle W, (Vf)X \rangle$ while the fifth term contributes $\langle W, (-Vf)X \rangle$ so these cancel.

The verification of the "Leibnitz rule"

$$\nabla_V(fW) = (Vf)W + f\nabla_V W$$

is similar as is the verification of the rule

$$\nabla_{fV} W = f\nabla_V W.$$

4.2. LEVI-CIVITA'S THEOREM.

So we obtain a well defined vector field, $\nabla_V W$. We must check that this definition of ∇ satisfies the conditions for a connection and is torsion free and isometric. Here are some of the details of this verification:

Interchange V and W in

$$V\langle W, X\rangle + W\langle X, V\rangle - X\langle V, W\rangle - \langle V, [W, X]\rangle + \langle W, [X, V]\rangle + \langle X, [V, W]\rangle$$

and subtract. All that is left is $2\langle X, [V, W]\rangle$. This shows that

$$\nabla_V W - \nabla_W V = [V, W], \quad \square$$

i.e. that the connection is torsion free.

Let us now verify that the connection defined by Koszul's formula is isometric:

Interchange W and X in

$$V\langle W, X\rangle + W\langle X, V\rangle - X\langle V, W\rangle - \langle V, [W, X]\rangle + \langle W, [X, V]\rangle + \langle X, [V, W]\rangle$$

and add. All that remains is $2V\langle X, W\rangle$. This shows that

$$\langle \nabla_V W, X\rangle + W\nabla_V X\rangle = V\langle W, X\rangle$$

which says that the connection given by Koszul's formula is isometric. $\square$

Tullio Levi-Civita

Born: 29 March 1873 in Padua, Italy
Died: 29 Dec. 1941 in Rome, Italy

J. L. Koszul

(1921-)

4.3 The Christoffel symbols of the Levi-Civita connection.

We can use the Koszul identity to derive a formula for the Christoffel symbols in terms of the metric. First some standard notations: We will use the symbol **g** to stand for the metric, so **g** is just another notation for $\langle\ ,\ \rangle$. In a local coordinate system we write

$$g_{ij} := \langle \partial_i, \partial_j \rangle$$

so

$$\mathbf{g} = \sum_{ij} g_{ij} dx^i \otimes dx^j.$$

Here the g_{ij} are functions on the coordinate neighborhood, but we are suppressing the functional dependence on the points in the notation. The metric **g** is a (symmetric) tensor of type (0,2). It induces an isomorphism (at each point) of the tangent space with the cotangent space, each tangent vector ξ going into the linear function $\langle \xi, \cdot \rangle$ consisting of scalar product by ξ. By the above formula the map is given by

$$\partial_i \mapsto \sum_j g_{ij} dx^j.$$

This isomorphism induces a scalar product on the cotangent space at each point, and so a tensor of type (2,0) which we shall denote by $\hat{\mathbf{g}}$ or sometimes by $\mathbf{g}\uparrow\uparrow$. We write

$$g^{ij} := \langle dx^i, dx^j \rangle$$

4.4. GEODESICS IN ORTHOGONAL COORDINATES.

so
$$\hat{\mathbf{g}} = \sum_{ij} g^{ij} \partial_i \otimes \partial_j.$$

(The transition from the two lower indices to the two upper indices is the reason for the vertical arrows notation.) The metric on the cotangent spaces induces a map into its dual space which is the tangent space given by

$$dx^i \mapsto \sum g^{ij} \partial_j$$

and the two maps - from tangent spaces to cotangent spaces and vice versa - are inverses of one another so

$$\sum_k g^{ik} g_{kj} = \delta^i_j,$$

the "matrices" (g_{ij}) and $(g^{k\ell})$ are inverses.

Now let us substitute $X = \partial_m, V = \partial_i, W = \partial_j$ into the Koszul formula (4.2). All brackets on the right vanish and we get

$$2\langle \nabla_i \partial_j, \partial_m \rangle = \partial_i(g_{jm}) + \partial_j(g_{im}) - \partial_m(g_{ij}).$$

Since
$$\nabla_i \partial_j = \sum_k \Gamma^k_{ij} \partial_k$$

is the definition of the Christoffel symbols, the preceding equation becomes

$$2 \sum_a \Gamma^a_{ij} g_{am} = \partial_i(g_{jm}) + \partial_j(g_{im}) - \partial_m(g_{ij}).$$

Multiplying this equation by g^{mk} and summing over m gives

$$\Gamma^k_{ij} = \frac{1}{2} \sum_m g^{km} \left\{ \frac{\partial g_{jm}}{\partial x^i} + \frac{\partial g_{im}}{\partial x^j} - \frac{\partial g_{ij}}{\partial x^m} \right\}. \tag{4.3}$$

In principle, we should substitute this formula into (3.11) and solve to obtain the geodesics. In practice this is a mess for a general coordinate system and so we will spend a good bit of time developing other means (usually group theoretical) for finding geodesics. However the equations are manageable in orthogonal coordinates.

4.4 Geodesics in orthogonal coordinates.

A coordinate system is called **orthogonal** if

$$g_{ij} = 0, \quad i \neq j.$$

If we are lucky enough to have an orthogonal coordinate system the equations for geodesics take on a somewhat simpler form. First notice that (4.3) becomes

$$\Gamma_{ij}^k = \frac{1}{2}g^{kk}\left\{\frac{\partial g_{jk}}{\partial x^i} + \frac{\partial g_{ik}}{\partial x^j} - \frac{\partial g_{ij}}{\partial x^k}\right\}$$

So (3.11) becomes

$$\frac{d^2 x^k}{dt^2} + g^{kk}\sum_i \frac{\partial g_{kk}}{\partial x^i}\frac{dx^k}{dt}\frac{dx^i}{dt} - \frac{1}{2}g^{kk}\sum_i \frac{\partial g_{ii}}{\partial x^k}\frac{dx^i}{dt}\frac{dx^i}{dt} = 0.$$

If we multiply this equation by g_{kk} and bring the negative term to the other side we obtain

$$\frac{d}{dt}\left(g_{kk}\frac{dx^k}{dt}\right) = \frac{1}{2}\sum_i \frac{\partial g_{ii}}{\partial x^k}\left(\frac{dx^i}{dt}\right)^2 \tag{4.4}$$

as the equations for geodesics in orthogonal coordinates.

4.5 The hereditary character of the Levi-Civita connection.

Suppose that $S \subset M$ is a submanifold which is itself a semi-Riemannian manifold under the induced metric from M. That is, the scalar product of two vectors in T_pS is the same as their scalar product in T_pM. (Since the restriction of the scalar product to such a subspace may not be non-degenerate, we have to impose the condition of non-degeneracy as a hypothesis. In the Riemannian case this is automatic.)

Let X and Y be vector fields on S and extend them locally to vector fields on M (which we can do by the implicit function theorem).

At each $p \in S$ we have the orthogonal decomposition $T_pM = T_pS + (T_pS)^\perp$. We will denote the decomposition of any $\xi \in T_pM$ according to this direct sum as

$$\xi = \xi^{tan} + \xi^{nor}.$$

Now we can consider the covariant derivative $\nabla_X Y$ where X and Y are thought of as vector fields on M, or we can consider their covariant derivative given by the Levi-Civita connection corresponding to the induced metric on S - let us denote this covariant derivative by $\nabla_X^S Y$.

It follows immediately from Koszul's formula that

$$\nabla_X^S Y = (\nabla_X Y)^{tan}. \tag{4.5}$$

So if C is a non-singular curve on S and Z is a vector field on S tangent to S, then the derivative of Z along this curve relative to the Levi-Civita connection of S in its induced metric is obtained by computing is derivative relative to the Levi-Civita metric on M and then projecting onto $T_{C(t)}S$.

4.6. BACK TO THE ISOMETRIC CONDITION $\nabla \mathbf{G} = 0$.

In particular, C will be a geodesic on S if and only if, at each t, $C''(t)$ is orthogonal to S where C'' is computed relative to the Levi-Civita metric of M.

For example, if $M = \mathbb{R}^n$ with its standard metric, then Koszul tells us that $\nabla_{\partial_i} \partial_j = 0$ for all i and j where the ∂_k are the standard constant vector fields in the coordinate directions, and hence the Christoffel symbols vanish and so the derivative of any vector field along any curve is given by $Z'(t) = \frac{d}{dt} Z(t)$.

Hence a curve on S will be a geodesic if and only if its acceleration (in the sense of freshman calculus) is orthogonal to S.

For example, as we have seen, the great circles on a sphere in Euclidean space have this property, hence they are the geodesics on the sphere.

4.6 Back to the isometric condition $\nabla \mathbf{g} = 0$.

Since $\nabla_X \mathbf{g} = 0$, any object that we can build algebraically out of the metric $\mathbf{g}$ will also be sent into zero by ∇_X for all X. For example, the metric $\mathbf{g}$ determines a scalar product on each of the spaces $\wedge^k TM$ and these will be sent to zero by ∇_X. In particular, $\mathbf{g}$ determines a density (a notion of volume on each tangent space) which we will sometimes denote by g, and we we have $\nabla_X g = 0$ for every X, i.e. that

$$\nabla g = 0.$$

4.7 Problems: Geodesics in the Schwarzschild exterior.

The purpose of this problem set is to go through the details of two of the famous results of general relativity, the explanation of the advance of the perihelion of Mercury and the deflection of light passing near the sun. (Einstein, 1915).

One of the postulates of general relativity is that a "small" particle will move along a geodesic γ in a four dimensional Lorentzian manifold of signature (-,+,+,+) whose metric is determined by the matter distribution over the manifold. Here the word "small" is taken to mean that the effect of the mass of the particle on the metric itself can be ignored. We can ignore the mass of a planet when the metric is determined by the mass distribution of the stars. This notion of "small" or "passive" is similar to that involved in the equations of motion of a charged particle in an electromagnetic field. General electromagnetic theory says that the particle itself affects the electromagnetic field, but for "small" particles we ignore this and treat the particles as passively responding to the field. Similarly here. We will have a lot to say about the philosophical underpinnings of the postulate "small particles move along geodesics" when we have enough mathematical machinery. The theory also specifies that if the particle is massive then the geodesic is timelike, i.e $\langle \gamma', \gamma' \rangle < 0$, while if the particle has mass zero then the geodesic is a null geodesic, i.e. lightlike: $\langle \gamma', \gamma' \rangle \equiv 0$.

The second component of the theory is how the distribution of matter determines the metric. This is given by the Einstein field equations: Matter distribution is described by a (possibly degenerate) symmetric bilinear form on the tangent space at each point called the *stress energy tensor*, T. The Einstein equations take the form $\mathcal{G} = 8\pi T$ where $\mathcal{G}$ is related to the Ricci curvature which we will define later on. In particular, in empty space, the Einstein equations become $\mathcal{G} = 0$.

4.7.1 The Schwarzschild solution.

Although the study of the Einstein field equations for determining the metric in space-time is a huge enterprise, the solution for the equations $\mathcal{G} = 0$ in the exterior of a star of mass M which is "spherically symmetric", "stationary" and tends to the Minkowski metric , i.e. the flat metric on $\mathbb{R}^{1,3}$, at large distances was found almost immediately by Schwarzschild. (The words in quotes need to be more carefully defined.) This is the metric

$$ds^2 := -h dt^2 + h^{-1} dr^2 + r^2 d\sigma^2 \tag{4.6}$$

where

$$h(r) := 1 - \frac{2GM}{r} \tag{4.7}$$

where G is Newton's gravitational constant and $d\sigma^2$ is the invariant metric on the ordinary unit sphere,

$$d\sigma^2 = d\theta^2 + \sin^2\theta d\phi^2. \tag{4.8}$$

To be more precise, let $P_I \subset \mathbb{R}^2$ consist of those pairs, (t, r) with

$$r > 2GM.$$

Let

$$N = P_I \times S^2,$$

the set of all (t, r, q), $r > 2GM$, $q \in S^2$. The coordinates (θ, ϕ) can be used on the sphere with the north and south pole removed, and (4.8) is the local expression for the invariant metric of the unit sphere in terms of these coordinates. Then the metric we are considering on N is given by (4.6) as above.

Notice that the structure of N is like that of a surface of revolution, with the interval on the z-axis replaced by the two dimensional region, N, the circle replaced by the sphere, and the radius of revolution, f, replaced by r^2.

The variables t, r, θ, ϕ form an orthogonal coordinate system for the Schwarzschild metric. Indeed, if we set $x^0 := t, x^1 := r, x^2 := \theta, x^3 := \phi$ then

$$g_{ij} = 0, \ i \neq j$$

while

$$g_{00} = -h, g_{11} = h^{-1}, g_{22} = r^2, g_{33} = r^2 \sin^2\theta.$$

4.7. PROBLEMS: GEODESICS IN THE SCHWARZSCHILD EXTERIOR.

We repeat that in an orthogonal coordinate system the equations for geodesics are given by

$$\frac{d}{ds}\left(g_{kk}\frac{dx^k}{ds}\right) = \frac{1}{2}\sum_j \frac{\partial g_{jj}}{\partial x^k}\left(\frac{dx^j}{ds}\right)^2. \tag{4.4}$$

1. Show that for the Schwarzschild metric, (4.6), the equation involving g_{22} on the left is

$$\frac{d}{ds}\left[r^2\frac{d\theta}{ds}\right] = r^2 \sin\theta \cos\theta \left(\frac{d\phi}{ds}\right)^2.$$

Conclude from the uniqueness theorem for solutions of differential equations that if $\theta(0) = \pi/2, \dot\theta(0) = 0$ then $\theta(s) \equiv \pi/2$ along the whole geodesic. Conclude from rotational invariance that all geodesics must lie in a plane, i.e. by suitable choice of poles of the sphere we can arrange that $\theta \equiv \pi/2$.

2. With the above choice of spherical coordinates along the geodesic, show that the g_{00} and g_{33} equations become

$$h\frac{dt}{ds} = E$$

$$r^2\frac{d\phi}{ds} = L$$

where E and L are constants. These constants are called the "energy" and the "angular momentum". Notice that for $L > 0$, as we shall assume, $d\phi/ds > 0$, so we can use ϕ as a parameter on the orbit if we like.

Later on we will show that general principles imply that there is a "constant of motion" associated to every one parameter group of symmetries of the system. The Schwarzschild metric is invariant under time translations $t \mapsto t + c$ and under rotations $\phi \mapsto \phi + \alpha$. We will find that E corresponds to time translation and that L corresponds to $\phi \mapsto \phi + \alpha$.

We now consider separately the case of a massive particle where we can choose the parameter s so that $\langle \gamma'(s), \gamma'(s)\rangle \equiv -1$ and massless particles for which $\langle \gamma'(s), \gamma'(s)\rangle \equiv 0$.

4.7.2 Massive particles.

We can write the tangent vector, $\gamma'(s)$ to the geodesic γ at the point s as

$$\gamma'(s) = \dot x^0(s)\left(\frac{\partial}{\partial x^0}\right)_{\gamma(s)} + \dot x^1(s)\left(\frac{\partial}{\partial x^1}\right)_{\gamma(s)} + \dot x^2(s)\left(\frac{\partial}{\partial x^2}\right)_{\gamma(s)} + \dot x^3(s)\left(\frac{\partial}{\partial x^3}\right)_{\gamma(s)}.$$

Let us assume that we use proper time as the parameterization of our geodesic so that

$$\langle \gamma'(s), \gamma'(s)\rangle_{\gamma(s)} \equiv -1.$$

3. Using this last equation and the results of problem 2, show that

$$E^2 = \left(\frac{dr}{ds}\right)^2 + (1 + \frac{L^2}{r^2})h(r) \qquad (4.9)$$

along any geodesic.

4.7.3 Orbit Types.

We can write (4.9) as

$$E^2 = \left(\frac{dr}{ds}\right)^2 + V(r) \qquad (4.10)$$

where the *effective potential* V is given as

$$V(r) := 1 - \frac{2GM}{r} + \frac{L^2}{r^2} - \frac{2GML^2}{r^3}.$$

The behavior of the orbit depends on the the relative size of L and GM. In particular, (4.10) implies that on any orbit, r is restricted to the interval

$$I \subset \{r : V(r) \leq E^2\} \qquad r(0) \in I.$$

If we differentiate (4.10) we get

$$2\left(\frac{d^2r}{ds^2}\right)\left(\frac{dr}{ds}\right) = -V'(r)\left(\frac{dr}{ds}\right). \qquad (4.11)$$

In particular, a critical point of V, i.e. a point r_0 for which $V'(r_0) = 0$, gives rise to a circular orbit $r \equiv r_0$. If R is a non-critical point of V for which $V(R) = E^2$, then R is a turning point - the orbit reaches the end point R of the interval I and then turns around to move along I in the opposite direction. We compute

$$V'(r) = \frac{2}{r^4}\left(GMr^2 - L^2r + 3GML^2\right) \qquad (4.12)$$

and the quadratic expression in parenthesis has discriminant

$$L^2(L^2 - 12G^2M^2).$$

If this discriminant is negative, there are no critical points. If this discriminant is positive there are two critical points, $r_1 < r_2$ with r_1 a local maximum and r_2 a local minimum. (We will ignore the exceptional case of discriminant zero.) In the positive discriminant case we must distinguish between the cases where the local maximum at r_1 is not a global maximum, and when it is. Since $V(r) \to 1$ as $r \to \infty$ these two cases are distinguished by $V(r_1) < 1$ and $V(r_1) > 1$. Ignoring non-generic cases we thus can classify the behavior of $r(s)$ as:

- $L^2 < 12G^2M^2$ so V has no critical points and hence is monotone increasing on the interval $[2GM, \infty)$. The behavior of $r(s)$ for $s \geq 0$ subdivides into four cases, all leading to "crashing" (i.e. reaching the Schwarzschild boundary $2GM$ in finite s) or escape to infinity. The four possibilities have to do with the sign of $\dot{r}(0)$ and whether $E^2 < 1$ or $E^2 > 1$.

4.7. PROBLEMS: GEODESICS IN THE SCHWARZSCHILD EXTERIOR.

1. $E^2 < 1$, $\dot{r}(0) < 0$. The particle crashes into the barrier at $2GM$.
2. $E^2 < 1$, $\dot{r} > 0$. The orbit initially moves in the direction of increasing r, reaches its maximum value where $V(r) = E^2$, turns around and crashes.
3. $E^2 > 1$, $\dot{r} > 0$. The particle escapes to infinity.
4. $E^2 > 1$, $\dot{r} < 0$. The particle crashes.

- $12G^2M^2 < L^2 < 16G^2M^2$. Here there are two critical points, but the maximum value at r_1 is < 1. There are now four types of intervals I, depending on the value of E:

  1. $E^2 < V(r_1)$, $r < r_1$. Here the interval I lies below the local maximum and to the left of the graph of V. The behavior will be like the first two cases above - "crash" if $\dot{r}(0) < 0$ and turn around then crash if $\dot{r} > 0$.
  2. $E^2 < V(r_1)$, $r > r_1$. the interval I now lies in a well to the right of r_1, and so the value of r has two turning points corresponding to the end points of this interval. In other words the value of r is bounded along the entire orbit. We call this a *bound orbit*. In the "non-relativistic" approximation, this corresponds to Kepler's ellipses. In problems **4** and **5** below we will examine more closely how this approximation works and derive Einstein's famous calculation of the advance of the perihelion of Mercury.
  3. $V(r_1) < E^2 < 1$. The interval I is bounded on the right by the curve and extends all the way to the left (up to the barrier at $2GM$). The behavior is again either direct crash or turn around and then crash according to the sign of $\dot{r}(0)$,
  4. $E^2 > 1$. Now the possible behaviors are "crash" if $\dot{r}(0) < 0$ of escape to infinity if $\dot{r}(0) > 0$.

- $L > 4GM$. Now $V(r_1) > 1$. Again there will be four possible intervals:

  1. $E^2 < V(r_1)$, $r(0) < r_1$. This is an interval lying to the left of the "potential barrier" and so yield either a crash or turn around then crash orbit.
  2. $1 < E^2 < V(r_1)$. Now I lies to the right of the barrier, but below its peak, extending out to ∞ on the right. The orbit will escape to infinity if $\dot{r}(0) > 0$ or turn around and then escape if $\dot{r}(0) < 0$.
  3. $E^2 > V(r_1)$. The interval I extends from $2GM$ to infinity and the orbit is either crash or escape depending on the sign of $\dot{r}(0)$.
  4. $V(r_2) < E^2 < 1$. The interval now lies in a "well" to the right of the peak at r_1. We have again a bound orbit.

We are interested in the bound orbits with $L > 0$. According to problem **2** we can use ϕ as a parameter on such an orbit and by the second equation in that problem we have

$$\dot{r} := \frac{dr}{ds} = \frac{dr/d\phi}{ds/d\phi} = \frac{L}{r^2}\frac{dr}{d\phi}.$$

Substituting this and the definition of h into (4.9) we get

$$E^2 = \frac{L^2}{r^4}\left(\frac{dr}{d\phi}\right)^2 + \left(1 + \frac{L^2}{r^2}\right)\left(1 - \frac{2GM}{r}\right).$$

It is now convenient to introduce the variable

$$u := \frac{1}{r}$$

instead of r. We have

$$\frac{du}{d\phi} = -\frac{1}{r^2}\left(\frac{dr}{d\phi}\right) = -u^2\frac{dr}{d\phi}$$

so

$$E^2 = L^2\left(\frac{du}{d\phi}\right)^2 + (1 + L^2 u^2)(1 - 2GMu). \tag{4.13}$$

We can rewrite this as

$$\left(\frac{du}{d\phi}\right)^2 = 2GMQ, \quad Q := u^3 - \frac{1}{2GM}u^2 + \beta_1 u + \beta_0 \tag{4.14}$$

where β_0 and β_1 are constants, combinations of E, L, and GM:

$$\beta_1 = \frac{1}{L^2}, \quad \beta_0 = \frac{E^2 - 1}{2GML^2}.$$

4.7.4 Perihelion advance.

We will be interested in the case of bound orbits. In this case, a maximum value, u_1 along the orbit must be a root of the cubic polynomial, Q, as must be a minimum, u_2, since these are turning points where the left hand side of (4.14) vanishes. Notice that these values are then independent of ϕ, being roots of a given polynomial with constant coefficients. Since two of the roots of Q are real, so is the third, and all three roots must add up to $\frac{1}{2GM}$, the negative of the coefficient of u^2. Thus the third root is

$$\frac{1}{2GM} - u_1 - u_2.$$

We thus have

4.7. PROBLEMS: GEODESICS IN THE SCHWARZSCHILD EXTERIOR.

$$\left(\frac{du}{d\phi}\right)^2 = 2GM(u-u_1)(u-u_2)\left(u - \frac{1}{2GM} + u_1 + u_2\right).$$

Since the first factor on the right is non-positive and the second non-negative, the third is non-positive as the product must equal the non-negative expression on the left. Furthermore, we will be interested in the region where $r \gg 2GM$ so

$$2GM(u + u_1 + u_2) < 6GMu_1 \ll 1.$$

We therefore have the following expressions for $|d\phi/du|$:

$$\left|\frac{d\phi}{du}\right| = \frac{1}{\sqrt{(u_1-u)(u-u_2)}} \cdot [1 - 2GM(u+u_1+u_2)]^{-\frac{1}{2}} \quad (4.15)$$

$$\dot{=} \frac{1 + GM(u+u_1+u_2)}{\sqrt{(u_1-u)(u-u_2)}} \quad (4.16)$$

$$\dot{=} \frac{1}{\sqrt{(u_1-u)(u-u_2)}} \quad (4.17)$$

Here (4.16) is obtained from (4.15) by ignoring terms which are quadratic in $2GM(u+u_1+u_2)$ and (4.17) is obtained from (4.15) by ignoring terms which are linear in $2GM(u+u_1+u_2)$.

The strategy now is to observe that (4.17) is really the equation of an ellipse, whose Appolonian parameters, the latus rectum and the eccentricity, are expressed in terms of u_1 and u_2. Then (4.16) is used to approximate the advance in the perihelion of Keplerian motion associate to this ellipse.

4. Show that the ellipse

$$u = \frac{1}{\ell}(1 + e\cos\phi)$$

is a solution of (4.17) where e and ℓ are determined from

$$u_1 = \frac{1}{\ell}(1+e), \quad u_2 = \frac{1}{\ell}(1-e)$$

so that the mean distance

$$a := \frac{1}{2}\left(\frac{1}{u_1} + \frac{1}{u_2}\right) = \frac{\ell}{1-e^2}.$$

This is the approximating ellipse with the same maximum and minimum distance to the sun as the true orbit, if we choose our angular coordinate ϕ so that the $x-$ axis is aligned with the axis of the ellipse.

In principle (4.15) is in solved form; if we integrate the right hand side from u_1 to u_2 and then back again, we will get the total change in ϕ across a complete cycle. Instead, we will approximate this integral by replacing (4.15) by (4.16) and then also make the approximate change of variables $u = \ell^{-1}(1 + e\cos\phi)$.

5. By making these approximations and substitutions show that the integral becomes
$$\int_0^{2\pi} [1 + GM\ell^{-1}(3 + \cos\phi)] \, d\phi = 2\pi + \frac{6\pi GM}{\ell}$$
so the perihelion advance in one revolution is
$$\frac{6\pi GM}{a(1-e^2)}.$$

We have done these computations in units where the speed of light is one. If we are given the various constants in conventional units, say
$$G = 6.67 \times 10^{-11} \mathrm{m}^3/\mathrm{kg\ sec},$$
and the mass of the sun in kilograms
$$M = 1.99 \times 10^{30} \mathrm{kg}$$
we must replace G by G/c^2 where c is the speed of light, $c = 3 \times 10^8$ m/sec. Then $2GM/c^2 \doteq 1.5$km. We may divide by the period of the planet to get the rate of advance as
$$\frac{6\pi GM}{a(1-e^2)T}.$$
If we substitute, for Mercury, the mean distance $a = 5.768 \times 10^{10}$m, eccentricity $e = 0.206$ and period $T = 88$ days, and use the conversions
$$\begin{aligned} \text{century} &= 36524 \text{ days} \\ \text{radian} &= [360/2\pi] \text{degrees} \\ \text{degree} &= 3600'' \end{aligned}$$
we get the famous value of 43.1"/century for the advance of the perihelion of Mercury. This advance had been observed in the middle of the last century.

Up until recently, this observational verification of general relativity was not conclusive. The reason is that Newton's theory is based on the assumption that the mass of the sun is concentrated at a point. A famous theorem of Newton says that the attraction due to a homogeneous ball (on a particle outside) is the same as if all the mass is concentrated at a point. But if the sun is not a perfect sphere, or if its mass is not uniformly distributed, one would expect some deviation from Kepler's laws. The small effect of the advance of the perihelion of Mercury might have an explanation in terms of Newtonian mechanics. In the recent years, measurements from pulsars indicate large perihelion advances of the order of degrees per year (instead of arc seconds per century) yielding a striking confirmation of Einstein's theory.

4.7.5 Massless particles.

We now have

$$\gamma'(s) = \dot{x}^0(s)\left(\frac{\partial}{\partial x^0}\right)_{\gamma(s)} + \dot{x}^1(s)\left(\frac{\partial}{\partial x^1}\right)_{\gamma(s)} + \dot{x}^2(s)\left(\frac{\partial}{\partial x^2}\right)_{\gamma(s)} + \dot{x}^3(s)\left(\frac{\partial}{\partial x^3}\right)_{\gamma(s)}.$$

$$\langle\gamma'(s),\gamma'(s)\rangle_{\gamma(s)} \equiv 0.$$

6. Using problem **2** verify that

$$E^2 = \left(\frac{dr}{ds}\right)^2 + \left(\frac{L^2}{r^2}\right)h$$

and then

$$\frac{d^2u}{d\phi^2} + u = 3GMu^2. \tag{4.18}$$

We will be interested in orbits which go out to infinity in both directions. For large values of r, the right hand side is negligibly small, so we should compare (4.18) with

$$\frac{d^2u_0}{d\phi^2} + u_0 = 0$$

whose solutions are

$$u_0 = a\cos\phi + b\sin\phi$$

or

$$1 = ax + by, \quad x = r\cos\phi, \ y = r\sin\phi,$$

in other words straight lines. We might as well choose our angular coordinate ϕ so that this straight line is parallel to the $y-$ axis, i.e.

$$u_0 = r_0^{-1}\cos\phi$$

where r_0 is the distance of closest approach to the origin. Suppose we are interested in light rays passing the sun. The radius of the sun is about 7×10^5 km while $2GM$ is about 1.5km. Hence in units where r_0 is of order 1 the expression $3GM$ is a very small quantity, call it ϵ. So write our equation as

$$u'' + u = \epsilon u^2, \quad \epsilon = 3GM. \tag{4.19}$$

We solve this by the method of perturbation theory: look for a solution of the form

$$u = u_0 + \epsilon v + \cdots$$

where the error is of order ϵ^2. We choose u_0 as above to solve the equation obtained by equating the zero-th order terms in ϵ.

7. Compare coefficients of ϵ to obtain the equation

$$v'' + v = \frac{1}{2r_0^2}(1 + \cos 2\phi)$$

and try a solution of the form $v = a + b\cos 2\phi$ to find the solution of this equation and so obtain the first order approximation

$$u = \frac{1}{r_0}\cos\phi - \frac{\epsilon}{3r_0^2}\cos^2\phi + \frac{2\epsilon}{3r_0^2} \qquad (4.20)$$

to (4.19).

The asymptotes as $r \to \infty$ or $u \to 0$ will be straight lines with angles obtained by setting $u = 0$ in (4.20). This gives a quadratic equation for $\cos\phi$.

8. Remembering that cosine must be ≤ 1 show that up through order ϵ we have

$$\cos\phi = -\frac{2\epsilon}{3r_0} = -\frac{2GM}{r_0}.$$

Writing $\phi = \pi/2 + \delta$ this gives $\sin\delta = 2GM/r_0$ or approximately $\delta = 2GM/r_0$. This was for one asymptote. The same calculation gives the same result for the other asymptote. Adding the two and passing to conventional units gives

$$\Delta = \frac{4GM}{c^2 r_0} \qquad (4.21)$$

for the deflection. For light just grazing the sun this predicts a deflection of 1.75". This was approximately observed in the expedition to the solar eclipse of 1919.

Recent, remarkable, photographs from the Hubble space telescope have given strong confirmation to Einstein's theory from the deflection of light by dark matter.

4.7.6 Kerr-Schild form.

We will show that by making a change of variables that the Schwarzschild metric is the sum of a flat metric, and a multiple of the square of a linear differential form α where $\|\alpha\|^2 = 0$ in the flat metric. A metric of this type is said to be of **Kerr-Schild** form. The generalization of this construction is important in the case of rotating black holes.

We make the change of variables in the metric in two stages: Let

$$u = t + T(r)$$

where T is any function of r (determined up to additive constant) such that

$$T' = \frac{1}{h}, \quad h = 1 - \frac{2M}{r}.$$

4.7. PROBLEMS: GEODESICS IN THE SCHWARZSCHILD EXTERIOR.

Then $du = dt + \frac{1}{h}dr$ so

$$\begin{aligned} hdt^2 &= hdu^2 - 2dudr + \frac{1}{h}dr^2 \text{ so} \\ -hdt^2 + \frac{1}{h}dr^2 &= -hdu^2 + 2dudr \\ &= -(du - dr)^2 + dr^2 + \frac{2M}{r}du^2. \end{aligned}$$

If we set $x^0 := u - r$ this beomes

$$-d(x^0)^2 + dr^2 - \frac{2M}{r}\left[dx^0 + dr\right]^2.$$

The linear differential form $dx^0 + dr$ has square length zero in the flat metric

$$-d(x^0)^2 + dx^2 + dy^2 + dz^2, \quad r^2 = x^2 + y^2 + z^2$$

so the Schwarzschild metric is given by

$$d(x^0)^2 + dr^2 + r^2 d\sigma^2 - \frac{2M}{r}\left[dx^0 + dr\right]^2$$

which is the desired Kerr-Schild form.

4.7.7 A brief biography of Schwarzschild culled from wikipedia and St. Andrews.

Karl Schwarzschild

(October 9, 1873 – May 11, 1916)

Karl Schwarzschild's parents were Henrietta Sabel and Moses Martin Schwarzschild. The family was Jewish, with Karl's father being a well-off member of the business community in Frankfurt.

He attended a Jewish primary school in Frankfurt up to the age of eleven, then he entered the Gymnasium there. It was at this stage that he became interested in astronomy and saved his pocket money to buy himself materials such as a lens from which he could construct a telescope. Karl's father was friendly with Professor J Epstein, who was professor at the Philanthropin Academy and had his own private observatory. Their friendship arose through a common interest in music. Professor Epstein has a son, Paul Epstein, who was two years older than Karl and the two boys became good friends. They shared an interest in astronomy, and Karl learnt how to use a telescope and also learnt some advanced mathematics from his friend Paul Epstein.

It was in large part what he learnt through his friendship with Epstein which led to Schwarzschild mastering celestial mechanics by the age of sixteen. Such was this mastery that he wrote his first two papers on the theory of orbits of double stars at this age (!) while still at the Frankfurt Gymnasium. The papers were published in Astronomische Nachrichten in 1890.

Schwarzschild studied at the University of Strasbourg during the two years 1891-93 where he learnt a great deal of practical astronomy, then at the University of Munich where he obtained his doctorate. His dissertation, on an application of Poincaré's theory of stable configurations of rotating bodies to tidal deformation of moons and to Laplace's origin of the solar system, was supervised by Hugo von Seeliger. Schwarzschild found great inspiration from Seeliger's teaching which influenced him throughout his life.

After the award of his doctorate, Schwarzschild was appointed as an assistant at the Von Kuffner Observatory in Ottakring which is a suburb of Vienna. He took up his appointment in October 1896 and held it until June 1899. While at the Observatory he worked on ways to determine the apparent brightness of stars using photographic plates.

From 1901 until 1909 he was Extraordinary Professor at Göttingen and also director of the Observatory there. In Göttingen he collaborated with Klein, Hilbert and Minkowski. In less than a year he had been promoted to Ordinary Professor.

Schwarzschild published on electrodynamics and geometrical optics during his time at Göttingen. He carried out a large survey of stellar magnitudes while at the Göttingen Observatory, publishing Aktinometrie (the first part in 1910, the second in 1912). In 1906 he studied the transport of energy through a star by radiation and published an important paper on radiative equilibrium of the atmosphere of the sun.

He married Else Posenbach, the daughter of a professor of surgery at Göttingen, on 22 October 1909. They had three children, Agathe, Martin who was born on 31 May 1912 and went on to became a professor of astronomy at Princeton, and Alfred.

After his marriage, near the end of 1909, Schwarzschild left Göttingen to take up an appointment as director of the Astrophysical Observatory in Potsdam. This was the most prestigious post available for an astronomer in Germany and he filled the position with great success. He had the opportunity to study photographs of the return of Halley's comet in 1910 taken by a Potsdam expedition

4.8. CURVATURE IDENTITIES.

to Tenerife. He also made major contributions to spectroscopy which became a topic of great interest to him around this time.

On the outbreak of war in August 1914 Schwarzschild volunteered for military service. He served in Belgium where he was put in charge of a weather station, France where he was assigned to an artillery unit and given the task of calculating missile trajectories, and then Russia.

He contracted an illness while in Russia called pemphigus, which is a rare autoimmune blistering disease of the skin, with a more common frequency among Ashkenazic Jews. For people with this disease the immune system mistakes the cells in the skin as foreign and attacks them causing painful blisters. In Schwarzschild's time there was no known treatment and, after being invalided home in March 1916, he died two months later.

Today, he is best known for providing the first exact solution to the Einstein field equations of general relativity, for the limited case of a single spherical non-rotating mass, which he accomplished in 1915, the same year that Einstein first introduced general relativity. The Schwarzschild solution, which makes use of Schwarzschild coordinates and the Schwarzschild metric, leads to the well-known Schwarzschild radius, which is the size of the event horizon of a non-rotating black hole. He wrote this paper while suffering from a painful disease and with full knowledge of his impending death.

4.8 Curvature identities.

The curvature of the Levi-Civita connection satisfies several additional identities beyond the two curvature identities that we have already discussed. Let us choose vector fields X, Y, V with vanishing brackets. We have

$$\begin{aligned}
-\langle R_{XY}V, V\rangle &= -\langle \nabla_X \nabla_Y V, V\rangle + \langle \nabla_Y \nabla_X V, V\rangle \\
&= Y\langle \nabla_X V, V\rangle - \langle \nabla_X V, \nabla_Y V\rangle - X\langle \nabla_Y V, V\rangle + \langle \nabla_Y V, \nabla_X V\rangle \\
&= \frac{1}{2}YX\langle V, V\rangle - \frac{1}{2}XY\langle V, V\rangle \\
&= \frac{1}{2}[X, Y]\langle V, V\rangle \\
&= 0.
\end{aligned}$$

This implies that for any three tangent vectors we have

$$\langle R_{\xi\eta}\zeta, \zeta\rangle = 0$$

and hence, by polarization, that for any four tangent vectors we have

$$\langle R_{\xi\eta}v, \zeta\rangle = -\langle v, R_{\xi\eta}\zeta\rangle. \tag{4.22}$$

This equation says that the curvature operator $R_{\xi\eta}$ acts as an infinitesimal orthogonal transformation on the tangent space.

The last identity we want to discuss is the symmetry property

$$\langle R_{\xi\eta}v, \zeta\rangle = \langle R_{v\zeta}\xi, \eta\rangle. \tag{4.23}$$

The proof consists of starting with the first Bianchi identity
$$\mathcal{C}yc R_{\eta,v}\xi = 0$$
and taking the scalar product with ζ to obtain
$$\langle \mathcal{C}yc\, R_{\eta,v}\xi, \zeta \rangle = 0.$$
This is an equation involving three terms. Take the cyclic permutation of the four vectors to obtain four equations like this involving twelve terms in all. When we add the four equations eight of the terms cancel in pairs and the remaining terms give (4.23). We summarize the symmetry properties of the Riemann curvature:

- $R_{\xi\eta} = -R_{\eta\xi}$
- $\langle R_{\xi\eta}v, \zeta \rangle = -\langle v, R_{\xi\eta}\zeta \rangle$
- $R_{\xi\eta}\zeta + R_{\eta\zeta}\xi + R_{\zeta\xi}\eta = 0$
- $\langle R_{\xi\eta}v, \zeta \rangle = \langle R_{v\zeta}\xi, \eta \rangle.$

Consider $R(X, Y, W, V)$. This is anti-symmetric in X, Y and in V, W and symmetric under the exchange of the pairs X, Y with W, V. So it defines a symmetric bilinear form on $\wedge^2(TM)$ by

$$\mathcal{R}(X \wedge Y, V \wedge W) = R(X, Y, W, V). \tag{4.24}$$

Notice the reversal of V and W in this definition. This will bring our definitions in what follows into line with O'Neill.

This bilinear form in turn defines a linear operator $\mathfrak{R} : \wedge^2 TM \to \wedge^2 TM$ by

$$\langle \mathfrak{R}(X \wedge Y), V \wedge W \rangle = \mathcal{R}(X \wedge Y, V \wedge W). \tag{4.25}$$

The $\langle \, , \, \rangle$ occurring on the left hand side of (4.25) is the "scalar product" induced on $\wedge^2(TM)$: If we are given a a bilinear form $\langle \, , \, \rangle$ on a vector space V it induces a $\langle \, , \, \rangle$ on each of the exterior powers $\wedge^k V$. See for example, the discussion at the beginning of Chapter 12. For the case $k = 2$, for example, this "scalar product" on decomposable vectors is given by

$$\langle v_1 \wedge v_2, w_1 \wedge w_2 \rangle = \langle v_1, w_1 \rangle \langle v_2, w_2 \rangle - \langle v_1, w_2 \rangle \langle v_2, w_1 \rangle.$$

4.9 Sectional curvature.

4.9.1 Degenerate and non-degenerate planes.

For a pair of vectors in a vector space with a non-degenerate symmetric bilinear form $\langle \, , \, \rangle$ define
$$Q(v, w) := \langle v, v \rangle \langle w, w \rangle - \langle v, w \rangle^2.$$

4.9. SECTIONAL CURVATURE.

In the positive definite case this is the square of the area of the parallelogram spanned by v and w. Assuming that v and w are independent, if $Q(v,w) = 0$, the same is true for any two vectors in the plane that they span. Such a plane is called degenerate because the bilinear form restricted to that subspace is degenerate. Of course, in the positive definite case, every two dimensional subspace is non-degenerate. In terms of the induced non-degenerate form on $\wedge^2 V$

$$Q(v,w) = \langle v \wedge w, v \wedge w \rangle.$$

We will let $\Pi = \Pi(v,w)$ denote the plane spanned by v and w.

In what follows we will assume unless otherwise stated that $\Pi(v,w)$ is non-degenerate. Notice that a slight perturbation will carry a degenerate plane into a non-degenerate one.

4.9.2 Definition of the sectional curvature.

Recall the directional curvature operator R_v defined by $R_v w = R(v,w)v$ and the operator $\mathfrak{R}$ we introduced above. Define

$$\begin{aligned} \sec(v,w) &= -\frac{\langle R_v(w), w \rangle}{Q(v,w)} \quad (4.26) \\ &= \frac{\langle R(w,v)v, w \rangle}{Q(v,w)} \\ &= \frac{\langle \mathfrak{R}(v \wedge w), v \wedge w \rangle}{Q(v,w)}. \end{aligned}$$

From the last expression it is clear that $\sec(v,w)$ depends only on the plane $\Pi(v,w)$ so we will also write $\sec(\Pi)$. So $\Pi \mapsto \sec(\Pi)$ is a real valued function on the space of non-degenerate planes.

4.9.3 The sectional curvature determines the Riemann curvature tensor.

We wish to prove that sec (together with the metric) determines R. This is a purely algebraic result for which we will need the following definitions and a lemma:

- A multilinear function F on a vector space is called **curvaturelike** if it satisfies the identities listed above for the curvature R. The space of curvaturelike multilinear functions is clearly a vector space.

- If F is a curvaturelike multilinear function on a vector space with a non-degenerate symmetric bilinear form we define

$$K_F(v,w) = \frac{F(w,v,v,w)}{Q(v,w)}, \quad \text{if } Q(v,w) \neq 0$$

just as in the definition of the sectional curvature. And just as in the definition of the sectional curvature $K_F(v,w)$ depends only on the plane Π spanned by v and w and so we may write $K_F(\Pi)$.

Lemma 5. *If F is a curvaturelike multilinear function and $K_F = 0$ then $F = 0$. As a consequence, if F_1 and F_2 are curvaturelike multilinear functions with $K_{F_1} = K_{F_2}$ then $F_1 = F_2$.*

The consequence follows from the first statement in the lemma since the evaluation map $F \mapsto F(v,w,v,w)$ is linear in F.

Proof. To begin the proof observe that the hypothesis is that $F(v,w,w,v) = 0$ for all non-degenerate pairs. But since these are dense and F is continuous $F(v,w,w,v) = 0$ for all v and w.

As a second step we will prove that $F(w,v,v,x) \equiv 0$. Indeed we know from multilinearity that
$$0 = F(w+x, v, v, w+x)$$
$$= F(w,v,v,w) + F(x,v,v,w) + F(w,v,v,x) + F(x,v,v,x).$$
The first and last term on the right vanish, from our assumption. The two middle terms are equal by one of the curvature-like identities. So we get $2F(w,v,v,x) = 0$ for all v, w, x.

We know that $F(w,v,v,x) \equiv 0$. So $F(w, v+y, v+y, x) \equiv 0$ and therefore
$$F(w,v,y,x) = -F(w,y,v,x) = F(y,w,v,x).$$
So $F(w,v,y,x)$ is unchanged by a cyclic permutation of the first three variables, But then the analogue of the first Bianchi identity for F implies that $3F(w,v,y,x) = 0$ so $F = 0$.

□

4.9.4 Constant curvature spaces.

M is said to have **constant curvature** k if $\sec(\Pi) = k$ (at a point p) for all non-degenerate two planes Π at p. This will certainly be the case if for any three vectors x, y, z at p we have
$$R(x,y)z = k(\langle y,z \rangle x - \langle x,z \rangle y) \tag{4.27}$$
for then
$$\langle R(w,v)v, w \rangle = kQ(v,w).$$
A famous result of Riemann says the converse - that if the $\sec(\Pi) = k$ for all Π then the curvature has the form given by (4.27). Indeed, the function given by the right hand side of (4.27) is curvaturelike (as you can check) and the result follows from the lemma.

Later on, when we learn effective and simple methods for computing curvature, will find that (in Riemannian geometry) all possible values of k can occur

4.10. RICCI CURVATURE.

as the constant curvature. Indeed spheres have constant positive curvature, Euclidean spaces have $k = 0$ and $k < 0$ in hyperbolic geometry.

In all of these cases, the sectional curvature is not only constant on the two planes at a point, but k is the same at all points of M. We will prove a theorem of F. Schur which says that if $\dim M \geq 3$, and if the Riemann metric on M has constant curvature $f(p)$ at each point p of M, then f is a constant. I will postpone the proof of this theorem until we will have developed the Cartan formalism in Riemannian geometry.

4.10 Ricci curvature.

If we hold $\xi \in TM_x$ and $\eta \in TM_x$ fixed in $R_{v\xi}\eta$ then the map

$$v \mapsto R_{v\xi}\eta \quad v \in TM_x$$

is a linear map of TM_x into itself. Its trace (which is biinear in ξ and η) is known as the **Ricci curvature tensor**.

$$\text{Ric}(\xi, \eta) := \text{tr}[v \mapsto R_{v\xi}\eta]. \tag{4.28}$$

Ricci curvature plays a key role in general relativity because it is the Ricci curvature rather than than the full Riemann curvature which enters into the Einstein field equations.

We will give a geometric interpretation of the sectional and Ricci curvatures in Chapter 10. In the next chapter, we will compute some of these these objects in a group theoretic setting.

But observe the following useful property of the Ricci curvature : By polarization and scalar multiplication the Ricci tensor is determined at a point by its values $\text{Ric}(u, u)$ on "unit vectors", i.e. vectors satisfying $|\langle u, u \rangle| = 1$. On such vectors, it follows from the definition that

$$\text{Ric}(u, u) = \langle u, u \rangle \sum_2^n \sec(u, e_i), \tag{4.29}$$

where $e_2, \ldots, e_n$ is an "orthonormal" basis of $u^\perp$. In other words, $\text{Ric}(u, u) = \langle u, u \rangle \times$ the sum of the sectional curvatures of any $n-1$ orthogonal non-degenerate planes through u.

The (Ricci) **scalar curvature S** is the contraction of the Ricci curvature. So in terms of an "orthonormal" basis of TM_m

$$\mathbf{S}(m) = \sum_{i \neq j} \sec(e_i, e_j). \tag{4.30}$$

In particular, if $(M, \mathbf{g})$ is a semi-Riemannian manifold with constant sectional curvature, k, then it follows from the above that

$$\text{Ric} = (n-1)k\mathbf{g} \quad \text{and} \quad \mathbf{S} = n(n-1)k. \tag{4.31}$$

4.11 Locally symmetric semi-Riemannian manifolds.

In this section we will discuss some information about locally symmetric Levi-Civita connections. We begin by using polar maps associated with the Levi-Civita connection to show that a local isometry of a semi-Riemannian manifold is determined by its value and its differential at a single point. More precisely: Let M and N be semi-Riemannian manifolds with M connected and let $\Phi, \Psi : M \to N$ be local isometries.

Proposition 13. *If there is a point $p \in M$ such that $\Phi(p) = \Psi(p)$ and $d\Phi_p = d\Psi_p$ then $\Phi = \Psi$.*

Proof. Let $A := \{q \in M | \Phi(q) = \Psi(q) \text{ and } d\Phi_q = d\Psi_q\}$. By continuity, A is closed. Since $p \in A$ it is non-empty. If we show that A is open then it must be all of M since M is connected. But on any normal neighborhood U of $q \in A$ we must have $\Phi = \phi_L$ where $L = d\phi_q$. Since $d\Psi_q = L$ as well, we must have $\Phi = \Psi$ on U and hence their differentials agree there as well. □

Since the orthogonal group of an n-dimensional space has dimension $n(n-1)/2$, it follows from Proposition 13 that

Proposition 14. *The maximal dimension of the group of isometries of an n-dimensional semi-Riemmanian manifold is $n(n+1)/2$.*

We can improve on Proposition 9 in the semi-Riemannian case by adding a third equivalent condition:

Proposition 15. *The following conditions on a semi-Riemmannian manifold M are equivalent:*

1. $\nabla R = 0$.

2. *If X, Y, Z are parallel vector fields on a non-singular curve α then $R(X,Y)Z$ is also parallel on α.*

3. *The sectional curvature $\sec(\Pi)$ of any non-degenerate tangent plane remains constant as Π is parallel translated along α.*

*If M satisfies any one, hence all, of these conditions then M is called **locally symmetric**.*

We already know that conditions 1 and 2 are equivalent. We will show that 2 and 3 are equivalent:

Proof that 2 ⇒ 3. Let Π be a non-degenerate tangent plane at $\alpha(0)$ and let X, Y be parallel vector fields along α with $X(0), Y(0)$ a basis for Π. Thus the parallel planes $\Pi(t)$ are spanned by $X(t), Y(t)$. Since parallel translation is an isometry, $Q(X(t), Y(t))$ is constant along α and 2 tells us that $R(Y(t), X(t))X(t)$

4.11. LOCALLY SYMMETRIC SEMI-RIEMANNIAN MANIFOLDS.

is parallel along α and hence $\langle R(Y(t), X(t))X(t), Y(t)\rangle$ is constant along α. Hence both the numerator and the denominator in the definition of $\sec(\Pi)(t)$ are constant along α. $\square$

Proof that 3 $\Rightarrow$ 2. This will be a variant of the fact that the sectional curvature determines the curvature. Let X, Y, Z, W be any parallel vector fields along α. It is enough to prove that $\langle R(X,Y)Z, W\rangle$ is constant for we may let W range over a parallel basis of $T_{\alpha(t)}M$. So define the multilinear function $A(t) : (T_{\gamma(t)}M)^4 \to \mathbb{R}$ by

$$A(t)(x, y, z, w) = \langle R(X, Y)Z, W\rangle(t)$$

where X, Y, Z, W are parallel vector fields extending $x, y, z, w \in T_{\gamma(t)}M$. This function is clearly curvaturelike at each t. Since $Q(X(t), Y(t))$ is constant and $\sec(\Pi)$ is constant we see that $A(t)$ is constant. $\square$

We now (following Cartan) derive some important consequences of the above improved proposition:

Theorem 7. *Let M and N be locally symmetric n dimensional semi-Riemannian manifolds and let $o \in M$ and $p \in N$. Let $L : T_oM \to T_pN$ be a linear isometry that preserves curvature, Then if $\mathcal{U}$ is a sufficiently small normal neighborhood of o there exists a unique isometry Φ of $\mathcal{U}$ onto a normal neighborhood of p such that $d\Phi_o = L$.*

The uniqueness follows from the fact that an isometry is determined by its value, and the value of its differential at a single point. In fact, we know that at any point such a map must be given by the polar map about that point. So for the existence, it is enough for us to prove that the polar map $\phi = \phi_L$ is an isometry.

Proof. Let $q \in \mathcal{U}$ and $v \in T_qM$. We wish to prove that

$$\langle d\phi_q v, d\phi_q v\rangle = \langle v, v\rangle \tag{4.32}$$

where $\phi = \phi_L$.

Let $\tilde{\mathcal{U}} = \exp_o^{-1}(\mathcal{U})$. There is a unique $x \in \tilde{\mathcal{U}}$ such that $\exp_o(x) = q$ and a unique $y_x \in T_x(T_oM)$ such that $d(\exp_o)_x(y_x) = v$. Let y be the vector in T_oM which corresponds to y_x under the identification of $T_x(T_oM)$ with T_oM. We know that

$$v = Y(1)$$

where Y is the Jacobi field on γ_x such that $Y(0) = 0$ and $Y'(0) = y$.

The polar map $\phi = \phi_L$ is given by $\phi = \exp_p \circ L \circ (\exp_o)^{-1}$ and hence

$$d\phi_q(v) = d(\exp_p)_{Lx}(dL_{Lx}(y_x)).$$

Under the identification of the tangent space to a vector space with that vector space, the differential of any linear function is identified with that linear function. So we can write the above as

$$d\phi_q(v) = d(\exp_p)_{Lx}(Ly)_{Lx}.$$

$$d\phi_q(v) = d(\exp_p)_{Lx}(Ly)_{Lx}.$$

By applying our theorem relating the Jacobi equations to the exponential map again, this time to N, we see that

$v = Y(1)$ where Y is the Jacobi field on γ_x such that $Y(0) = 0$ and $Y'(0) = y$ and $d\phi_q(v) = Z(1)$ where Z is the Jacobi field on γ_{Lx} such that $Z(0) = 0$ and $Z'(0) = Ly$.

Let $E_1, \ldots, E_n$ be a parallel orthonormal frame field on γ_x and let $F_1, \ldots F_n$ be a parallel orthonormal frame field on γ_{Lx} such that $L(E_i(0)) = F_i(0)$ for all i. The coefficients of the curvature tensors relative to these frames are constant, since we are assuming that both M and N are locally symmetric. Also they are equal to one another at 0, since we are assuming that L preserves curvature. Hence the Jacobi equations expressed as second order linear differential equations are the same on γ_{Lx} as on γ_x. Also, the coefficients of Ly in terms of the F-basis are the same as the coefficients of y in terms of the E-basis.

So these coefficients, call them $y(t)$ and $z(t)$ satisfy the same differential equations and the same initial conditions so are equal. But

$$\langle Y(1), Y(1) \rangle = \sum_i \epsilon_i y_i(1)^2 \text{ and } \langle Z(1), Z(1) \rangle = \sum_i \epsilon_i z_i(1)^2$$

so

$$\langle d\phi_q(v), d\phi_q(v) \rangle = \langle v, v \rangle.$$

$\square$

Corollary 2. *Let M and N be two Riemannian manifolds of the same dimension and the same (constant) constant curvature. Then for any $o \in M$ and $p \in N$ there is an isometry of a neighborhood of o with a neighborhood of p.*

Indeed, by Sylvester's theorem, all vector spaces of the same dimension with positive definite scalar product are isomorphic. Since the sectional curvatures are assumed to be the same at all points of M and of N they are both locally symmetric spaces and any linear isometry between any two of their tangent spaces preserves the sectional curvature and hence the curvature. $\square$

Again, by Sylvester's theorem, we can let M and N be semi-Riemannian manifolds provided that we assume in addition that the metrics have the same signature.

In two dimensions this is a theorem of Minding, (1839),

Proposition 16. *Any two two dimensional Riemannian manifolds with the same constant Gaussian curvature are locally isometric.*

4.11.1 Why the word "symmetric" in locally symmetric?

Let M be any semi-Riemannian manifold and let $o \in M$. The map $-I_o$, i.e. the map $v \mapsto -v$ on T_oM is an isometry. We can construct the polar map (defined in some normal neighborhood of o):

$$\zeta_o := \phi_{-I_o}.$$

It sends $\gamma_x(t)$ into $\gamma_x(-t)$. Let us call this map the **local geodesic symmetry** at o. Suppose we know that this is an isometry for every $o \in M$. Since it is an isometry, it preserves curvature and covariant derivative. In other words, for every four tangent vectors v, x, y, z at o we have

$$-(\nabla_v R)(x,y)z = \nabla_{-v} R(-x,-y)(-z).$$

But

$$(\nabla_{-v} R)(-x,-y)(-z) = (\nabla_v R)(x,y)z.$$

So $\nabla R \equiv 0$ and M is locally symmetric according to our definition. On the other hand the theorem asserts that if M is locally symmetric (according to our definition) then any linear map $L: T_pM \to T_qM$ which preserves curvature and is a linear isometry extends to a local isometry.

Now $-I_o$ actually preserves every two plane Π so automatically is a linear isometry which preserves sectional curvature and hence preserves curvature. So we have proved the following theorem due to Cartan:

Theorem 8. *The following conditions on a semi-Riemannian manifold are equivalent:*

- *M is locally symmetric according to any one of our previously stated equivalent conditions. (For example $\nabla R \equiv 0$.)*

- *If $L: T_pM \to T_qM$ is a linear isometry that preserves (sectional) curvature then there is an isometry ϕ of normal neighborhoods of p and q such that $\phi(p) = q$ and $d\phi_p = L$.*

- *At each $p \in M$ the local geodesic symmetry at p is a local isometry.*

4.12 Curvature of the induced metric of a submanifold.

We saw in Section 4.5 that the Levi-Civita connection of the metric induced on a submanifold is the tangential component of the Levi-Civita connection of M. See equation (4.5).

Let us now consider the normal component $(\nabla_X Y)^{nor}$ where X and Y are vector fields on S (extended so as to be vector fields on M). This assigns to every point $s \in S$ an element of TM_s which is orthogonal to TS_s. It is a "vector field of M along S".

Since $\nabla_{fX}Y = f\nabla_X Y$ for $f \in \mathfrak{F}(M)$ the value of $(\nabla_X Y)^{nor}$ at a point $s \in S$ depends only on $X(s)$. Let $g \in \mathfrak{F}(S)$, extended to be a smooth function on M. Then
$$\nabla_X(gY) = (Xg)Y + g\nabla_X Y.$$
Upon taking the normal component, the first term disappears, since Y is tangential. So $(\nabla_X Y)^{nor}(s)$ depends only on $X(s)$ and $Y(s)$. To shorten notation we will set
$$\mathcal{N}(X,Y) := (\nabla_X Y)^{nor}$$
for any $X, Y \in \mathcal{V}(S)$.

Let V, W, X, Y be vector fields on S. We can consider
$$\langle R_{VW} X, Y \rangle$$
where R is the curvature tensor of M, and also
$$\langle R^S_{VW} X, Y \rangle$$
where R^S is the curvature tensor of the induced metric and its connection. The following formula gives the relation between the two:
$$\langle R^S_{VW} X, Y \rangle$$
$$= \langle R_{VW} X, Y \rangle - \langle \mathcal{N}(V,X), \mathcal{N}(W,Y) \rangle + \langle \mathcal{N}(V,Y), \mathcal{N}(W,X) \rangle. \tag{4.33}$$

Proof. With no loss of generality we may assume that $[V, W] = 0$ so that $R_{VW} = \nabla_V \nabla_W - \nabla_W \nabla_V$. We write $\nabla_W X = \nabla^S_W X + \mathcal{N}(W,X)$ so that
$$\langle \nabla_V \nabla_W X, Y \rangle = \langle \nabla_V \nabla^S_W X, Y \rangle + \langle \nabla_V (\mathcal{N}(W,X)), Y \rangle. \tag{$*$}$$
Since Y is tangent to S, the first term in $(*)$ is the same as would be obtained by replacing $\nabla_V \nabla^S_W X$ by its tangential component which is $\nabla^S_V \nabla^S_W X$. We massage the second term in $(*)$ by writing it as
$$V \langle \mathcal{N}(W,X), Y \rangle - \langle \mathcal{N}(W,X), \nabla_V Y \rangle. \tag{$**$}$$
Since Y is tangent to S the function $\langle \mathcal{N}(W,X), Y \rangle$ vanishes on S, and since V is tangent to S the first term in $(**)$ vanishes. Since $\mathcal{N}(W,X)$ is normal to S we may replace the $\nabla_V Y$ in the second term in $(**)$ by its normal component which is $\mathcal{N}(V,Y) \rangle$. So $(*)$ becomes
$$\langle \nabla_V \nabla_W X, Y \rangle = \langle \nabla^S_V \nabla^S_W X, Y \rangle - \langle \mathcal{N}(V,X), \mathcal{N}(W,Y) \rangle.$$
Interchanging V and W and subtracting yields (4.33).

□

4.13. THE DE SITTER UNIVERSE AND ITS RELATIVES.

The induced sectional curvature.

Let v and w span a non-degenerate plane in the tangent space TS_s at some point $s \in S$. Let $\sec^M(v,w)$ denote the sectional curvature of this plane considered as a plane in TM_s and $\sec^S(v,w)$ its sectional curvature considered as a plane in TS_s relative to the induced metric on S. Recalling the definition of the sectional curvature it follows from (4.33) that

$$\sec^S(v,w) = \sec^M(v,w) + \frac{\langle \mathcal{N}(v,v), \mathcal{N}(w,w)\rangle - \langle \mathcal{N}(v,w), \mathcal{N}(v,w)\rangle}{\langle v,v\rangle\langle w,w\rangle - \langle v,w\rangle^2}. \qquad (4.34)$$

4.12.1 The case of a hypersurface.

The above formulas simplify somewhat when S is a hypersurface. The assumption that S is non-degenerate implies that the line normal to S is non-degenerate. So if y is a non-zero vector normal to S, we have $\langle y,y\rangle \neq 0$. But (in the semi-Riemannian case) this might be positive or negative.

Let us choose, locally, a normal vector field, with $|\langle U,U\rangle| = 1$ and let (following O'Neill) $\epsilon = \langle U,U\rangle \equiv \pm 1$. For V tangent to S we have

$$0 = \frac{1}{2}V\langle U,U\rangle = \langle \nabla_V U, U\rangle.$$

So $\nabla_V U$ is tangent to S. Let W be another vector field tangent to S. Then since $\langle W,U\rangle \equiv 0$, we have

$$\langle \nabla_V U, W\rangle = -\langle \nabla_V W, U\rangle = -\langle \mathcal{N}(V,W), U\rangle.$$

So if we define the $(1,1)$ tensor $\mathfrak{S}$ on S which assigns to each vector field V the vector field $-\nabla_V U$ we have

$$\langle \mathfrak{S}(V), W\rangle = \langle \mathcal{N}(V,W), U\rangle.$$

So

$$\langle \mathcal{N}(V,W), \mathcal{N}(V',W')\rangle = \langle \mathfrak{S}(V), W\rangle \langle \mathfrak{S}(V'), W'\rangle \langle U,U\rangle = \epsilon \langle \mathfrak{S}(V), W\rangle \langle \mathfrak{S}(V'), W'\rangle$$

for any four vector fields $V, W, V'W'$ tangent to S. So (4.34) becomes

$$\sec^S(v,w) = \sec^M(v,w) + \epsilon \frac{\langle \mathfrak{S}v, v\rangle\langle \mathfrak{S}w, w\rangle - \langle \mathfrak{S}v, w\rangle^2}{\langle v,v\rangle\langle w,w\rangle - \langle v,w\rangle^2} \qquad (4.35)$$

for any pair of tangent vectors at a point of S which span a non-degenerate plane. (O'Neill calls the tensor field $\mathfrak{S}$ the "shape operator". See the discussion following the "warning" in Section 1.3 in Chapter I.)

4.13 The de Sitter universe and its relatives.

In 1917, de Sitter wrote down a solution of the Einstein field equations (see below in this section) for a universe devoid of matter provided that these equations contain a "cosmological constant".

His key papers are

Koninklijke Akademie van Wetenschappen te Amsterdam 26, Nr. 2, (1917), 222-236. Over de Kromming der ruimte. (On the curvature of space), Proc. Amsterdam 20 (1918) 229,

and

Koninklijke Akademie van Wetenschappen te Amsterdam 26, Nr. 10, (1918), 1472-1475. Nadere opmerkingen omtrent de oplossingen der veldverglijkingen van Einstein's gravitatie-theorie. (Further remarks on the solutions of the field-equations of Einstein's theory of gravitation), Proc. Amsterdam 20 (1918) 1309.

The de Sitter universe has a vey simple description: it is a quadratic hypersurface (a "pseudosphere") in the five dimensional space $\mathbb{R}^{1,4}$ - the space $\mathbb{R}^5$ equipped with the quadratic form

$$q(\mathbf{y}) = -y_1^2 + y_2^2 + y_3^2 + y_4^2 + y_5^2.$$

The de Sitter space with parameter $r > 0$ is just the hypersurface

$$S_r := q^{-1}(r^2).$$

We will find, using the method of the preceding section, that in the induced metric, this pseudosphere has constant curvature $1/r^2$. There is nothing special about the values $(1,4)$. The argument works for a pseudosphere in $\mathbb{R}^{p,q}$ for any p and q. The treatment here follows O'Neill pp. 108-114 with minor changes in notation.

On any vector space we have the identification of the tangent space at any point with the vector space itself. In particular, with have the vector field $\mathcal{E}$ which assigns to each point p itself, thought of as an element of the tangent space at p. In terms of a system $x_1, \ldots, x_n$ of linear coordinates we have

$$\mathcal{E} = \sum x_i \partial_i.$$

If v is a tangent vector at any point of our vector space, we have

$$\nabla_v \mathcal{E} = \sum v_i \partial_i = v$$

where ∇ is the covariant derivative of the (flat) metric. We have

$$\langle \mathcal{E}, \mathcal{E} \rangle = q.$$

So

$$vq = v\langle \mathcal{E}, \mathcal{E} \rangle = 2\langle \nabla_v \mathcal{E}, \mathcal{E} \rangle = 2\langle v, \mathcal{E} \rangle.$$

In particular, if v is tangent to S_r, the above expressions vanish. So $\mathcal{E}$ is normal to S_r. Therefore

$$U := \frac{1}{r}\mathcal{E}$$

4.13. THE DE SITTER UNIVERSE AND ITS RELATIVES.

is the "outward pointing" unit normal vector to S_r and the shape operator $\mathfrak{S}$ (from the preceding section) defined by

$$\mathfrak{S}(v) = -\nabla_v U, \quad v \text{ tangent to } S_r$$

is given by $\mathfrak{S}v = -v/r$, in other words,

$$\mathfrak{S} = -\frac{1}{r}\text{Id}.$$

Since the curvature of $\mathbb{R}^{(1,4)}$ vanishes, it follows from (4.35) that

Proposition 17. *The de Sitter space S_r has constant curvature $1/r^2$.*

Notice that the the group $O(1,4)$, the orthogonal group of $\mathbb{R}^{1,4}$ (which is ten dimensional) acts transitively as isometries of S_r (just as the orthogonal group in three dimensions acts transitively as isometries of the sphere). So by Proposition 14, S_r has maximal possible dimension of its isometry group.

Willem De Sitter

Born: 6 May 1872 in Sneek, the Netherlands
Died: 20 Nov. 1934 in Leiden, the Netherlands

4.13.1 The Einstein field equations with cosmological constant Λ.

In covariant form this is the equation

$$\text{Ric} - \frac{1}{2}\text{Sg} + \Lambda\mathbf{g} = T \tag{4.36}$$

as an equation for the Lorentzian metric **g** in terms of the "source term" T which is the stress energy tensor describing the distribution of matter and energy of the universe. I will "derive" these equations from the Hilbert functional in Chapter 14.

From (4.31) it follows that if S is a four dimensional manifold with constant (sectional) curvature k, then

$$\text{Ric} = 3k\mathbf{g} \quad \text{and} \quad \mathbf{S} = 12k.$$

So S is a solution of the Einstein field equations with $T = 0$ if $\Lambda = 3k$. So

Proposition 18. *The de Sitter universe S_r is a solution of the Einstein field equations with $T = 0$ and cosmological constant*

$$\Lambda = \frac{3}{r^2}.$$

I will leave it to the professionals to decide whether to move the Λ term to the other side of the Einstein equations and consider its contribution as consisting of "dark matter".

4.13.2 Cosmological considerations

In cosmology one "breaks" a fundamental principle of general relativity by singling out a special coordinate system - i.e. by a choice of cosmic time. More technically, one describes the universe in terms of a Friedmann-Robertson-Walker metric. See Section 6.7 below.

But the same space can be described (at least locally) as an FRW metric in different ways. In de Sitter's original papers cited above he chose as his cosmic time one that made its level sets spheres, and led to a static universe which (temporarily) pleased Einstein. In 1925 Lemaître described (part of) de Sitter space so that the hypersurfaces of constant time are copies of flat $\mathbb{R}^3$ leading to the idea of the expanding universe. We describe this computation using the Cartan calculus in Section 6.7.1 below.

Georges Henri Joseph Edouard Lemaitre

Born: 17 July 1894 in Charleroi, Belgium
Died: 20 June 1966 in Charleroi, Belgium

For an excellent account of the history of the "expanding universe", see the book *Discovering the Expanding Universe* by Harry Nussbaumer and Lydia Bieri where misattributions are corrected. For a good article with many enlightening pictures (but unfortunately different conventions from ours) see "The de Sitter and anti-de Sitter sightseeing tour" by U. Moschella, available on the web.

Chapter 5

Bi-invariant metrics on a Lie group.

The simplest example of a Riemman manifold is Euclidean space, where the geodesics are straight lines and all curvatures vanish. We may think of Euclidean space as a commutative Lie group under addition, and view the straight lines as translates of one parameter subgroups (lines through the origin). An easy but important generalization of this is when we consider bi-invariant metrics on a Lie group, a concept we shall explain below. In this case also, the geodesics are the translates of one parameter subgroups. Some of the material in this chapter is a repeat or review of the Problem set in Chapter 2.

5.1   The Lie algebra of a Lie group.

Let G be a Lie group. This means that G is a group, and is a smooth manifold such that the multiplication map $G \times G \to G$ is smooth, as is the map

$$\text{inv} : G \to G$$

sending every element into its inverse:

$$\text{inv} : a \mapsto a^{-1}, \quad a \in G.$$

Until now the Lie groups we studied were given as subgroups of $Gl(n)$. We can continue in this vein, or work with the more general definition just given. We have the left action of G on itself

$$L_a : G \to G, \quad b \mapsto ab$$

and the right action

$$R_a : G \to G, \quad b \mapsto ba^{-1}.$$

We let $\mathfrak{g}$ denote the tangent space to G at the identity:

$$\mathfrak{g} = TG_e.$$

We identify $\mathfrak{g}$ with the space of all left invariant vector fields on G, so $\xi \in \mathfrak{g}$ is identified with the vector field X which assigns to every $a \in G$ the tangent vector

$$d(L_a)_e \xi \in TG_a.$$

We will alternatively use the notation X, Y or ξ, η for elements of $\mathfrak{g}$.

The left invariant vector field X generates a one parameter group of transformations which commutes with all left multiplications and so must consist of a one parameter group of right multiplications. In the case of a subgroup of $Gl(n)$, where $\mathfrak{g}$ was identified with a subspace of of the space of all $n \times n$ matrices, we saw that this was the one parameter group of transformations

$$A \mapsto A \exp tX,$$

i.e. the one parameter group

$$R_{\exp -tX}.$$

So we might as well use this notation in general: $\exp tX$ denotes the one parameter subgroup of G obtained by looking at the solution curve through e of the left invariant vector field X, and then the one parameter group of transformations generated by the vector field X is $R_{\exp -tX}$.

Let X and Y be elements of $\mathfrak{g}$ thought of as left invariant vector fields, and let us compute their Lie bracket as vector fields. So let

$$\phi_t = R_{\exp -tX}$$

be the one parameter group of transformations generated by X. According to the general definition, the Lie bracket $[X, Y]$ is obtained by differentiating the time dependent vector field $\phi_t^* Y$ at $t = 0$. By definition, the pull-back $\phi_t^* Y$ is the vector field which assigns to the point a the tangent vector

$$(d\phi_t)_a^{-1} Y(\phi_t(a)) = (dR_{\exp tX})_{a(\exp tX)} Y(a(\exp tX)). \tag{5.1}$$

In the case that G is a subgroup of the general linear group, this is precisely the left invariant vector field

$$A \mapsto A(\exp tX) Y (\exp -tX).$$

Differentiating with respect to t and setting $t = 0$ shows that the vector field $[X, Y]$ is precisely the left invariant vector field corresponding to the commutator of the two matrices X and Y.

We can mimic this computation for a general Lie group, not necessarily given as a subgroup of $Gl(n)$: First let us record the special case of (5.1) when we take $a = e$:

$$(d\phi_t)_e^{-1} Y(\phi_t(e)) = (dR_{\exp tX})_{\exp tX} Y(\exp tX). \tag{5.2}$$

5.2. THE GENERAL MAURER-CARTAN FORM.

For any $a \in G$ we let A_a denote conjugation by the element $a \in G$, so

$$A_a : G \to G, A_a(b) = aba^{-1}.$$

We have $A_a(e) = e$ and A_a carries one-parameter subgroups into one parameter subgroups. In particular the differential of A_a at $TG_e = \mathfrak{g}$ is a linear transformation of $\mathfrak{g}$ which we shall denote by Ad_a:

$$d(A_a)_e =: \mathrm{Ad}_a : TG_e \to TG_e.$$

We have

$$A_a = L_a \circ R_a = R_a \circ L_a.$$

So if Y is the left invariant vector field on G corresponding to $\eta \in TG_e = \mathfrak{g}$, we have $dL_a(\eta) = Y(a)$ and so

$$d(A_a)_e \eta = d(R_a)_a \circ d(L_a)_e \eta = d(R_a)_a Y(a).$$

Set $a = \exp tX$, and compare this with (5.2). Differentiate with respect to t and set $t = 0$. We see that the left invariant vector field $[X, Y]$ corresponds to the element of TG_e obtained by differentiating $\mathrm{Ad}_{\exp tX}\, \eta$ with respect to t and setting $t = 0$. In symbols, we can write this as

$$\frac{d}{dt} \mathrm{Ad}_{\exp tX}{}_{|t=0} = \mathrm{ad}(X) \quad \text{where} \quad \mathrm{ad}(X) : \mathfrak{g} \to \mathfrak{g}, \quad \mathrm{ad}(X)Y = [X, Y]. \quad (5.3)$$

Now $\mathrm{ad}(X)$ as defined above is a linear transformation of $\mathfrak{g}$. So we can consider the corresponding one parameter group $\exp t\, \mathrm{ad}(X)$ of linear transformations of $\mathfrak{g}$ (using the usual formula for the exponential of a matrix). But (5.3) says that $\mathrm{Ad}_{\exp tX}$ is a one parameter group of linear transformations with the same derivative, $\mathrm{ad}(X)$ at $t = 0$. The uniqueness theorem for linear differential equations then implies the important formula

$$\exp(t\, \mathrm{ad}(X)) = \mathrm{Ad}_{\exp tX}. \quad (5.4)$$

5.2 The general Maurer-Cartan form.

For the convenience of the reader, I now review and expand upon the discussion in Section 2.14.16:

If $v \in TG_a$ is tangent vector at the point $a \in G$, there will be a unique left invariant vector field X such that $X(a) = v$. In other words, there is a linear map

$$\omega_a : TG_a \to \mathfrak{g}$$

sending the tangent vector v to the element $\xi = \omega_a(v) \in \mathfrak{g}$ where the left invariant vector field X corresponding to ξ satisfies $X(a) = v$. So we have defined a $\mathfrak{g}$ valued linear differential form ω identifying the tangent space at any $a \in G$ with $\mathfrak{g}$. If

$$dL_b v = w \in TG_{ba}$$

then $X(ba) = w$ since $X(a) = v$ and X is left invariant. In other words,

$$\omega_{L_b a} \circ dL_b = \omega_a,$$

or, what amounts to the same thing

$$L_b^* \omega = \omega$$

for all $b \in G$. The form ω is left invariant. When we proved this for a subgroup of $Gl(n)$ this was a computation. But in the general case, as we have just seen, it is a tautology. We now want to establish the generalization of the Maurer-Cartan equation (2.10) which said that for subgroups of $Gl(n)$ we have

$$d\omega + \omega \wedge \omega = 0.$$

Since we no longer have, in general, the notion of matrix multiplication which enters into the definition of $\omega \wedge \omega$, we must first must rewrite $\omega \wedge \omega$ in a form which generalizes to an arbitrary Lie group.

So let us temporarily consider the case of a subgroup of $Gl(n)$. Recall that for any two form τ and a pair of vector fields X and Y we write $\tau(X,Y) = i(Y)i(X)\tau$. Thus

$$(\omega \wedge \omega)(X,Y) = \omega(X)\omega(Y) - \omega(Y)\omega(X),$$

the commutator of the two matrix valued functions, $\omega(X)$ and $\omega(Y)$. Consider the commutator of two matrix valued one forms, ω and σ,

$$\omega \wedge \sigma + \sigma \wedge \omega$$

(according to our usual rules of superalgebra). We denote this by

$$[\omega \wedge, \sigma].$$

In particular we may take $\omega = \sigma$ to obtain

$$[\omega \wedge, \omega] = 2\omega \wedge \omega.$$

So we can rewrite the Maurer-Cartan equation for a subgroup of $Gl(n)$ as

$$d\omega + \frac{1}{2}[\omega \wedge, \omega] = 0. \tag{5.5}$$

Now for a general Lie group we *do* have the Lie bracket map

$$\mathfrak{g} \times \mathfrak{g} \to \mathfrak{g}.$$

So we can define the two form $[\omega \wedge, \omega]$. It is a $\mathfrak{g}$ valued two form which satisfies

$$i(X)[\omega \wedge, \omega] = [X, \omega] - [\omega, X]$$

5.3. LEFT INVARIANT AND BI-INVARIANT METRICS.

for any left invariant vector field X. Hence

$$[\omega\wedge,\omega](X,Y) := i(Y)i(X)[\omega\wedge,\omega] = i(Y)\left([X,\omega] - [\omega,X]\right)$$
$$= [X,Y] - [Y,X] = 2[X,Y]$$

for any pair of left invariant vector fields X and Y. So to prove (5.5) in general, we must verify that for any pair of left invariant vector fields we have

$$d\omega(X,Y) = -\omega([X,Y]).$$

But this is a consequence of our general formula (2.7) for the exterior derivative which in our case says that

$$d\omega(X,Y) = X\omega(Y) - Y\omega(X) - \omega([X,Y]).$$

In our situation the first two terms on the right vanish since, for example, $\omega(Y) = Y = \eta$ a constant element of $\mathfrak{g}$ so that $X\omega(Y) = 0$ and similarly $Y\omega(X) = 0$.

5.3 Left invariant and bi-invariant metrics.

Any non-degenerate scalar product, $\langle \, , \, \rangle$, on $\mathfrak{g}$ determines (and is equivalent to) a left invariant semi-Riemann metric on G via the left-identification $dL_a : \mathfrak{g} = TG_e \to TG_a$, $\forall\, a \in G$.

Since $A_a = L_a \circ R_a$, the left invariant metric, $\langle \, , \, \rangle$ is right invariant if and only if it is A_a invariant for all $a \in G$, which is the same as saying that $\langle \, , \, \rangle$ is invariant under the adjoint representation of G on $\mathfrak{g}$, i.e. that

$$\langle Ad_a Y, Ad_a Z\rangle = \langle Y, Z\rangle, \quad \forall Y, Z \in \mathfrak{g},\ a \in G.$$

Setting $a = \exp tX$, $X \in \mathfrak{g}$, differentiating with respect to t and setting $t = 0$ gives

$$\langle [X,Y], Z\rangle + \langle Y, [X,Z]\rangle = 0, \quad \forall X, Y, Z \in \mathfrak{g}. \tag{5.6}$$

If every element of G can be written as a product of elements of the form $\exp \xi$, $\xi \in \mathfrak{g}$ (which will be the case if G is connected), this condition implies that $\langle \, , \, \rangle$ is invariant under Ad and hence is invariant under right and left multiplication. Such a metric is called **bi-invariant**.

Recall that inv denotes the map sending every element into its inverse:

$$\text{inv} : a \mapsto a^{-1}, \quad a \in G.$$

Since $\text{inv}\exp tX = \exp(-tX)$ we see that

$$d\, \text{inv}_e = -\text{id}\,.$$

Also

$$\text{inv} = R_a \circ \text{inv} \circ L_{a^{-1}}$$

since the right hand side sends $b \in G$ into

$$b \mapsto a^{-1}b \mapsto b^{-1}a \mapsto b^{-1}.$$

Hence $d\,\mathrm{inv}_a : TG_a \to TG_{a^{-1}}$ is given, by the chain rule, as

$$dR_a \circ d\mathrm{inv}_e \circ dL_{a^{-1}} = -dR_a \circ dL_{a^{-1}}$$

implying that a bi-invariant metric is invariant under the map inv. Conversely, if a left invariant metric is invariant under inv then it is also right invariant, hence bi-invariant since

$$R_a = \mathrm{inv} \circ L_a \circ \mathrm{inv}.$$

5.4 Geodesics are cosets of one parameter subgroups.

The Koszul formula simplifies considerably when applied to left invariant vector fields and bi-invariant metrics since all scalar products are constant, so their derivatives vanish, and we are left with

$$2\langle \nabla_X Y, Z \rangle = -\langle X, [Y,Z] \rangle - \langle Y, [X,Z] \rangle + \langle Z, [X,Y] \rangle$$

and the first two terms cancel by (5.6). We are left with

$$\nabla_X Y = \frac{1}{2}[X,Y]. \tag{5.7}$$

Conversely, if $\langle\,,\,\rangle$ is a left invariant metric for which (5.7) holds, then

$$\begin{aligned}
\langle X, [Y,Z] \rangle &= 2\langle X, \nabla_Y Z \rangle \\
&= -2\langle \nabla_Y X, Z \rangle \\
&= -\langle [Y,X], Z \rangle \\
&= \langle [X,Y], Z \rangle
\end{aligned}$$

so the metric is bi-invariant.

Let α be an integral curve of the left invariant vector field X. Equation (5.7) implies that $\alpha'' = \nabla_X X = 0$ so α is a geodesic. Thus the one-parameter groups are the geodesics through the identity, and all geodesics are left cosets of one parameter groups. (This is the reason for the name exponential map in the theory of linear connections.)

In Chapter ?? we will study Riemannian submersions. It will emerge from this study that if we have a quotient space $B = G/H$ of a group with a bi-invariant metric (satisfying some mild conditions), then the geodesics on B in the induced metric are orbits of certain one parameter subgroups. For example, the geodesics on spheres are the great circles.

5.5 The Riemann curvature of a bi-invariant metric.

We compute the Riemann curvature of a bi-invariant metric by applying the definition (3.13) to left invariant vector fields:

$$R_{XY}Z = \frac{1}{4}[X,[Y,Z]] - \frac{1}{4}[Y,[X,Z]] - \frac{1}{2}[[X,Y],Z]$$

Jacobi's identity implies the first two terms add up to $\frac{1}{4}[[X,Y],Z]$ and so

$$R_{XY}Z = -\frac{1}{4}[[X,Y],Z]. \tag{5.8}$$

5.6 Sectional curvatures.

In particular

$$\langle R_{XY}X,Y\rangle = -\frac{1}{4}\langle[[X,Y],X],Y\rangle = -\frac{1}{4}\langle[X,Y],[X,Y]\rangle$$

so the sectional curvature is given by

$$K(X,Y) = \frac{1}{4}\frac{||[X,Y]||^2}{||X \wedge Y||^2}. \tag{5.9}$$

Notice that in the Riemannian case (but not in general in the semi-Riemannian case) this expression is non-negative.

5.7 The Ricci curvature and the Killing form.

Recall that for each $X \in \mathfrak{g}$ the linear transformation of $\mathfrak{g}$ consisting of bracketing on the left by X is called ad X. So

$$\text{ad } X : \mathfrak{g} \to \mathfrak{g}, \quad \text{ad } X(V) := [X,V].$$

We can thus write our formula for the curvature as

$$R_{XV}Y = \frac{1}{4}(\text{ad }Y)(\text{ad }X)V.$$

Now the Ricci curvature was defined as

$$\text{Ric }(X,Y) = \text{tr }[V \mapsto R_{XV}Y].$$

We thus see that for any bi-invariant metric, the Ricci curvature is always given by

$$\text{Ric } = \frac{1}{4}B \tag{5.10}$$

where B, the **Killing form**, is defined by

$$B(X,Y) := \text{tr } (\text{ad } X)(\text{ad } Y). \tag{5.11}$$

The Killing form is symmetric, since tr (AC) = tr CA for any pair of linear operators. It is also invariant. Indeed, let $\mu : \mathfrak{g} \to \mathfrak{g}$ be any automorphism of $\mathfrak{g}$, so $\mu([X,Y]) = [\mu(X), \mu(Y)]$ for all $X, Y \in \mathfrak{g}$. We can read this equation as saying
$$\text{ad } (\mu(X))(\mu(Y)) = \mu(\text{ad}(X)(Y))$$
or
$$\text{ad } (\mu(X)) = \mu \circ \text{ad } X \mu^{-1}.$$
Hence
$$\text{ad } (\mu(X))\text{ad } (\mu(Y)) = \mu \circ \text{ad } X \text{ad } Y \mu^{-1}.$$
Since trace is invariant under conjugation, it follows that
$$B(\mu(X), \mu(Y)) = B(X,Y).$$

Applied to $\mu = \exp(t\text{ad } Z)$ and differentiating at $t = 0$ shows that $B([Z,X],Y) + B(X,[Z,Y]) = 0$.

So the Killing form defines a bi-invariant symmetric bilinear form on G. Of course it need not, in general, be non-degenerate. For example, if the group is commutative, it vanishes identically. A group G is called *semi-simple* if its Killing form is non-degenerate. So on a semi-simple Lie group, we can always choose the Killing form as the bi-invariant metric. For such a choice, our formula above for the Ricci curvature then shows that the group manifold with this metric is **Einstein**, i.e. the Ricci curvature is a multiple of the scalar product.

Suppose that the adjoint representation of G on $\mathfrak{g}$ is irreducible. Then $\mathfrak{g}$ can not have two invariant non-degenerate scalar products unless one is a multiple of the other. In this case, we can also conclude from our formula that the group manifold is Einstein.

5.8 Bi-invariant forms from representations.

Here is a way to construct invariant scalar products on a Lie algebra $\mathfrak{g}$ of a Lie group G. Let ρ be a representation of G. This means that ρ is a smooth homomorphism of G into $Gl(n, \mathbb{R})$ or $Gl(n, \mathbb{C})$. This induces a representation $\dot\rho$ of $\mathfrak{g}$ by
$$\dot\rho(X) := \frac{d}{dt}\rho(\exp tX)_{|t=0}.$$
So
$$\dot\rho : \mathfrak{g} \to gl(n)$$
where $gl(n)$ is the Lie algebra of $Gl(n)$, and
$$\dot\rho([X,Y]) = [\dot\rho(X), \dot\rho(Y)]$$

5.8. BI-INVARIANT FORMS FROM REPRESENTATIONS.

where the bracket on the right is in $gl(n)$. More generally, a linear map $\dot\rho : \mathfrak{g} \to gl(n, \mathbf{C})$ or $gl(n, \mathbb{R})$ satisfying the above identity is called a representation of the Lie algebra $\mathfrak{g}$. Every representation of G gives rise to a representation of $\mathfrak{g}$ but not every representation of $\mathfrak{g}$ need come from a representation of G in general.

If $\dot\rho$ is a representation of $\mathfrak{g}$, with values in $gl(n, \mathbb{R})$, we may define

$$\langle X, Y\rangle_\mathfrak{g} := \operatorname{tr} \dot\rho(X)\dot\rho(Y).$$

This is real valued, symmetric in X and Y, and

$$\langle [X,Y], Z\rangle_\mathfrak{g} + \langle Y, X, Z\rangle_\mathfrak{g} =$$

$$\operatorname{tr}\left(\dot\rho(X)\dot\rho(Y)\dot\rho(Z) - \dot\rho(Y)\dot\rho(X)\dot\rho(Z) + \dot\rho(Y)\dot\rho(X)\dot\rho(Z) - \dot\rho(Y)\dot\rho(Z)\dot\rho(X)\right) = 0.$$

So this is invariant. Of course it need not be non-degenerate.

A case of particular interest is when the representation $\dot\rho$ takes values in $u(n)$, the Lie algebra of the unitary group. An element of $u(n)$ is a skew adjoint matrix, i.e. a matrix of the form iA where $A = A^*$ is self adjoint. If $A = A^*$ and $A = (a_{ij})$ then

$$\operatorname{tr} A^2 = \operatorname{tr} AA^* = \sum_{i,j} a_{ij}a_{ji} = \sum_{i,j} a_{ij}\overline{a_{ij}} = \sum_{ij}|a_{ij}|^2$$

which is positive unless $A = 0$. So

$$-\operatorname{tr}(iA)(iA)$$

is positive unless $A = 0$. This implies that if $\dot\rho : \mathfrak{g} \to u(n)$ is injective, then the form

$$\langle X, Y\rangle = -\operatorname{tr} \dot\rho(X)\dot\rho(Y)$$

is a positive definite invariant scalar product on $\mathfrak{g}$.

For example, let us consider the Lie algebra $\mathfrak{g} = u(2)$ and the representation $\dot\rho$ of $\mathfrak{g}$ on the exterior algebra of $\mathbf{C}^2$. We may decompose

$$\wedge(\mathbf{C}^2) = \wedge^0(\mathbf{C}^2) \oplus \wedge^1(\mathbf{C}^2) \oplus \wedge^2(\mathbf{C}^2)$$

and each of the summands is invariant under our representation. Every element of $u(2)$ acts trivially on $\wedge^0(\mathbf{C}^2)$ and acts in its standard fashion on $\wedge^1(\mathbf{C}^2) = \mathbf{C}^2$. Every element of $u(2)$ acts via multiplication by its trace on $\wedge^2(\mathbf{C}^2)$ so in particular all elements of $su(2)$ act trivially there. Thus restricted to $su(2)$, the induced scalar product is just

$$\langle X, Y\rangle = -\operatorname{tr} XY, \quad X, Y \in su(2),$$

while on scalar matrices, i.e. matrices of the form $S = riI$ we have

$$\langle S, S\rangle = -\operatorname{tr} \dot\rho(S)^2 = 2r^2 + (2r)^2 = 6r^2 = -3\operatorname{tr} S^2 = -\frac{3}{2}(\operatorname{tr} S)^2.$$

5.9 The Weinberg angle.

The preceding example illustrates the fact that if the adjoint representation of $\mathfrak{g}$ is not irreducible, there may be more than a one parameter family of invariant scalar products on $\mathfrak{g}$. Indeed the algebra $u(2)$ decomposes as a sum

$$u(2) = su(2) \oplus u(1)$$

of subalgebras, where $u(1)$ consists of the scalar matrices (which commute with all elements of $u(2)$). It follows from the invariance condition that $u(1)$ must be orthogonal to $su(2)$ under any invariant scalar product. Each of these summands is irreducible under the adjoint representation, so the restriction of any invariant scalar product to each summand is determined up to positive scalar multiple, but these multiples can be chosen independently for each summand. So there is a two parameter family of choices.

In the physics literature it is conventional to write the most general invariant scalar product on $u(2)$ as

$$\langle A, B \rangle = -\frac{2}{g_2^2} \operatorname{tr}\left(A - \frac{1}{2}(\operatorname{tr} A)I\right)\left(B - \frac{1}{2}(\operatorname{tr} B)I\right) + -\frac{1}{g_1^2} \operatorname{tr} A \operatorname{tr} B \quad (5.12)$$

where g_1 and g_2 are sometimes called "coupling strengths". The first summand vanishes on $u(1)$ and the second summand vanishes on $su(2)$. The **Weinberg angle** θ_W is defined by

$$\sin \theta_W := \frac{g_1}{\sqrt{g_1^2 + g_2^2}}$$

and plays a key role in Electro-Weak theory which unifies the electromagnetic and weak interactions. In the current state of knowledge, there is no broadly agreed theory that predicts the Weinberg angle. It is an input derived from experiment. The data as of 2010 from the Particle Data Group gives a value of

$$\sin^2 \theta_W = 0.23116... \ .$$

Notice that the computation that we did from the exterior algebra has

$$g_1^2 = \frac{2}{3} \quad \text{and} \quad g_2^2 = 2$$

so

$$\sin^2 \theta_W = \frac{\frac{2}{3}}{\frac{2}{3} + 2} = .25 \ .$$

Of course several quite different representations will give the same metric or Weinberg angle.

Chapter 6

Cartan calculations in semi-Riemannian geometry.

6.1 Frame fields and coframe fields.

By a **frame field** on a manifold M we mean an n-tuplet $E = (E_1, \ldots, E_n)$ of vector fields (defined on some neighborhood) whose values at every point form a basis of the tangent space at that point. These then define a dual collection of differential forms

$$\theta = \begin{pmatrix} \theta^1 \\ \vdots \\ \theta^n \end{pmatrix}$$

whose values at every point form the dual basis. For example, a coordinate system $x^1, \ldots, x^n$ provides the frame field $\partial_1, \ldots, \partial_n$ with dual forms $dx^1, \cdots, dx^n$. But the use of more general frame fields allows for flexibility in computation. This was another one of Cartan's great ideas, along with the exterior differential calculus which he invented.

A frame field on a semi-Riemannian manifold M is called **"orthonormal"** if $\langle E_i, E_j \rangle \equiv 0$ for $i \neq j$ and $\langle E_i, E_i \rangle \equiv \epsilon_i$ where $\epsilon_i = \pm 1$. For example, applying the Gram-Schmidt procedure to an arbitrary frame field for a positive definite metric yields an orthonormal one.

The reason for the quotation marks in the semi-Riemannian case is because we might have $\langle E_i, E_i \rangle \equiv -1$ for some i.

So for an "orthonormal" frame field, the $E_1(p), \ldots, E_n(p)$ form an "orthonormal" basis of the tangent space TM_p at each point p in the domain of definition of the frame field. The dual basis of the cotangent space then provides a family of linear differential forms, $\theta^1, \ldots, \theta^n$. It follows from the definition, that if

$v \in TM_p$ then
$$\langle v, v \rangle = \epsilon_1 \left(\theta^1(v)\right)^2 + \cdots + \epsilon_n \left(\theta^n(v)\right)^2.$$

This equation, true at all points in the domain of definition of the frame field, is usually written as

$$ds^2 = \epsilon_1(\theta^1)^2 + \cdots + \epsilon_n(\theta^n)^2. \tag{6.1}$$

Conversely, if $\theta^1, \ldots, \theta^n$ is a collection of linear differential forms satisfying (6.1) (defined on some open set) then the dual vector fields constitute an "orthonormal" frame field.

6.1.1 The tautological tensor.

On any manifold, we have the tautological tensor field of type (1,1) which assigns to each tangent space the identity linear transformation. We will denote this tautological tensor field by id. Thus for any $p \in M$ and any $v \in TM_p$,

$$\mathrm{id}(v) = v.$$

In terms of a frame field we have

$$\mathrm{id} = E_1 \otimes \theta^1 + \cdots E_n \otimes \theta^n$$

in the sense that both sides yield v when applied to any tangent vector v in the domain of definition of the frame field. We can say that the θ^i give the expression for id in terms of the frame field.

We introduce the "column vector of differential forms"

$$\theta := \begin{pmatrix} \theta^1 \\ \vdots \\ \theta^n \end{pmatrix}$$

as a shorthand for the collection of the θ^i and the "row vector" of frame fields $E = (E_1, \ldots E_n)$ and write the equation

$$\mathrm{id} = E_1 \otimes \theta^1 + \cdots E_n \otimes \theta^n$$

in shorthand as
$$\mathrm{id} = E \otimes \theta.$$

6.2 Connection and curvature forms in a frame field.

6.2.1 Connection forms.

For each i the Levi-Civita connection yields a tensor field ∇E_i, the covariant differential of E_i with respect to the connection, and hence linear differential

6.2. CONNECTION AND CURVATURE FORMS IN A FRAME FIELD.

forms ω_j^i defined by
$$\omega_j^i(\xi) = \theta^i(\nabla_\xi E_j). \tag{6.2}$$

So
$$\nabla_\xi E_j = \sum_m \omega_j^m(\xi) E_m.$$

6.2.2 Cartan's first structural equation.

This says that
$$d\theta^i = -\sum_m \omega_m^i \wedge \theta^m. \tag{6.3}$$

If we let $\omega := (\omega_j^i)$ (a matrix of linear differential forms) we can write (6.3) in shorthand as
$$d\theta = -\omega \wedge \theta.$$

Proof of Cartan's first structural equation.

Proof. To prove (6.3), we apply the formula
$$d\theta(X,Y) = X(\theta(Y)) - Y(\theta(X)) - \theta([X,Y]) \tag{2.7}$$

valid for any linear differential form θ and vector fields X and Y to θ^i, E_a, E_b to obtain
$$d\theta^i(E_a, E_b) = E_a \theta^i(E_b) - E_b \theta^i(E_a) - \theta^i([E_a, E_b]).$$

Since $\theta^i(E_b)$ and $\theta^i(E_a) = 0$ or 1 are constants, the first two terms vanish and so the left hand side of (6.3) when evaluated on (E_a, E_b) becomes
$$-\theta^i([E_a, E_b]).$$

As to the right hand side of (6.3) we have
$$\begin{aligned}
\left[-\sum \omega_m^i \wedge \theta^m\right](E_a, E_b) &= \left(i(E_a)\left[-\sum \omega_m^j \wedge \theta^m\right]\right)(E_b) \\
&= \left[-\sum \omega_m^i(E_a)\theta^m + \sum \theta^m(E_a)\omega_m^i\right](E_b) \\
&= -\omega_b^i(E_a) + \omega_a^i(E_b) \\
&= -\theta^i\left(\nabla_{E_a}(E_b) - \nabla_{E_b}(E_a)\right) \\
&= -\theta^i([E_a, E_b]).
\end{aligned}$$

$\square$

6.2.3 Symmetry properties of ω.

Apology - In some of the formulas below I use (,) for scalar product instead of $\langle\ ,\ \rangle$. I am afraid that it would introduce too many new errors in trying to be consistent.

Notice that
$$\omega^i_j(\xi) = \theta^i(\nabla_\xi E_j) = \epsilon_i(\nabla_\xi E_j, E_i).$$
Since
$$0 = d(E_i, E_j)$$
we have $0 = \xi(E_i, E_j) = (\nabla_\xi E_i, E_j) + (E_i, \nabla_\xi E_j)$ and hence
$$\epsilon_j \omega^j_i = -\epsilon_i \omega^i_j. \qquad (6.4)$$

In particular $\omega^i_i = 0$.

Doing a Cartan style computation.

The "detective" work in doing a Cartan style computation is to find ω from the conditions
$$d\theta = -\omega \wedge \theta \qquad (6.3)$$
and
$$\epsilon_j \omega^j_i = -\epsilon_i \omega^i_j \qquad (6.4).$$
I will illustrate later on.

6.2.4 Curvature forms in a frame field.

For tangent vectors $\xi, \eta \in TM_p$ let $(\Omega^i_j(\xi, \eta))$ be the matrix of the curvature operator $R_{\xi\eta}$ with respect to the basis $E_1(p), \ldots, E_n(p)$. So
$$R_{\xi\eta}(E_j)(p) = \sum_i \Omega^i_j(\xi, \eta) E_i(p).$$

Since $R_{\eta,\xi} = -R_{\xi,\eta}$, $\Omega^i_j(\xi, \eta) = -\Omega^i_j(\eta, \xi)$ so the Ω^i_j are exterior differential forms of degree two.

The way to do computations with the equation
$$R_{\xi\eta}(E_j)(p) = \sum_i \Omega^i_j(\xi, \eta) E_i(p)$$
is to remember that $(\Omega^i_j(\xi, \eta))$ is an ordinary matrix with numbers as entries so the above equation says to multiply the "row vector" $E = (E_1, \ldots, E_n)$ on the *right* by this matrix, i.e. compute
$$E \times (\Omega^i_j(\xi, \eta)).$$

The j-th entry of the resulting "row vector" is the image of E_j. (I have dropped the (p) to simplify the notation.)

6.2.5 Cartan's second structural equation.

This asserts that
$$\Omega_j^i = d\omega_j^i + \sum_m \omega_m^i \wedge \omega_j^m. \tag{6.5}$$

Proof of Cartan's second structural equation.

Proof. We have
$$R_{E_a E_b}(E_j) = \sum_i \Omega_j^i(E_a, E_b) E_i$$

by definition. We must show that the right hand side of (6.5) yields the same result when we substitute E_a, E_b into the differential forms, multiply by E_i and sum over i.
Write
$$R_{E_a E_b}(E_j) = \nabla_{E_a}(\nabla_{E_b} E_j) - \nabla_{E_b}(\nabla_{E_a} E_j) - \nabla_{[E_a, E_b]} E_j.$$

Since $\nabla_{E_b}(E_j) = \sum_i \omega_j^i(E_b) E_i$ we get

$$\begin{aligned}
\nabla_{E_a}(\nabla_{E_b} E_j) &= \sum E_a[\omega_j^i(E_b)] E_i + \sum \omega_j^m(E_b) \nabla_{E_a} E_m \\
&= \sum E_a[\omega_j^i(E_b)] E_i + \sum_{i,m} \omega_j^m(E_b) \omega_m^i(E_a) E_i \\
\nabla_{E_b}(\nabla_{E_a} E_j) &= \sum E_b[\omega_j^i(E_a)] E_i + \sum_{i,m} \omega_j^m(E_a) \omega_m^i(E_b) E_i \\
\nabla_{[E_a, E_b]} E_j &= \sum \omega_j^i([E_a, E_b]) E_j \quad \text{so} \\
R_{E_a E_b} E_j &= \sum_i \left[E_a \omega_j^i(E_b) - E_b \omega_j^i(E_a) - \omega_j^i([E_a, E_b]) \right] E_i \\
&\quad + \sum_{m,i} \left[\omega_m^i(E_a) \omega_j^m(E_b) - \omega_m^i(E_b) \omega_j^m(E_a) \right] E_i.
\end{aligned}$$

The first expression in square brackets is the value on E_a, E_b of $d\omega_j^i$ by (2.7) while the second expression in square brackets is the value on E_a, E_b of $\sum \omega_m^i \wedge \omega_j^m$. This proves Cartan's second structural equation. $\square$

6.2.6 Both structural equations in compact form.

$$d\theta + \omega \wedge \theta = 0 \tag{6.6}$$
$$d\omega + \omega \wedge \omega = \Omega \tag{6.7}$$

6.3 Cartan's lemma and Levi-Civita's theorem.

In this section we will show that the equations (6.6) and (6.4) determine the ω_j^i. First a result in exterior algebra:

6.3.1 Cartan's lemma in exterior algebra.

Lemma 6. *Let $x_1, \ldots, x_p$ be linearly independent elements of a vector space, V, and suppose that $y_1, \ldots y_p \in V$ satisfy*

$$x_1 \wedge y_1 + \cdots x_p \wedge y_p = 0.$$

Then

$$y_j = \sum_{k=1}^{p} A_{jk} x_k \quad \text{with } A_{jk} = A_{kj}.$$

Proof. Choose $x_{p+1}, \ldots, x_n$ if $p < n$ so as to obtain a basis of V and write

$$y_i = \sum_{j=1}^{p} A_{ij} x_j + \sum_{k=p+1}^{n} B_{ik} x_k.$$

Substituting into the equation of the lemma gives

$$\sum_{i<j\leq p} (A_{ij} - A_{ji}) x_i \wedge x_j + \sum_{i\leq p<k} B_{ik} x_i \wedge x_k = 0.$$

Since the $x_i \wedge x_\ell$, $i < \ell$ form a basis of $\wedge^2(V)$, we conclude that $B_{ik} = 0$ and $A_{ij} = A_{ji}$ which is the content of the lemma. $\square$

6.3.2 Using Cartan's lemma to prove Levi-Civita's theorem.

Suppose that ω and ω' are two matrices of one forms which satisfy (6.3). Then their difference, $\sigma := \omega - \omega'$ satisfies $\sigma \wedge \theta = 0$. Applying the lemma we conclude that

$$\sigma_k^i = \sum A_{jk}^i \theta^j, \quad A_{jk}^i = A_{kj}^i.$$

If we set

$$B_{jk}^i = \epsilon_i A_{jk}^i$$

and if both ω and ω' satisfy (6.4) so that σ does as well, then

$$B_{jk}^i = B_{kj}^i \quad \text{and} \quad B_{jk}^i = -B_{ki}^j.$$

We claim that these two equations imply that all the $B_{jk}^i = 0$ and hence that $\sigma = 0$. Indeed,

$$\begin{aligned} B_{jk}^i = B_{kj}^i &= -B_{ij}^k = B_{ki}^j \\ &= B_{ik}^j = -B_{jk}^i. \end{aligned}$$

The upshot is that if we have found ω satisfying (6.6) and (6.4) then we know that it is the matrix of connection forms.

This is useful in computations, and also gives us another proof of Levi-Civita's theorem.

6.4 Examples of Cartan style computations in semi-Riemannian geometry.

6.4.1 Polar coordinates in two dimensions.

Suppose we have coordinates (r, ϑ) in terms of which the metric has the form

$$ds^2 = dr^2 + \phi(r)^2 d\vartheta^2$$

so that

$$\theta = \begin{pmatrix} dr \\ \phi d\vartheta \end{pmatrix}$$

is a coframe field. Then

$$d\theta = \begin{pmatrix} 0 \\ \phi' dr \wedge d\vartheta \end{pmatrix}$$

and so

$$\omega = \begin{pmatrix} 0 & -\phi' d\vartheta \\ \phi' d\vartheta & 0 \end{pmatrix}.$$

In two dimensions we always have $\omega \wedge \omega = 0$ and therefore $\Omega = d\omega$. In our case this gives

$$\Omega = \begin{pmatrix} 0 & -\phi'' dr \wedge d\vartheta \\ \phi'' dr \wedge d\vartheta & 0 \end{pmatrix}.$$

It is convenient to express this in terms of the coframe field

$$\begin{pmatrix} \theta_1 \\ \theta_2 \end{pmatrix} = \begin{pmatrix} dr \\ \phi d\vartheta \end{pmatrix}$$

so

$$\Omega = \begin{pmatrix} 0 & -\frac{\phi''}{\phi} \theta_1 \wedge \theta_2 \\ \frac{\phi''}{\phi} \theta_1 \wedge \theta_2 & 0 \end{pmatrix}.$$

Let (E_1, E_2) be the corresponding frame field. Evaluating Ω on (E_1, E_2) gives the matrix of $R_{E_1 E_2}$ as

$$\begin{pmatrix} 0 & -\frac{\phi''}{\phi} \\ \frac{\phi''}{\phi} & 0 \end{pmatrix}.$$

Multiplying gives

$$(E_1, E_2) \times \begin{pmatrix} 0 & -\frac{\phi''}{\phi} \\ \frac{\phi''}{\phi} & 0 \end{pmatrix} = \frac{\phi''}{\phi} (E_2, -E_1).$$

Therefore,

$$\langle R(E_1, E_2) E_2, E_1 \rangle = -\frac{\phi''}{\phi}.$$

In two dimensions there is only one plane Π at each point, and E_1, E_2 give an orthonormal basis of Π.

So the sectional curvature, a.k.a. the Gaussian curvature is $-\phi''/\phi$. For example, polar coordinates centered about the north pole on the unit sphere have $\phi(r) = \sin r$. So the unit sphere has constant curvature 1. More generally, we can consider a surface of revolution where r is an arc length parameter on the generating curve and $\phi(r)$ is the distance from the axis of rotation.

6.4.2 Hyperbolic geometry.

Let M be the subset of $\mathbb{R}^n$ consisting of all points where $x_n > 0$. The hyperbolic metric is defined to be

$$\frac{1}{(x^n)^2}((dx^1)^2 + \cdots + (dx^n)^2)$$

so that

$$\theta := \frac{1}{x^n} \begin{pmatrix} dx^1 \\ dx^2 \\ \vdots \\ dx^n \end{pmatrix}$$

is a globally defined coframe field.

To simplify the notation we will do the calculations for $n = 3$. We determine the connection form as follows:

We have

$$\theta = \frac{1}{x^3} \begin{pmatrix} dx^1 \\ dx^2 \\ dx^3 \end{pmatrix}$$

so

$$d\theta = \frac{1}{(x^3)^2} \begin{pmatrix} dx^1 \wedge dx^3 \\ dx^2 \wedge dx^3 \\ 0 \end{pmatrix}.$$

Hence

$$\omega = \frac{1}{x^3} \begin{pmatrix} 0 & 0 & -dx^1 \\ 0 & 0 & -dx^2 \\ dx^1 & dx^2 & 0 \end{pmatrix}.$$

Next, we determine the curvature form: We have

$$d\omega = \frac{1}{(x_3)^2} \begin{pmatrix} 0 & 0 & -dx^1 \wedge dx^3 \\ 0 & 0 & -dx^2 \wedge dx^3 \\ dx^1 \wedge dx^3 & dx^2 \wedge dx^3 & 0 \end{pmatrix}$$

$$\omega \wedge \omega = \frac{1}{(x^3)^2} \begin{pmatrix} 0 & -dx^1 \wedge dx^2 & 0 \\ dx^1 \wedge dx^2 & 0 & 0 \\ 0 & 0 & 0 \end{pmatrix}$$

6.4. EXAMPLES OF CARTAN STYLE COMPUTATIONS.

so
$$\Omega = \frac{1}{(x^3)^2} \begin{pmatrix} 0 & -dx^1 \wedge dx^2 & -dx^1 \wedge dx^3 \\ dx^1 \wedge dx^2 & 0 & -dx^2 \wedge dx^3 \\ dx^1 \wedge dx^3 & dx^2 \wedge dx^3 & 0 \end{pmatrix}.$$

From the curvature form we can determine the Riemann curvature tensor (relative to our frame field): Let $E = (E_1, E_2, E_3)$ be the dual frame field. Consider, for example $R(E_1, E_2)$. Evaluating Ω on E_1, E_2 gives

$$\begin{pmatrix} 0 & -1 & 0 \\ 1 & 0 & 0 \\ 0 & 0 & 0 \end{pmatrix}.$$

Multiplying E on the right by this matrix gives $(E_2, -E_1, 0)$. So

$$R(E_1, E_2)E_1 = E_2, \quad R(E_1, E_2)E_2 = -E_1 \quad R(E_1, E_2)E_3 = 0$$

so

$$(R(E_1, E_2)E_1, E_2) = 1 \quad \text{and} \quad (R(E_1, E_2)E_1, E_3) = 0.$$

From this we get the curvature operator: We have

$$(R(E_1, E_2)E_2, E_1) = -1 \quad \text{and} \quad (R(E_1, E_2)E_3, E_i) = 0, \ i = 1, 2.$$

In other words,

$$\mathfrak{R}(E_1 \wedge E_2) = -E_1 \wedge E_2$$

with a similar result for any other pair of elements of the frame field. Thus

$$\mathfrak{R} = -I.$$

We have shown that **Hyperbolic space has constant curvature** -1.

In general, a Riemannian manifold has constant sectional curvature (at each point) if the curvature operator $\mathfrak{R} = kI$ (where $k = k(p)$. So in terms of a coframe field, this says that

$$\Omega = k\theta \wedge \theta.$$

This says that for any three tangent vectors at any point

$$R(x, y)z = k\left\{\langle z, y \rangle x - \langle z, x \rangle y\right\}.$$

In fact, the function $F(x, y, v, w) := k\left(\langle v, y \rangle \langle x, w \rangle - \langle v, x \rangle \langle y, w \rangle\right)$ defines a "curvaturelike" function whose corresponding sectional curvature is k, so by a theorem we proved earlier, any Riemann metric with (pointwise) constant sectional curvature must have its curvature of the above form, or in Cartan's notation $\Omega = k\theta \wedge \theta$. The above calculation was in the Riemannian case. I will leave to the reader to develop a corresponding formula in the non-positive definite case.

6.4.3 The Schwarzschild metric.

On $\mathbb{R}^2 \times \mathcal{S}^2$ set $\quad ds^2 = -h dt^2 + (1/h) dr^2 + r^2 (d\vartheta^2 + S^2 d\phi^2)$ where

$$h := 1 - \frac{2M}{r} \quad \text{and} \quad S = \sin \vartheta.$$

This does not define a metric when $r = 2M$ or when $r = 0$ or when $\vartheta = 0$ or π. These last two points are not really serious problems, but there is some real trouble at $r = 2M$. For simplicity I will only consider the "Schwarzschild exterior", the set where $r > 2M$ and so $h > 0$. So an "orthonormal" coframe field is given by

$$\theta = \begin{pmatrix} \theta^0 \\ \theta^1 \\ \theta^2 \\ \theta^3 \end{pmatrix} = \begin{pmatrix} \sqrt{h}\, dt \\ \frac{1}{\sqrt{h}} dr \\ r\, d\vartheta \\ rS\, d\phi \end{pmatrix}.$$

Computing $d\theta$.

$d\left(\frac{2M}{r}\right) = -\frac{2M\, dr}{r^2}$ and hence

$$d\sqrt{h} = \frac{M}{r^2 \sqrt{h}} dr.$$

Therefore

$$d\theta^0 = \frac{M}{r^2 \sqrt{h}} dr \wedge dt = -\frac{M}{r^2 \sqrt{h}} \theta^0 \wedge \theta^1, \qquad d\theta^1 = 0$$

$$d\theta^2 = dr \wedge d\vartheta = -\frac{\sqrt{h}}{r} \theta^2 \wedge \theta^1,$$

$$d\theta^3 = S\, dr \wedge d\phi + rC\, d\vartheta \wedge d\phi$$

$$= -\frac{\sqrt{h}}{r} \theta^3 \wedge \theta^1 - \frac{C}{rS} \theta^3 \wedge \theta^2$$

where $C := \cos \vartheta$. In short

$$d\theta = \begin{pmatrix} -\frac{M}{r^2 \sqrt{h}} \theta^0 \wedge \theta^1 \\ 0 \\ -\frac{\sqrt{h}}{r} \theta^2 \wedge \theta^1 \\ -\frac{\sqrt{h}}{r} \theta^3 \wedge \theta^1 - \frac{C}{rS} \theta^3 \wedge \theta^2 \end{pmatrix}.$$

Finding the connection form.

We seek ω satisfying

$$d\theta = -\omega \wedge \theta$$

and the symmetry-anti-symmetry condition

$$\epsilon_j \omega_i^j = -\epsilon_i \omega_j^i.$$

6.4. EXAMPLES OF CARTAN STYLE COMPUTATIONS.

Now $d\theta^0 = d(\sqrt{h}) \wedge dt$. Since anti-symmetry requires that $\omega_0^0 = 0$, and h is function of r alone we know without any computation that ω_1^0 is of the form $f(r)dt$. This is consistent with the fact that $d\theta^1 = 0$. Since the coefficients in the definition of θ^2 and θ^3 do not involve t, we know from the structure equation that ω_0^2 and ω_0^3 must each be some function multiple of dt, and hence the same is true of ω_2^0 and ω_3^0, and then the structure equation implies that these must vanish. So we find that

$$\omega_1^0 = \omega_0^1 = \frac{M}{r^2\sqrt{h}}\theta^0, \quad \omega_i^0 = \omega_0^i = 0, \quad i = 2, 3.$$

We have computed the top row and left hand column of ω. From $d\theta^1 = 0$ we conclude that ω_2^1 is some multiple of $d\vartheta$ and that ω_3^1 is some multiple of $d\phi$. From $d\theta^2 = dr \wedge d\vartheta$ we see that $\omega_1^2 = \sqrt{h}d\vartheta$ and that ω_3^2 is some multiple of $d\phi$.
From

$$d\theta^3 = -S\sqrt{h}d\phi \wedge \theta^1 - Cd\phi \wedge \theta^2$$

we then deduce that $\omega_1^3 = S\sqrt{h}d\phi$ and that $\omega_2^3 = \frac{rC}{S}d\phi$.

Putting it all together gives

$$\omega = \begin{pmatrix} 0 & \frac{M}{r^2}dt & 0 & 0 \\ \frac{M}{r^2}dt & 0 & -\sqrt{h}d\vartheta & -S\sqrt{h}d\phi \\ 0 & \sqrt{h}d\vartheta & 0 & -Cd\phi \\ 0 & S\sqrt{h}d\phi & Cd\phi & 0 \end{pmatrix}$$

A check shows that

$$d\theta = -\omega \wedge \theta.$$

Computing the curvature form.

Using our formula for $d\sqrt{h}$ derived above we get $d\omega =$

$$\begin{pmatrix} 0 & -\frac{2M}{r^3}dr \wedge dt & 0 & 0 \\ -\frac{2Mdr\wedge dt}{r^3} & 0 & -\frac{Mdr\wedge d\vartheta}{r^2\sqrt{h}} & -\frac{SMdr\wedge d\phi}{r^2\sqrt{h}} - C\sqrt{h}d\vartheta \wedge d\phi \\ 0 & \frac{M}{r^2\sqrt{h}}dr \wedge d\vartheta & 0 & Sd\vartheta \wedge d\phi \\ 0 & \frac{SMdr\wedge d\phi}{r^2\sqrt{h}} + C\sqrt{h}d\vartheta \wedge d\phi & -Sd\vartheta \wedge d\phi & 0 \end{pmatrix}$$

and $\omega \wedge \omega =$

$$\begin{pmatrix} 0 & 0 & -\frac{M\sqrt{h}}{r^2}dt \wedge d\vartheta & -\frac{SM\sqrt{h}}{r^2}dt \wedge d\phi \\ 0 & 0 & 0 & C\sqrt{h}d\vartheta \wedge d\phi \\ \frac{M\sqrt{h}}{r^2}d\vartheta \wedge dt & 0 & 0 & -Shd\vartheta \wedge d\phi \\ \frac{SM\sqrt{h}}{r^2}d\phi \wedge dt & C\sqrt{h}d\phi \wedge d\vartheta & -Shd\phi \wedge d\vartheta & 0 \end{pmatrix}.$$

We thus obtain (using the definition of the θ^i and the explicit expression for h):

$$\Omega = d\omega + \omega \wedge \omega = \frac{M}{r^3} \begin{pmatrix} 0 & 2\theta^0 \wedge \theta^1 & -\theta^0 \wedge \theta^2 & -\theta^0 \wedge \theta^3 \\ 2\theta^0 \wedge \theta^1 & 0 & -\theta^1 \wedge \theta^2 & -\theta^1 \wedge \theta^3 \\ -\theta^0 \wedge \theta^2 & \theta^1 \wedge \theta^2 & 0 & 2\theta^2 \wedge \theta^3 \\ -\theta^0 \wedge \theta^3 & \theta^1 \wedge \theta^3 & -2\theta^2 \wedge \theta^3 & 0 \end{pmatrix}.$$

The Schwarzschild metric is Ricci flat.

The form of the curvature matrix shows that all the sectional curvatures of the "coordinate planes" are either $\pm Mr^{-3}$ or $\pm 2Mr^{-3}$. The curvature tensor in the frame field E_0, E_1, E_2, E_3 dual to the θ^i is given by

$$R^i_{jk\ell} = \Omega^i_j(E_k, E_\ell).$$

From the entries of Ω we see that $R^m_{imj} = 0$ if $i \neq j$. Since

$$\mathrm{Ric}_{ij} = \sum R^m_{imj}$$

we see that $\mathrm{Ric}_{ij} = 0$ if $i \neq j$. So we must evaluate Ric_{ii} for $i = 0, 1, 2, 3$. Basically this just involves looking at the appropriate columns of Ω. For example, for $i = 0$ we find from the leftmost column of Ω that

$$R^1_{010} = 2\frac{M}{r^3}$$

$$R^2_{020} = -\frac{M}{r^3}$$

$$R^3_{030} = -\frac{M}{r^3}$$

so $\mathrm{Ric}_{00} = 0$. Similarly for $i = 1, 2, 3$.

Conclusion: **The Schwarzschild metric is Ricci flat** in the sense that its Ricci tensor vanishes everywhere. In particular, it is a solution of the Einstein field equations with zero on the right hand side.

6.5 The second Bianchi identity.

Take the exterior derivative of Cartan's second equation

$$d\omega + \omega \wedge \omega = \Omega. \qquad (6.7)$$

We obtain

$$d\omega \wedge \omega - \omega \wedge d\omega = d\Omega.$$

Substitute $d\omega = \Omega - \omega \wedge \omega$ in the left hand side. The terms involving $\omega \wedge \omega \wedge \omega$ cancel, and we are left with

$$d\Omega = \Omega \wedge \omega - \omega \wedge \Omega \qquad (6.8)$$

which is known as the **second Bianchi identity**.

6.6 A theorem of F. Schur.

Theorem 9. *Let M be a connected Riemannian manifold with the property that at every point $p \in M$ the sectional curvature $\sec(\Pi)$ is independent of Π. Then if the dimension of M is ≥ 3 this sectional curvature is independent of p.*

Proof. Let $K(p)$ denote the sectional curvature at p. So K is a function on M and, in terms of a coframe field, we are assuming that

$$\Omega^i_j = K \theta^i \wedge \theta^j.$$

Take d of both sides. On the left we get

$$\sum_k (\Omega^i_k \wedge \omega^k_j - \omega^i_k \wedge \Omega^k_j)$$

by the second Bianchi identity. On the right we get

$$dK \wedge \theta^i \wedge \theta^j + K d\theta^i \wedge \theta^j - K \theta^i \wedge d\theta^j.$$

$$= dK \wedge \theta^i \wedge \theta^j - K \sum_k (\omega^i_k \wedge \theta^k \wedge \theta^j - \theta^i \wedge \omega^j_k \wedge \theta^k)$$

by the first structural equation.

Now

$$K \sum_k \omega^i_k \wedge \theta^k \wedge \theta^j = \sum_k \omega^i_k \wedge \Omega^k_j$$

and

$$K \sum_k \theta^i \wedge \omega^j_k \wedge \theta^k = -\sum_k \Omega^i_k \wedge \omega^j_k = \sum_k \Omega^i_k \wedge \omega^k_j$$

by the antisymmetry of ω. So we arrive at the conclusion that

$$dK \wedge \theta^i \wedge \theta^j = 0$$

for all i and j. Expand $dK = \sum_k a_k \theta^k$. If $\dim M \geq 3$ we can, for each k find i and j both $\neq k$ so $a_k = 0$ and hence $dK = 0$. $\square$

6.7 Friedmann Robertson Walker metrics in general relativity.

These are metrics on the product M of an interval $I \subset \mathbb{R}$ (with a negative definite metric $-dt^2$) and a three dimensional Riemannian manifold S of constant curvature. The metric has the form

$$-dt^2 + f^2 d\sigma^2$$

where f is a function on I and $d\sigma^2$ is the metric on S.

The reason for the constant curvature assumption has to do with Schur's theorem. If we believe that space time is of the form $I \times S$, and if we believe that at all points of S "all directions are the are the same", a sort of democracy principle, then if follows from Schur's theorem that the curvature must be constant. One of the consequences of the computations we will do here are that the Einstein field equations reduce to an ordinary differential equation, from which we can deduce the "big bang" under plausible additional assumptions. I will postpone some of the discussion of these issues until after I will have introduced the Einstein gravitational equations, but you can look in O'Neill for a lucid exposition.

Over the next few paragraphs I will do Cartan calculations without assuming that $d\sigma^2$ has constant curvature, and then specialize to the constant curvature case. I will keep the condition that S be three dimensional in order that the formulas fit on a page. But there is no difference between 3 and $n \geq 3$.

The coframe field.

Let ϕ^1, ϕ^2, ϕ^3 be a coframe field on S so that

$$\theta^0 = dt, \ \theta^1 = f\phi^1, \ \theta^2 = f\phi^2, \ \theta^3 = f\phi^3$$

is a coframe field on M and the metric is given by

$$-(\theta^0)^2 + (\theta^1)^2 + (\theta^2)^2 + (\theta^3)^2.$$

Searching for ω.

We seek ω satisfying $d\theta + \omega \wedge \theta = 0$ and the symmetry-anti-symmetry conditions

$$\omega_i^0 = \omega_0^i, \quad \omega_0^0 = 0, \quad \omega_j^i = -\omega_i^j, \quad i, j = 1, 2, 3.$$

Since $d\theta^0 = 0$, we know that ω_i^0 must be a multiple of θ^i, $i = 1, 2, 3$. In fact, we can be more precise: For $i = 1, 2, 3$ we have $\theta^i = f\phi^i$ so

$$d\theta^i = df \wedge \phi^i + f d\phi^i = f' dt \wedge \phi^i + f d\phi^i = \frac{f'}{f} \theta^0 \wedge \theta^i + f d\phi^i.$$

From this we draw the conclusions that $\omega_0^i = \omega_i^0 = \frac{f'}{f}\theta^i$, $i = 1, 2, 3$ and that the ω_j^i on M coincide with the ω_j^i of S (pulled back to M). So, by abuse of notation (not putting in notation for the pullback):

$$\omega = \begin{pmatrix} 0 & \frac{f'}{f}\theta^1 & \frac{f'}{f}\theta^2 & \frac{f'}{f}\theta^3 \\ \frac{f'}{f}\theta^1 & 0 & \omega_2^1 & \omega_3^1 \\ \frac{f'}{f}\theta^2 & -\omega_2^1 & 0 & \omega_3^2 \\ \frac{f'}{f}\theta^3 & -\omega_3^1 & -\omega_3^2 & 0 \end{pmatrix}$$

where the ω_j^i are the connection forms on S.

6.7. FRIEDMANN ROBERTSON WALKER METRICS. 175

Computing the curvature.

We begin by computing $d\omega$. In the lower 3×3 block we will simply write $d\omega^i_j$ (coming from S) so let us look at the top row: We have $d\left(\frac{f'}{f}\theta^i\right) = d\left(\frac{f'}{f}\right) \wedge \theta^i + \frac{f'}{f}d\theta^i$,

$$d\left(\frac{f'}{f}\right) = \frac{f''f - (f')^2}{f^2}\theta^0,$$

and

$$d\theta^i = -\frac{f'}{f}\theta^i \wedge \theta^0 - \sum_{j=1}^{3} \omega^i_j \wedge \theta^j.$$

So the terms involving $(f')^2/f^2$ cancel, and the entries in the top row of $d\omega$ are:

$$\sigma^0_i := \frac{f''}{f}\theta^0 \wedge \theta^i - \frac{f'}{f}\sum_j \omega^i_j \wedge \theta^j.$$

So

$$d\omega = \begin{pmatrix} 0 & \sigma^0_1 & \sigma^0_2 & \sigma^0_3 \\ \sigma^0_1 & 0 & d\omega^1_2 & d\omega^1_3 \\ \sigma^0_2 & -d\omega^1_2 & 0 & d\omega^2_3 \\ \sigma^0_3 & -d\omega^1_3 & -d\omega^2_3 & 0 \end{pmatrix}$$

where $\sigma^0_i := \frac{f''}{f}\theta^0 \wedge \theta^i - \frac{f'}{f}\sum_j \omega^i_j \wedge \theta^j$. Let's look at the top entry in the second column of $\omega \wedge \omega$. It is $-\frac{f'}{f}\left[\theta^2 \wedge \omega^1_2 + \theta^3 \wedge \omega^1_3\right]$. This cancels the term involving f'/f in σ^0_1. So the expression for the curvature is:

$$\Omega = d\omega + \omega \wedge \omega =$$

$$\begin{pmatrix} 0 & \frac{f''}{f}\theta^0 \wedge \theta^1 & \frac{f''}{f}\theta^0 \wedge \theta^2 & \frac{f''}{f}\theta^0 \wedge \theta^3 \\ \frac{f''}{f}\theta^0 \wedge \theta^1 & 0 & \left(\frac{f'}{f}\right)^2\theta^1 \wedge \theta^2 + \Omega^1_2 & \left(\frac{f'}{f}\right)^2\theta^1 \wedge \theta^3 + \Omega^1_3 \\ \frac{f''}{f}\theta^0 \wedge \theta^2 & \left(\frac{f'}{f}\right)^2\theta^2 \wedge \theta^1 + \Omega^2_1 & 0 & \left(\frac{f'}{f}\right)^2\theta^2 \wedge \theta^3 + \Omega^2_3 \\ \frac{f''}{f}\theta^0 \wedge \theta^3 & \left(\frac{f'}{f}\right)^2\theta^3 \wedge \theta^1 + \Omega^3_1 & \left(\frac{f'}{f}\right)^2\theta^3 \wedge \theta^2 + \Omega^3_2 & 0 \end{pmatrix}$$

where the Ω^i_j are the curvature forms of S.

Now let us make the assumption that S has constant curvature k so $\Omega^i_j = k\phi^i \wedge \phi^j = \frac{k}{f^2}\theta^i \wedge \theta^j$. We get $\Omega =$

$$\begin{pmatrix} 0 & \frac{f''}{f}\theta^0 \wedge \theta^1 & \frac{f''}{f}\theta^0 \wedge \theta^2 & \frac{f''}{f}\theta^0 \wedge \theta^3 \\ \frac{f''}{f}\theta^0 \wedge \theta^1 & 0 & \left[\left(\frac{f'}{f}\right)^2 + \frac{k}{f^2}\right]\theta^1 \wedge \theta^2 & \left[\left(\frac{f'}{f}\right)^2 + \frac{k}{f^2}\right]\theta^1 \wedge \theta^3 \\ \frac{f''}{f}\theta^0 \wedge \theta^2 & \left[\left(\frac{f'}{f}\right)^2 + \frac{k}{f^2}\right]\theta^2 \wedge \theta^1 & 0 & \left[\left(\frac{f'}{f}\right)^2 + \frac{k}{f^2}\right]\theta^2 \wedge \theta^3 \\ \frac{f''}{f}\theta^0 \wedge \theta^3 & \left[\left(\frac{f'}{f}\right)^2 + \frac{k}{f^2}\right]\theta^3 \wedge \theta^1 & \left[\left(\frac{f'}{f}\right)^2 + \frac{k}{f^2}\right]\theta^3 \wedge \theta^2 & 0 \end{pmatrix}$$

6.7.1 The expanding universe and the big bang.

From this it is easy to compute the Ricci curvature and the scalar curvature. See the discussion in O'Neill pp.341-353. Let us look at the special case of our formula for Ω where $k = 0$ and $f = e^{t/r}$. Then the coefficients of all the $\theta^i \wedge \theta^j$ are equal to $1/r^2$. In other words, the space with this Robertson-Walker metric has constant curvature $1/r^2$. So it is locally symmetric. In fact, on general principles, it must be isometric to an open subset of the deSitter space-time that we studied in Section 4.13. Thus the Friedmann-Robertson-Walker metric

$$-dt^2 + e^{2t/r}(dx^2 + dy^2 + dz^2) \tag{6.9}$$

(so with flat space) describes an open subset of deSitter space. To see this explicitly, consider the map $(t, x, y, z) \mapsto (W, V, X, Y, Z)$ given by

$$\begin{aligned} V + W &= e^{t/r} \\ V - W &= -(x^2 + y^2 + z^2)e^{t/r} + r^2 e^{-t/r} \\ X &= e^{t/r}x \\ Y &= e^{t/r}y \\ Z &= e^{t/r}z. \end{aligned}$$

Then $-W^2 + V^2 + X^2 + Y^2 + Z^2 = r^2$ and the metric $-dW^2 + dV^2 + dX^2 + dY^2 + dZ^2$ pulls back, under this map to the metric (6.9) given above, as can easily be checked. Notice that the hypersurfaces $t = $ const. go over under this map to the intersection of the quadric with the hyperplanes $V + W = $ constant. Since $e^{t/r} > 0$, we must have $V + W > 0$ so only "half" of deSitter space is covered by this map.

It is this model (and modifications thereof) which are currently used in cosmology.

6.7. FRIEDMANN ROBERTSON WALKER METRICS. 177

In deSitter's original model, he chose his "cosmic time" so the the quadric was sliced by "horizontal" hyperplanes, rather than the diagonal ones as in the figure which was Lemaitre's choice of "cosmic time".

Back to more general Robertson-Walker metrics.

As O'Neill points out on page 347, according to Hubble (1929) all distant galaxies are moving away from us at a rate proportional to their distance. The distance between $\gamma_p(t)$ and $\gamma_q(t)$ in $S(t)$ is $f(t)d(p,q)$ where d is the Riemannian distance in S. Hubble's constant

$$H_0 = \frac{f'(t)}{f(t)}$$

is estimated as

$$\frac{1}{(18 \pm 2) \times 10^9 \text{yr}}.$$

In particular f' is positive so the universe is "expanding".

The distribution of energy and "pressure" in the universe, together with the Einstein field equations (which become ordinary differential equations) show that $f'' < 0$. This, together with the value of Hubble's constant, shows that there must have been a singularity between ten and twenty billion years ago. In other words, that the universe as we know it had a definite beginning somewhere between ten and twenty billion years ago.

6.8 The rotating black hole.

Scwarzschild discovered his solution to the Einstein field equations in 1916. The next major black hole solution was not found until 1963! This was the rotating black hole found by Kerr (discovered by using a theoretical classification of possible space-times by Petrov in 1954). The purpose of this problem set is to walk through the part of the computation which shows that the Kerr metric is Ricci flat. The material in this problem set follows (with slight changes in notation) the excellent book *The Geometry of Kerr Black Holes* by Barrett O'Neill.

6.8.1 Killing fields and Noether's theorem.

Before writing down the Kerr metric, I want to go over some elementary facts concerning continuous symmetries of a semi-Riemannian manifold M with semi-Riemannian metric **g**.

A vector field X on M is called a **Killing** vector field (named after the mathematician Wilhelm Killing (1847 - 1923)) if

$$L_X \mathbf{g} = 0,$$

which is the same as saying that the flow generated by X preserves **g**. Another way of saying the same thing is that for any pair of vector fields V and W

$$X\langle V, W \rangle = \langle [X, V], W \rangle + \langle V, [X, W] \rangle.$$

6.8. THE ROTATING BLACK HOLE.

For any three vector fields we have

$$X\langle V, W\rangle = \langle \nabla_X V, W\rangle + \langle V, \nabla_X W\rangle,$$

and

$$\nabla_X V = \nabla_V X + [X, V], \quad \nabla_X W = \nabla_W X + [X, W].$$

So

Proposition 19. *X is Killing vector field if and only if for any pair of vector fields V and W we have*

$$\langle \nabla_V X, W\rangle + \langle V, \nabla_W X\rangle = 0.$$

Another way of saying this is that the linear transformation ∇X (which sends every vector field V to $\nabla_V X$) is skew symmetric at all points.

The following is a very special case of Noether's theorem which says that continuous symmetries of variational equations give rise to conserved quantities:

Proposition 20. *Let X be a Killing vector field and let γ be a geodesic. Then $\langle X, \gamma'\rangle$ is constant along γ.*

Proof. We may work locally and so assume that the flow ψ_s generated by X is defined near γ for some $0 \leq s \leq \epsilon$. Since ψ_s are isometries, the two parameter family

$$(t, s) \mapsto \psi_s(\gamma(t))$$

is a geodesic variation of γ. For fixed t, the curve $s \mapsto \psi_s(\gamma(t))$ is an integral curve of X. In other words, the variational vector field of this two parameter family is just the restriction of X to γ. Now

$$\frac{d}{dt}\langle X, \gamma'\rangle = \langle \nabla_{\gamma'} X, \gamma'\rangle$$

since $\gamma'' = 0$. But by the preceding Proposition, $\langle \nabla_{\gamma'} X, \gamma'\rangle = 0$. □

For a fascinating and in depth study of Noether's theorem, its history, and the history of its subsequent developments, see *The Noether theorems* by Yvette Kosmann Schwarzbach, Springer, (2011).

6.8.2 The definition of the Kerr metric and some of its elementary properties.

We will begin by expressing the metric in terms of t, r, ϑ, ϕ where (in the far exterior, where we live) t is time and r, ϑ, ϕ are spherical coordinates on three dimensional space, and where the conventions in general relativity are that ϑ denotes the co-latitude (the angle from the north pole) and ϕ denotes the longitude (the angle from the x-axis).

The parameters $M > 0$ (called the "mass") and a (called the "angular momentum per unit mass") fix the metric as we will soon see. We will use the following notations to shorten the formulas:

$$\begin{aligned} S &:= \sin\vartheta \\ C &:= \cos\vartheta \\ \rho^2 &:= r^2 + a^2 C^2 \\ \Delta &:= r^2 - 2Mr + a^2. \end{aligned}$$

The g_{ij} in terms of these coordinates are defined to be

$$\begin{aligned} g_{tt} &= -1 + 2Mr/\rho^2 \\ g_{rr} &= \rho^2/\Delta \\ g_{\vartheta\vartheta} &= \rho^2 \\ g_{\phi\phi} &= \left[r^2 + a^2 + \frac{2Mra^2 S^2}{\rho^2}\right] S^2 \\ g_{ij}(i \neq j) &= 0 \quad \text{except for} \\ g_{\phi t} = g_{t\phi} &= -\frac{2MraS^2}{\rho^2}. \end{aligned}$$

We still have to check that this is a metric, but assuming that it is, the following facts are immediate:

- Since the g_{ij} do not depend explicitly on t and ϕ it follows that $\partial_t := \frac{\partial}{\partial t}$ and $\partial_\phi := \frac{\partial}{\partial \phi}$ are Killing fields, and hence we get conserved quantities E and L just as in the Schwarzschild case.

- When $a = 0$ we get the Schwarzschild metric and when a and M are both zero we get the flat Minkowski metric.

- For large r the metric is asymptotically Minkowskian.

6.8.3 Checking that we do have a semi-Riemannian metric.

We want to check that $\det(g_{ij}) \neq 0$. Since the only non-vanishing cross-terms are $g_{\phi t} = g_{t\phi}$, the first step is to compute the two by two determinant $g_{tt}g_{\phi\phi} - g_{t\phi}^2$ and then to multiply by $g_{rr}g_{\vartheta\vartheta}$.

1. Verify that

$$g_{tt}g_{\phi\phi} - g_{t\phi}^2 = -g_{\phi\phi} - a^{-1}(r^2 + a^2)g_{t\phi}.$$

6.8. THE ROTATING BLACK HOLE.

We now simplify the right hand side as follows:

$$ag_{\phi\phi} = a\left[r^2 + a^2 + \frac{2MraS^2}{\rho^2}\right]S^2$$

$$(r^2 + a^2)g_{t\phi} = -\frac{2MraS^2}{\rho^2}(r^2 + a^2)$$

$$\frac{2Mra^2S^2}{\rho^2}(S^2 - 1) = -\frac{2Mra^2S^2}{\rho^2}C^2$$

$$= -\frac{2Mra^2S^2C^2}{r^2 + a^2C^2}$$

$$= (2MrS^2)\left(\frac{r^2}{\rho^2} - 1\right) \quad \text{so}$$

$$ag_{\phi\phi} + (r^2 + a^2)g_{t\phi} = \Delta aS^2.$$

Substituting this into the right hand side of Problem **1** gives $-\Delta S^2$. Multiplying by $g_{rr}g_{\vartheta\vartheta}$ gives

$$\det(g_{ij}) = -\rho^4 S^2.$$

So as long as we stay away from the problematic sets where

$$\Delta = 0, \ \rho = 0, \ \text{or} \ \sin\vartheta = 0$$

we have a semi-Riemannian metric.

To check the signature it is convenient to introduce the vector fields

$$V := (r^2 + a^2)\partial_t + a\partial_\phi$$

and

$$W := \partial_\phi + aS^2\partial_t.$$

Repeat apology - In some of the formulas below I use (,) for scalar product instead of ⟨ , ⟩. I am afraid that it would introduce too many new errors in trying to be consistent.

2. Verify the following formulas for the scalar products

$$(V, \partial_\phi) = \Delta a S^2$$
$$(V, \partial_t) = -\Delta$$
$$(W, \partial_\phi) = (r^2 + a^2)S^2$$
$$(W, \partial_t) = -aS^2$$

and then that

$$(V, V) = -\Delta \rho^2 \tag{6.10}$$
$$(V, W) = 0 \tag{6.11}$$
$$(W, W) = \rho^2 S^2 \tag{6.12}$$

Thus $V, \partial_r, \partial_\vartheta, W$ form an orthogonal frame field (where the metric is non-singular). Later on we will normalize these vector fields and then pass to the coframe field to compute the connection forms and the curvature.

6.8.4 The domains and the signature.

Notice that if $a \neq 0$, the value $r = 0$ does not present a singularity in the expression for the metric, it is the value $\rho = 0$ which is problematic (and in fact represents an actual singularity - not just a failure of the coordinate system).

So the professionals in this subject allow r to range over the entire real line. We now turn to the issue of the zeros of Δ: The case where really interesting things happen is where
$$0 < a^2 < M^2$$
(known as the "slowly rotating black hole"). This is the case we will consider. In this case Δ has two positive roots
$$r_{\pm} = M \pm (M^2 - a^2)^{\frac{1}{2}} < 2M.$$

So there are three regions where the coordinates introduced above are ok (excluding $\rho = 0$ and $S = 0$):

- Region I (where we live) where $r > r_+$.
- Region II where $r_- < r < r_+$.
- Region III where $r < r_-$.

It turns out (although we will not prove it here) that the hypersurfaces $\Delta = 0$ represent failures of the coordinate system but not of the metric: that there exists a four dimensional semi-Riemannian manifold which includes all of the region $\rho \neq 0$. In this manifold the hypersurfaces $\Delta = 0$ are *horizons* in the sense that a forward timelike path (to be defined in the next paragraph) can cross from Region I into Region II but not from Region II to Region I and can cross from Region II to Region III but not from Region III back to Region II.

In all three regions we have
$$(W, W) > 0 \quad \text{and} \quad (\partial_\vartheta, \partial_\vartheta) > 0.$$

For large values of r we have $(\partial_t, \partial_t) < 0$, so we can orient all time-like paths to coincide with increasing t for $r \gg 0$. So we now have a notion of "forward".

$(V, V) < 0$ in Regions I and III but $(V, V) > 0$ in Region II. On the other hand, $(\partial_r, \partial_r) > 0$ in Regions I and III but $(\partial_r, \partial_r) < 0$ in Region II.

So in all three regions the metric is Lorentzian. But notice that ∂_t is spacelike in Region II (since it is orthogonal to ∂_r and the metric is Lorentzian). So it is not a good idea to think of t as time unless we are in Region I and we also have $r > 2M$.

The ergosphere.

The region $\mathcal{D}$ where the vector field ∂_t is spacelike includes all of Region II but also a region $\mathcal{E}$ lying in Region I called the **ergosphere**, and also a region $\mathcal{E}'$ in Region III. In other words, $\mathcal{D}$ is the region where $g_{tt} > 0$, and we live in a region

6.8. THE ROTATING BLACK HOLE.

where $g_{tt} < 0$. So the boundary L between where we live and the ergosphere will be given by the equation $g_{tt} = 0$. To see that this boundary is in fact a smooth hypersurface, write $g_{tt} = -1 + 2Mr/\rho^2$ as

$$g_{tt} = \frac{\ell}{\rho^2} \quad \text{where} \quad \ell := r((2M - r) - a^2 C^2.$$

Since ρ^2 is positive (away from the singularity) we see that L is given by $\ell = 0$. Since

$$d\ell = 2(M - r)dr + 2a^2 SC d\vartheta$$

we see that $d\ell = 0$ implies that $M = r$ so at points where $d\ell = 0$ we have $\ell = M^2 - a^2 C^2$. We are assuming that $a^2 < M^2$ so we see that $d\ell$ does not vanish on L. The implicit function theorem then guarantees that L is a smooth hypersurface and that g_{tt} does indeed change sign when we cross L.

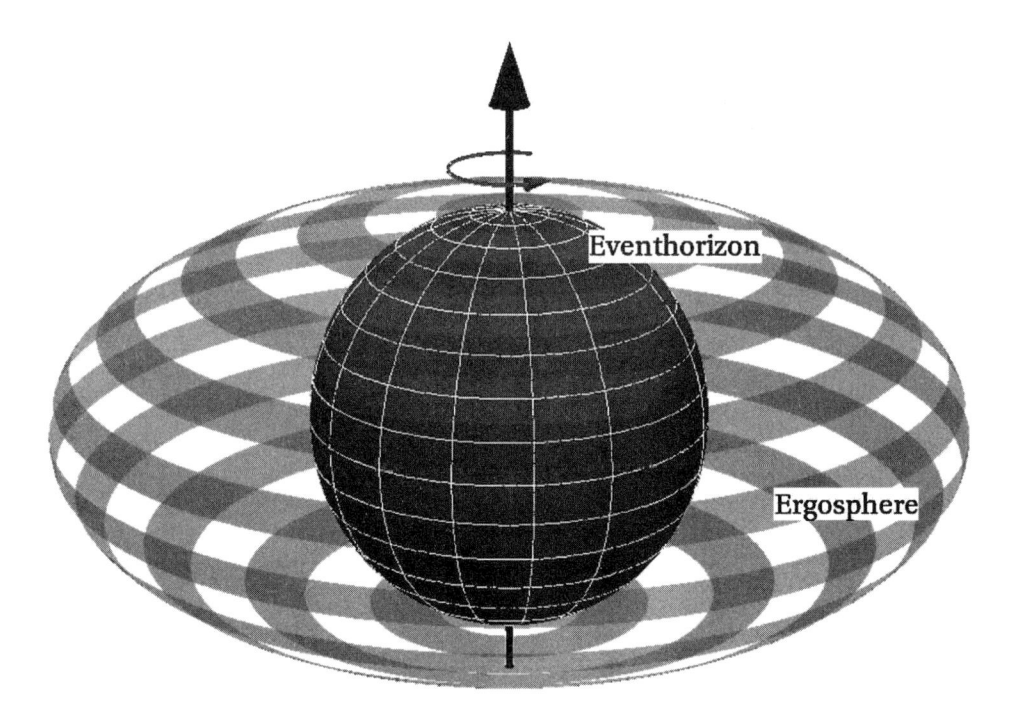

Penrose energy extraction.

The existence of the ergosphere led Roger Penrose (in 1969) to invent the following device to extract energy from a slowly rotating black hole:

Recall that we proved above, that if X is a Killing vector field, then $\langle \gamma, X \rangle$ is constant along any geodesic γ. We know that ∂_t is a Killing vector field.

If γ is a geodesic we let

$$E = E(\gamma) := -\langle \gamma', \partial_t \rangle$$

and call E the energy of the geodesic. The sign is chosen so that for a forward pointing time-like geodesic in our region (outside the ergosphere) $E > 0$. In fact, for $r \gg 0$ the energy E as defined above is approximately the sum of the kinetic energy and the mass energy together with a term coming from the potential energy of attraction. I will not prove this here.

Let α be a particle which falls from our region (where $r \gg 0$), and so has positive energy across L into the ergosphere. Arrange that inside $\mathcal{E}$ the particle α splits into two particles, α_+ and α_- where $E(\alpha_-) < 0$. This is possible because ∂_t is spacelike in $\mathcal{E}$. Conservation of energy-momentum at the splitting implies that
$$\dot{\alpha} = \dot{\alpha}_+ + \dot{\alpha}_-$$
at the splitting. Taking the scalar product of this equation with ∂_t gives
$$E(\alpha) = E(\alpha_+) + E(\alpha_-)$$
so $E(\alpha_+) > E(\alpha)$, in fact $E(\alpha_+) = E(\alpha) + |E(\alpha_-)|$.

Penrose arranges that α_- heads across the horizon into Region II never to be seen again, while α_+ returns to us with its increased energy. For more details see Wald, *General Relativity*, Chapter 12.

The Carter time machine.

It is not advisable for you to cross into Region II and certainly not into Region III. But if you do, you will find that near the singularity $\rho = 0$ the vector field ∂_ϕ is time-like. Since the integral curves of ∂_ϕ are circles, this means that someone who moves along one of these integral curves experiences an endless repetition of the same events, like in the movie "Groundhog Day".

The region $\mathfrak{T}$ in Region III where ∂_ϕ is timelike is called the Carter time machine. The reason is that one can prove that if p and q are any two points in Region III one can join p to q by a future pointing timelike curve. (Of course this means that we can also go from q to p by a future pointing timelike curve.) The method of proof is to go from p to $\mathfrak{T}$ and then circle around an appropriate number of times and then head to q. For the details I refer to O'Neill pp. 76-78.

All of the above discussion in this section is posited on the assumption that the metric we are using is in fact a solution to the Einstein field equations, in fact that its Ricci tensor vanishes. The computation here is much more complicated than in the Schwarzschild case, but not impossibly hard. We will now get onto this computation.

6.8.5 An orthonormal frame field and its coframe field.

Let ϵ be the function which is $+1$ on Regions I and III and is -1 on Region II, and let ρ denote the positive square root of $\rho^2 = r^2 + a^2 C^2$. Now normalize the orthogonal vector fields introduced above so define E_0 to be the normalized

6.8. THE ROTATING BLACK HOLE.

vector field corresponding to V, define E_1 to be the normalized vector field corresponding to ∂_r, define E_2 to be the normalized vector field corresponding to ∂_ϑ and E_3 the normalized vector field corresponding to W so

$$E_0 = \frac{1}{\rho\sqrt{\epsilon\Delta}}\left((r^2+a^2)\partial_t + a\partial_\phi\right)$$

$$E_1 = \frac{\sqrt{\epsilon\Delta}}{\rho}\partial_r$$

$$E_2 = \frac{1}{\rho}\partial_\vartheta$$

$$E_3 = \frac{1}{rS}(\partial_\phi + aS^2\partial_t)$$

and we have

$$(E_0, E_0) = -\epsilon$$
$$(E_1, E_1) = \epsilon$$
$$(E_2, E_2) = 1$$
$$(E_3, E_3) = 1.$$

3. Verify that the following give the dual coframe basis expressed in terms of the t, r, ϑ, ϕ coordinates:

$$\theta^0 = \frac{\sqrt{\epsilon\Delta}}{\rho}(dt - aS^2 d\phi), \quad \theta^1 = \frac{\rho}{\sqrt{\epsilon\Delta}}dr,$$

$$\theta^2 = \rho d\vartheta, \quad \theta^3 = \frac{S}{\rho}\left((r^2+a^2)d\phi - a dt\right).$$

4. Solve for $dt, dr, d\vartheta, d\phi$ in terms of $\theta^0, \theta^1, \theta^2, \theta^3$.

Our next step is to express the $d\theta^i$ in terms of $\theta^0, \theta^1, \theta^2, \theta^3$ so as to find the connection forms. For this the following collection of partial derivatives (which you might want to privately check) will be useful:

$$\frac{\partial \rho}{\partial r} = \frac{r}{\rho}, \quad \frac{\partial}{\partial r}\frac{1}{\rho} = -\frac{r}{\rho^3}, \quad \frac{\partial \rho}{\partial \vartheta} = -\frac{a^2 SC}{\rho}, \quad \frac{\partial}{\partial \vartheta}\frac{1}{\rho} = \frac{a^2 SC}{\rho^3}$$

$$\frac{\partial \Delta}{\partial r} = 2(r-M), \quad \frac{\partial}{\partial r}\sqrt{\epsilon\Delta} = \epsilon\frac{r-M}{\sqrt{\epsilon\Delta}}, \quad \frac{\partial}{\partial r}\frac{1}{\sqrt{\epsilon\Delta}} = -\frac{r-M}{\Delta\sqrt{\epsilon\Delta}}.$$

The following function enters into the computation of the connection forms:

$$F := \frac{\partial}{\partial r}\frac{\sqrt{\epsilon\Delta}}{\rho} = \epsilon\frac{(r-M)\rho^2 - r\Delta}{\rho^3\sqrt{\epsilon\Delta}}$$

and the functions

$$\mathbf{I} := \frac{Mr}{\rho^6}(r^2 - 3a^2 C^2), \quad \mathbf{J} := \frac{MaC}{\rho^6}(3r^2 - a^2 C^2)$$

enter into the computation of the curvature.

6.8.6 The connection forms.

5. Compute $d\theta^0, d\theta^1, d\theta^2, d\theta^3$ in terms of $dt, dr, d\vartheta, d\phi$

You can then use the results of Problem 4 to express $d\theta^0, d\theta^1, d\theta^2, d\theta^3$ in terms of the exterior products of the $\theta^0, \theta^1, \theta^2, \theta^3$. You should get:

$$d\theta^0 = F\theta^1 \wedge \theta^0 - \frac{a^2 SC}{\rho^3}\theta^2 \wedge \theta^0 - \frac{2aC\sqrt{\epsilon\Delta}}{\rho^3}\theta^2 \wedge \theta^3$$

$$d\theta^1 = \frac{a^2 SC}{\rho^3}\theta^1 \wedge \theta^2$$

$$d\theta^2 = \frac{r\sqrt{\epsilon\Delta}}{\rho^3}\theta^1 \wedge \theta^2$$

$$d\theta^3 = \frac{2arS}{\rho^3}\theta^1 \wedge \theta^0 + \frac{r\sqrt{\epsilon\Delta}}{\rho^3}\theta^1 \wedge \theta^3 + \frac{C}{S}\frac{r^2 + a^2}{\rho^3}\theta^2 \wedge \theta^3.$$

In principle, we now have to do some detective work together with Cartan's lemma to determine the forms ω^i_j so that $d\theta + \omega \wedge \theta = 0$. That is, we want all four equations

$$d\theta^i + \sum_j \omega^i_j \wedge \theta^j = 0, \quad i = 0, 1, 2, 3$$

to hold.

I will write down the answer. But first we have to write out the relations between ω^i_j and ω^j_i. Since E_2 and E_3 are spacelike throughout, we have

$$\omega^2_3 = -\omega^3_2.$$

Since E_0 and E_1 have opposite signatures throughout, we have we have

$$\omega^1_0 = \omega^0_1.$$

The remaining relations will involve ϵ:

$$\omega^1_2 = -\epsilon\omega^2_1, \quad \omega^1_3 = -\epsilon\omega^3_1, \quad \omega^0_2 = \epsilon\omega^2_0, \quad \omega^0_3 = \epsilon\omega^3_0.$$

6.8. THE ROTATING BLACK HOLE.

The claim is that the connection forms are then given by

$$\omega_1^0 = F\theta^0 - \frac{\epsilon a r S}{\rho^3}\theta^3$$

$$\omega_2^0 = -\frac{a^2 SC}{\rho^3}\theta^0 - \frac{aC\sqrt{\epsilon\Delta}}{\rho^3}\theta^3$$

$$\omega_3^0 = -\epsilon\frac{arS}{\rho^3}\theta^1 + \frac{aC\sqrt{\epsilon\Delta}}{\rho^3}\theta^2$$

$$\omega_2^1 = -\frac{a^2 SC}{\rho^3}\theta^1 - \frac{\epsilon r\sqrt{\epsilon\Delta}}{\rho^3}\theta^2$$

$$\omega_3^1 = -\epsilon\frac{arS}{\rho^3}\theta^0 - \epsilon\frac{r\sqrt{\epsilon\Delta}}{\rho^3}\theta^3$$

$$\omega_3^1 = -\epsilon\frac{aC\sqrt{\epsilon\Delta}}{\rho^3}\theta^0 - \frac{C(r^2+a^2)}{S\rho^3}\theta^3.$$

6. Check that the first Cartan structural equations $d\theta + \omega \wedge \theta = 0$, i.e

$$d\theta^i + \sum_j \omega_j^i \wedge \theta^j = 0$$

are satisfied.

6.8.7 The curvature.

I am not going to torture you by requesting that you compute the curvature. Instead, I refer you to pages 96 - 100 of O'Neil's *The Geometry of Kerr Black Holes*. O'Neil's notation is slightly different from the notation we have been using in that he writes ω^i for our θ^i. So in his notation the first Cartan structure equations read

$$d\omega^i + \sum_m \omega_m^i \wedge \omega^m = 0.$$

With this change in notation you should be able to read his computation of the curvature and the proof that the Kerr metric is Ricci flat.

Roy Patrick Kerr

(1934-)

Chapter 7

Gauss's lemma and some of its consequences.

We have defined geodesics as being curves which are self parallel. But there are several other characterizations of geodesics which are just as important: for example, in a Riemann manifold geodesics locally minimize arc length: "a straight line is the shortest distance between two points". We want to give one explanation of this fact here, using the exponential map for the Levi-Civita connection of a Riemannian manifold.

The main idea introduced by al Biruni (973-1048) went unappreciated for about 1000 years. The key result, known as Gauss' lemma asserts that radial geodesics are orthogonal to the images of spheres under the exponential map, and this will allow us to relate geodesics to extremal properties of arc length.

7.1 Geodesics locally minimize arc length in a Riemannian manifold.

Let M be a Riemannian manifold. Suppose that $\sigma : [0, b] \to M$ is a geodesic with the property that there are no conjugate points to 0 along σ at any t with $0 < t \leq b$. Let $o = \sigma(0)$. According to Theorem 6 this is equivalent to the assertion that $d(\exp_o)_{t\sigma'(0)}$ is non-singular as a map from

$$T_{t\sigma'(0)}(T_oM)) \to T_{\sigma(t)}M$$

for all $0 \leq t < b$. Then

Theorem 10. *Under these conditions the geodesic σ gives a strict local minimum for arc length among all curves joining $o := \sigma(0)$ to $q = \sigma(b)$.*

The proof makes use of Gauss's lemma, which we will state and prove below. I want to be more explicit about the statement of the theorem. If $\alpha : [0, b] \to M$

is a curve close to σ then by lifting we can write $\alpha = \exp_o \circ \tau$ where $\tau : [0,b] \to T_oM$ and $d\exp_o$ is non-singular at all points in the image of τ and

$$\tau(b) = v = \sigma'(0).$$

So we reformulate a more precise version of the theorem as follows: Let $\sigma : [0,1] \to M$ be the geodesic joining $o = \sigma(0)$ to $q = \sigma(1)$ with $\sigma'(0) = v$ so that $L(\sigma) = \|v\|$ (and σ is the image under $\exp_0$ of the line segment joining 0 to v). Let $\alpha = \exp_o \circ \tau$ where $\tau : [0,b] \to T_oM$ and $d\exp_o$ is non-singular at all points in the image of τ and with $\tau(b) = v$ so that $\alpha(b) = q$. Then:

Theorem 11. $L(\alpha) \geq L(\sigma)$. Furthermore, if $L(\alpha) = L(\sigma)$ then α is a monotone reparametrization of σ.

7.2 Gauss's lemma.

For the proof we state and prove Gauss's lemma (valid for any semi-Riemannian manifold): Let $o \in M$, $0 \neq x \in T_oM$. We have (as in any vector space) an identification of $T_x(T_o(M))$ with T_oM. If v, w are elements of T_oM we will denote the corresponding elements of $T_x(T_oM))$ by v_x, w_x. The vector v_x is called **radial** if v is a scalar multiple of x.

Lemma 7. *If v_x is radial then*

$$\langle d(\exp_o)_x v_x, d(\exp_o)_x w_x \rangle_{\exp_o(x)} = \langle v, w \rangle.$$

If the lemma is true for v it is true for any scalar multiple of v so in the proof we may assume that $v = x$.

Proof. Let $\mathbf{y} : \mathbb{R}^2 \to T_oM$ be given by $\mathbf{y}(t,s) = t(v + sw)$ and let

$$\mathbf{x} : \mathbb{R}^2 \to M, \quad \mathbf{x} := \exp_o \circ \mathbf{y}.$$

We have $\mathbf{y}_t(1,0) = v$ and $\mathbf{y}_s(1,0) = w$. So

$$\mathbf{x}_t(1,0) = d(\exp_o)_x(v_x) \quad \text{and} \quad \mathbf{x}_s(1,0) = d(\exp_o)_x(w_x).$$

So we must prove that

$$\langle \mathbf{x}_t(1,0), \mathbf{x}_s(1,0) \rangle = \langle v, w \rangle.$$

So we want to study the function $\langle \mathbf{x}_t, \mathbf{x}_s \rangle$ on $\mathbb{R}^2$. The longitudinal curve $t \mapsto \mathbf{x}(t,s)$ is a geodesic with initial velocity $v + sw$ by the definition of the exponential map. So $\langle \mathbf{x}_t, \mathbf{x}_t \rangle = \langle v + sw, v + sw \rangle$ and $\mathbf{x}_{tt} = 0$. So

$$\frac{\partial}{\partial t} \langle \mathbf{x}_t, \mathbf{x}_s \rangle = \langle \mathbf{x}_t, \mathbf{x}_{st} \rangle.$$

7.2. GAUSS'S LEMMA.

A general theorem about two parameter families that we proved earlier says that $\mathbf{x}_{st} = \mathbf{x}_{ts}$. So

$$\langle \mathbf{x}_t, \mathbf{x}_{st}\rangle = \langle \mathbf{x}_t, \mathbf{x}_{ts}\rangle = \frac{1}{2}\frac{\partial}{\partial s}\langle \mathbf{x}_t, \mathbf{x}_t\rangle = \frac{1}{2}\frac{\partial}{\partial s}\langle v+sw, v+sw\rangle.$$

In other words, we have shown that

$$\frac{\partial}{\partial t}\langle \mathbf{x}_t, \mathbf{x}_s\rangle = \langle v, w\rangle + s\langle w, w\rangle.$$

Setting $s = 0$ gives

$$\frac{\partial}{\partial t}\langle \mathbf{x}_t, \mathbf{x}_s\rangle(t,0) = \langle v, w\rangle.$$

Now $\mathbf{x}(0,s) \equiv o$ so $\langle \mathbf{x}_t, \mathbf{x}_s\rangle(0,0) = 0$. So from the preceding equation we get

$$\langle \mathbf{x}_t, \mathbf{x}_s\rangle(t,0) = t\langle v, w\rangle$$

and setting $t = 1$ gives Gauss's lemma. □

As special cases we have

- If v_x is radial then $\langle d(\exp_o)_x v_x, d(\exp_o)_x v_x\rangle_{\exp_o(x)} = \langle v, v\rangle$.

- If w_x is perpendicular to the radial vector $v_x \neq 0$ then $d(\exp_o)_x w_x$ is perpendicular to $d(\exp_o)_x v_x$.

Back to the proof of the theorem that geodesics locally minimize up to a conjugate point.

We are assuming that $\alpha(t) = \exp_o \circ \tau(t)$. Without loss of generality we may assume that $\tau(t) \neq 0$ for any $t > 0$ for otherwise we simply ignore the portion of τ going from 0 to 0. Let $u(t)$ denote the radial unit vector at $\tau(t)$ for $t > 0$. So we can write

$$\tau'(t) = \langle \tau'(t), u(t)\rangle u(t) + n(t)$$

where $n(t) \in T_{\tau(t)}(T_o(M))$ is perpendicular to the radial vectors.

Let

$$U(t) := d(\exp_o)_{\tau(t)} u(t) \quad \text{and} \quad N(t) := d(\exp_o)_{\tau(t)} n(t).$$

We have $\alpha'(t) = d(\exp_o)_{\tau(t)} \tau'(t)$ so by the two special cases of Gauss's lemma we have

$$\alpha'(t) = \langle \tau'(t), u(t)\rangle U(t) + N(t)$$

where $U(t)$ is a unit vector and $N(t)$ is perpendicular to $U(t)$. So

$$\|\alpha'(t)\| = \sqrt{\langle \tau'(t), u(t)\rangle^2 + \|N(t)\|^2} \geq \langle \tau'(t), u(t)\rangle.$$

But by Euclidean geometry, $u(t) = (\text{grad} r)(\tau(t))$ so

$$\langle \tau'(t), u(t)\rangle = \frac{d}{dt} r(\tau(t)).$$

Thus we have $\|\alpha'(t)\| \geq \frac{d}{dt}r(\tau(t))$ so

$$L(\alpha) = \int_0^b \|\alpha'(t)\| dt \geq r(\tau(b)).$$

But $\tau(b) = \sigma'(0)$ so $r(\tau(b)) = \|\sigma'(0)\| = L(\sigma)$. This shows that

$$L(\alpha) \geq L(\sigma)$$

for any curve α joining o to q satisfying the conditions of the theorem.

If we had equality, then the inequalities in the above argument would have to be equalities, so $N(t) \equiv 0$ and $\frac{dr}{dt} > 0$, so $\alpha' = \frac{dr}{dt} \cdot U$ and hence α travels monotonically along the radial geodesic from o to q and so is a monotone reparametrization of σ. $\square$

7.3 Short enough geodesics give an absolute minimum for arc length.

Here is a variation on the above argument which gives another important consequence of Gauss's lemma:

Let M be a Riemannian manifold. Suppose that the exponential map $\exp_0 : T_oM \to M$ is a diffeomorphism on a neighborhood U of 0 in T_oM which contains the ball of radius r about the origin. Let $v \in T_o(M)$ with $\|v\| = r$ so that the radial geodesic $\sigma : [0,1] \to M$ with $\sigma(0) = o$ and $\sigma'(0) = v$ has length r. Let $q = \sigma(1)$. Then

Theorem 12. *Any curve in M joining o to q has length $\geq r$ with equality if and only if it is a monotone reparametrization of σ.*

Proof. We may assume that any such curve lies entirely in the image of the ball of radius r under the exponential map, since the portion of the curve which does lie in the image already has length $\geq r$ by the theorem we have just proved. For this sort of curve we may now apply the theorem to conclude the inequality and the assertion about monotone reparametrization if we have equality. $\square$

Chapter 8

Variational formulas.

8.1 Jacobi fields in semi-Riemannian geometry.

Recall that in Chapter 3 we defined the Jacobi equations, Jacobi fields and conjugate points in the context of a general torsion free connection. We now study these objects in more detail in the case of the Levi-Civita connection arising from a semi-Riemannian metric.

8.1.1 Tangential Jacobi fields.

If $\alpha : I \to M$ is a smooth curve, then a vector field V on α is called **tangential** to α iff $V = f \cdot \alpha'$ for some function f on I.

Suppose that α is a geodesic and V is a tangential Jacobi field along α. Since $R(f\alpha', \alpha') \equiv 0$, the Jacobi equation becomes $V'' \equiv 0$. But

$$V'' = (f'\alpha')' = f''\alpha'$$

since $\alpha'' \equiv 0$. Thus $f'' \equiv 0$ so $f(t) = at + b$ for some constants a and b. In other words a tangential Jacobi vector field corresponds to reparametrization of the geodesic by an affine transformation of the line (and hence is not very interesting).

Notice that for a tangential Jacobi field, $V(0) = 0$ if and only if $f(t) = at$.

Also notice that if V is a tangential vector field along a geodesic, so is V' since $\alpha'' = 0$ and hence $V' = (f\alpha')' = f'\alpha'$.

8.1.2 Perpendicular Jacobi fields.

Suppose that V is perpendicular to the geodesic α, i.e.

$$\langle V, \alpha' \rangle \equiv 0.$$

Differentiating this equation gives $\langle V', \alpha' \rangle \equiv 0$ since $\alpha'' \equiv 0$. So the derivative of a perpendicular vector field along a geodesic is again perpendicular. Suppose

that V is a Jacobi vector field along the geodesic α, so that
$$\langle V, \alpha' \rangle'' = \langle V'', \alpha' \rangle = -\langle R(V, \alpha')\alpha', \alpha' \rangle = 0.$$
In other words, if V is a Jacobi vector field, then
$$\langle V, \alpha' \rangle(t) = At + B$$
for some constants A and B. So

Proposition 21. *A Jacobi vector field is perpendicular to the geodesic α $\Leftrightarrow$ $V(t)$ is perpendicular to $\alpha'(t)$ at two distinct points t_1 and t_2 $\Leftrightarrow$ $\langle V(t), \alpha'(t) \rangle = \langle V'(t), \alpha'(t) \rangle = 0$ for some t.*

8.1.3 Decomposition of a Jacobi field into its tangential and perpendicular components.

Suppose that the curve α is non-null in the sense that $\|\alpha'(t)\|^2 \neq 0$ at any t. At any t, the tangent space decomposes into a direct sum of the one dimensional space spanned by α' and the $(n-1)$-dimensional space $\alpha'(t)^\perp$. So for any vector field Y along α we get a decomposition
$$Y = Y^\top + Y^\perp$$
where $Y^\top$ is tangent to α and $Y^\perp$ is perpendicular to α.

Since $Y^\top = f\alpha'$ for some f, we have
$$R(Y^\top, \alpha') \equiv 0. \qquad (*)$$

Now suppose that α is a geodesic and consider $(*)$.
If Y is a Jacobi field along α, then by $(*)$, $Y'' = -R(Y, \alpha')\alpha' = -R(Y^\perp, \alpha')\alpha'$ and
$$\langle R(Y^\perp, \alpha')\alpha', \alpha' \rangle = 0.$$

So $R(Y^\perp, \alpha')\alpha'$ is perpendicular to α. Hence decomposing the right hand side of the equation $Y'' = R(Y, \alpha')\alpha'$ into its tangential and perpendicular components gives
$$\begin{aligned}(Y^\top)'' &= 0 \\ (Y^\perp)'' &= -R(Y^\perp, \alpha')\alpha'.\end{aligned}$$

We know that the first equation says that $Y^\top$ is a Jacobi field, and the second is the Jacobi equation for $Y^\perp$. So

Proposition 22. *Y is a Jacobi vector field along a non-null geodesic if and only if its tangential and perpendicular components are Jacobi vector fields. Furthermore*
$$Y' = (Y^\top)' + (Y^\perp)'$$
is the decomposition of Y' into tangential and perpendicular components.

8.2. VARIATIONS OF ARC LENGTH.

(The second assertion in the Proposition follows from that fact that the derivative of a tangential vector field along a geodesic is tangential, and the derivative of a perpendicular vector field along a geodesic is perpendicular, as we have already observed.)

8.2 Variations of arc length.

Let $\mathbf{x} : \mathcal{D} \to M$ where $\mathcal{D}$ is an open set in the plane containing the rectangle $[a,b] \times (-\delta, \delta)$, $\delta > 0$, and let $L = L_{\mathbf{x}}(v)$ be the length of the longitudinal curve

$$u \mapsto \mathbf{x}(u,v), \quad a \leq u \leq b.$$

If α is the curve $u \mapsto \mathbf{x}(u,0)$ then $L(0)$ is the arc length of α, and we are going to be interested in

$$L'(0) = \left.\frac{dL}{dv}\right|_{v=0} \quad \text{and} \quad L''(0) = \left.\frac{d^2 L}{dv^2}\right|_{v=0}$$

called the **first** and **second variations** of the arc length of α along the variation $\mathbf{x}$.

By definition, the arc length $L(v) = \int_a^b |\mathbf{x}_u(u,v)| du$. We will be interested in curves for which $|\alpha'|$ does not vanish (so either "space-like" or "time-like") so (for sufficiently small δ) we may assume that

$$L(v) = \int_a^b (\epsilon \langle \mathbf{x}_u, \mathbf{x}_u \rangle)^{\frac{1}{2}} du, \quad \epsilon = \pm 1.$$

8.2.1 The first variation.

We may differentiate under the integral sign and the derivative of the integrand with respect to v is

$$\frac{1}{2} (\epsilon \langle \mathbf{x}_u, \mathbf{x}_u \rangle)^{-\frac{1}{2}} 2\epsilon \langle \mathbf{x}_u, \mathbf{x}_{uv} \rangle = \epsilon \frac{\langle \mathbf{x}_u, \mathbf{x}_{vu} \rangle}{|\langle \mathbf{x}_u, \mathbf{x}_u \rangle|^{\frac{1}{2}}}.$$

Let us write $|\xi|$ for $|\langle \xi, \xi \rangle|^{\frac{1}{2}}$ so that the denominator of the right hand side of the above equation can be written simply as $|\mathbf{x}_u|$. If we set $h(u,v) := |\mathbf{x}_u|(u,v)$ then the above equation can be written as

$$\frac{\partial h}{\partial v} = \frac{\epsilon}{h} \langle \mathbf{x}_u, \mathbf{x}_{vu} \rangle.$$

So we have

$$L'(v) = \epsilon \int_a^b \frac{\langle \mathbf{x}_u, \mathbf{x}_{vu} \rangle}{|\mathbf{x}_u|} du.$$

When $v = 0$ we have $\mathbf{x}_u = \alpha'$ and $\mathbf{x}_v = V$ so $\mathbf{x}_{vu} = V'$. So

$$L'(0) = \epsilon \int_a^b \left\langle \frac{\alpha'}{|\alpha'|}, V' \right\rangle du.$$

Suppose that α is parametrized proportional to arc length, so that $|\alpha'|$ is constant; call this constant c. Then the above expression simplifies a bit to

$$L'(0) = \frac{\epsilon}{c} \int_a^b \langle \alpha', V' \rangle du.$$

Since $\langle \alpha', V \rangle' = \langle \alpha'', V \rangle + \langle \alpha', V' \rangle$ or

$$\langle \alpha', V' \rangle = \langle \alpha', V \rangle' - \langle \alpha'', V \rangle$$

we obtain

$$L'(0) = -\frac{\epsilon}{c} \int_a^b \langle \alpha'', V \rangle du + \frac{\epsilon}{c} \langle \alpha', V \rangle \Big|_a^b. \tag{8.1}$$

We can now draw a number of very important conclusions from this formula:

8.3 Geodesics are stationary for arc length.

Suppose that α has the property that $L'(0) = 0$ for all variations which fix the end points, i.e. such that $\mathbf{x}(a, v) \equiv \alpha(a)$ and $\mathbf{x}(b, v) \equiv \alpha(b)$. For such variations, $V(a) = 0$ and $V(b) = 0$ so the second summand in (8.1) disappears. But otherwise V can be arbitrary, and the only way that $\int_a^b \langle \alpha'', V \rangle du$ can vanish for all V vanishing at the end points is for $\alpha'' \equiv 0$ which says that α is a geodesic. So we have proved:

Theorem 13. *A twice differentiable non-null curve is stationary for all variations with fixed end points if and only if it is a geodesic.*

Suppose that instead of fixing the end points we allow variations for which $\mathbf{x}(a, v)$ lies on a submanifold S_a and $\mathbf{x}(b, v)$ lies on a submanifold S_b. If α is an extremal for all such variations, it certainly is an extremal for the fixed end point variational problem, hence must be a geodesic. So now the integral in (8.1) vanishes and we are left with the second term, where $V(a)$ can vary over the tangent space $T(S_a)_a$ and $V(b)$ can vary over the tangent space $T(S_b)_b$. So we have

Theorem 14. *A twice differentiable non-null curve is stationary for all variations with end points constrained to lie on submanifolds if and only if it is a geodesic which is orthogonal at the end points to these submanifolds.*

8.3.1 Piecewise smooth variations.

A variation $\mathbf{x}$ of a piecewise smooth α is called a *piecewise smooth* variation if $\mathbf{x}$ is continuous and for breaks

$$u_0 = a < u_1 < \cdots < u_k < u_{k+1} = b$$

8.4. THE SECOND VARIATION.

the restriction of u to each $[u_{i-1}, u_i] \times (-\delta, \delta)$ is smooth. At each $i = 1, \ldots k$ the right and left hand derivatives $\alpha'(u_i^+)$ and $\alpha'(u_i^-)$ exist, and we define

$$\Delta \alpha'(u_i) = \alpha'(u_i^+) - \alpha'(u_i^-) \in T_{\alpha(u_i)}M.$$

We can apply the formula (8.1) for the first variation to each $[u_{i-1}, u_i] \times (-\delta, \delta)$ and assemble the pieces to obtain:

$$L'(0) = -\frac{\epsilon}{c}\int_a^b \langle \alpha'', V \rangle du - \frac{\epsilon}{c}\sum_i \langle \Delta\alpha'(u_i), V(u_i)\rangle + \frac{\epsilon}{c}\langle \alpha', V\rangle \Big|_a^b.$$

So we can conclude

Proposition 23. *A piecewise smooth curve which is stationary for piecewise smooth variations must in fact be smooth.*

We next investigate when a geodesic in a Riemannian manifold is in fact a minimum for local variations of arc length. For this we need to study the second variation. We begin our discussion in the general semi-Riemannian case, but eventually we will restrict to the Riemannian case.

8.4 The second variation.

In what follows $\sigma : [a, b] \to M$ will be a non-null geodesic. If $\mathbf{x} : [a, b] \times (-\delta, \delta) \to M$ is a variation of σ then

$$V(u) = \mathbf{x}_v(u, 0)$$

is the **transverse velocity**. We call the vector field A along σ given by

$$A(u) = \mathbf{x}_{vv}(u, 0)$$

the **transverse acceleration**.

Recall that any vector field Y along a non-null curve splits into tangential and perpendicular pieces $Y = Y^\top + Y^\perp$ and along a non-null geodesic we have

$$(Y^\perp)' = (Y')^\perp.$$

We will denote this common value by $Y^{\perp'}$.

8.4.1 Synge's formula for the second variation.

Let σ be a geodesic of speed $c > 0$ and sign ϵ. Let $\mathbf{x}$ be a variation of σ. Then

Theorem 15. **Synge's formula for the second variation.**

$$L''(0) = \frac{\epsilon}{c}\int_a^b \left\{ \langle V^{\perp'}, V^{\perp'}\rangle + \langle R(V, \sigma')V, \sigma'\rangle \right\} du + \frac{\epsilon}{c}\langle \sigma', A\rangle \Big|_a^b.$$

Proof. In the proof of this formula we let $h(u,v) := |\mathbf{x}_u(u,v)|$ so that $L''(v) = \int_a^b \frac{\partial^2 h}{\partial v^2} du$, and recall that we have already computed

$$\frac{\partial h}{\partial v} = \frac{\epsilon}{h}\langle \mathbf{x}_u, \mathbf{x}_{uv}\rangle.$$

So

$$\frac{\partial^2 h}{\partial v^2} = \frac{\epsilon}{h}\left\{\langle \mathbf{x}_{uv}, \mathbf{x}_{uv}\rangle + \langle \mathbf{x}_u, \mathbf{x}_{uvv}\rangle - \frac{\epsilon}{h^2}\langle \mathbf{x}_u, \mathbf{x}_{uv}\rangle^2\right\}.$$

Using $\mathbf{x}_{uv} = \mathbf{x}_{vu}$ and $\mathbf{x}_{uvv} = \mathbf{x}_{vuv} = R(\mathbf{x}_v, \mathbf{x}_u)\mathbf{x}_v + \mathbf{x}_{vvu}$ this becomes

$$\frac{\epsilon}{h} \times \left[\langle \mathbf{x}_{vu}, \mathbf{x}_{vu}\rangle + \langle \mathbf{x}_u, R(\mathbf{x}_v, \mathbf{x}_u)\mathbf{x}_v\rangle + \langle \mathbf{x}_u, \mathbf{x}_{vvu}\rangle - \frac{\epsilon}{h^2}\langle \mathbf{x}_u, \mathbf{x}_{vu}\rangle^2\right].$$

Setting $v = 0$ gives the substitutions $h \mapsto c$, $\mathbf{x}_u \mapsto \sigma'$, $\mathbf{x}_v \mapsto V$, $\mathbf{x}_{vu} \mapsto V'$, $\mathbf{x}_{vv} \mapsto A$ and $\mathbf{x}_{vvu} \mapsto A'$ yielding

$$\frac{\epsilon}{c} \times \left[\langle V', V'\rangle + \langle R(V, \sigma')V, \sigma'\rangle + \langle \sigma', A'\rangle - \left(\frac{\epsilon}{c^2}\right)\langle \sigma', V'\rangle^2\right].$$

In the above expression we can write $\langle \sigma', A'\rangle = \langle \sigma', A\rangle'$ since $\sigma'' = 0$. Also, since $\frac{\sigma'}{c}$ is a unit vector field, the tangential component of V' is

$$\epsilon\left\langle V', \frac{\sigma'}{c}\right\rangle \frac{\sigma'}{c}.$$

So the contribution to $\langle V', V'\rangle$ coming from the tangential component $V'^\top$ cancels the last term; the above expression simplifies to

$$\langle V^{\perp'}, V^{\perp'}\rangle + \langle R(V, \sigma')V, \sigma'\rangle + \langle \sigma', A\rangle'.$$

Integrating from a to b then gives Synge's formula. □

Synge's formula for variations with fixed end points.

If the variation $\mathbf{x}$ is such that $\mathbf{x}(a, v) \equiv \sigma(a)$ and $\mathbf{x}(b, v) \equiv \sigma(b)$ then $A(a) = 0$ and $A(b) = 0$. In this case Synge's formula becomes

$$L''(0) = \frac{\epsilon}{c}\int_a^b \left\{\langle V^{\perp'}, V^{\perp'}\rangle + \langle R(V, \sigma')V, \sigma'\rangle\right\} du.$$

We can consider the right hand side of the above expression as a quadratic form on the space of all (continuous) piecewise differentiable vector fields along σ. The corresponding bilinear form is

$$I_\sigma(V, W) := \frac{\epsilon}{c}\int_a^b \left\{\langle V^{\perp'}, W^{\perp'}\rangle + \langle R(V, \sigma')W, \sigma'\rangle\right\} du.$$

The symmetry property $\langle R(V, \sigma')W, \sigma'\rangle = \langle R(W, \sigma')V, \sigma'\rangle$ of the curvature guarantees that I_σ is symmetric.

At all points where V and W are differentiable we have

$$\langle V^{\perp'}, W^{\perp'}\rangle = \langle V^\perp, W\rangle = \langle V^{\perp'}, W\rangle' - \langle (V^\perp)'', W\rangle$$

so

$$\int_a^b \langle V^{\perp'}, W^{\perp'}\rangle du = -\int_a^b \langle (V^\perp)'', W\rangle du - \sum_i \langle \Delta V^{\perp'}, W\rangle(u_i).$$

Also $\langle R(V, \sigma')W, \sigma'\rangle = -\langle R(V, \sigma')\sigma', W\rangle = -\langle R(V, \sigma')\sigma', W^\perp\rangle.$

Substituting these into the above expression for $I_\sigma(V, W)$ gives:

$$I_\sigma(V, W) = -\frac{\epsilon}{c}\int_a^b \langle (V^\perp)'' + R(V^\perp, \sigma')\sigma', \ W^\perp\rangle du - \frac{\epsilon}{c}\sum_i \langle \Delta V^{\perp'}, W^\perp\rangle(u_i).$$

The Riemannian case.

If M is a Riemannian manifold and σ gives a minimum for arc length among all curves joining $\sigma(a)$ to $\sigma(b)$ then $L''(0) \geq 0$.

For simplicity, assume that we are parametrizing by arc length (so $c = 1$) and also $\epsilon = 1$ since we are in the Riemannian case. The formula for I_σ becomes

$$I_\sigma(V, W) = \int_a^b \left\{ \langle V^{\perp'}, W^{\perp'}\rangle + \langle R(V, \sigma')W, \sigma'\rangle \right\} du.$$

Notice that if V is tangential to σ then $V^{\perp'} = 0$ and $R(V, \sigma') = 0$ so

$$I_\sigma(V, W) = 0 \quad \forall \ W.$$

So there is no loss of information in restricting I_σ to $\{V | V \perp \sigma\}$.

8.5 Conjugate points and the Morse index.

If $\sigma : [a, b] \to M$ is a geodesic, recall that we say that the points a and b are conjugate along σ if there is a non-zero Jacobi field J along σ such that $J(a) = 0$ and $J(b) = 0$.

One also says that the points $p = \sigma(a)$ and $q = \sigma(b)$ are conjugate along σ if the above holds.

For simplicity I will now restrict to the Riemannian case (with curves usually parametrized by arc length), referring to O'Neill for the semi-Riemannian generalizations.

Recall (Theorem 10) a geodesic locally minimizes arc length up to a conjugate point. Using the quadratic form I we can prove

Theorem 16. Jacobi. *Geodesics do not locally minimize past a conjugate point.*

Proof. Let $\sigma : [0, b] \to M$ be the geodesic in question. The hypothesis of the theorem is that there is a non-zero Jacobi field J on σ restricted to $[0, r]$, $r < b$ with $J(0) = J(r) = 0$. Since $J \neq 0$, we know that $J'(r) \neq 0$ and that J is everywhere perpendicular to σ. Let Y be the vector field along σ which equals J on $[0, r)$ and is 0 on $[r, b]$. So Y' has a jump at r, i.e. $\Delta Y' \neq 0$. Let W be a vector field along σ with $W(r) = \Delta Y'(r)$.

Now recall the formula $I_\sigma(V, Z)$

$$= -\frac{\epsilon}{c} \int_a^b \langle (V^\perp)'' + R(V^\perp, \sigma')\sigma', \ Z^\perp \rangle \, du - \frac{\epsilon}{c} \sum_i \langle \Delta V^{\perp'}, Z^\perp \rangle(u_i).$$

Taking $V = Y$, the integral vanishes since Y satisfies the Jacobi equation on $[0, r)$ and on $[r, b]$. So (with $\epsilon = c = 1$) we see that $I(Y, Z) = -\langle \Delta Y^{\perp'}(r), Z(r) \rangle$. Since $Y(r) = 0$ we get $I(Y, Y) = 0$. Since $W(r) = \Delta Y'(r) \neq 0$, we see that

$$I(Y, W) < 0.$$

But then
$$I(Y + sW, Y + sW) = 2sI(Y, W) + s^2 I(W, W)$$

is < 0 for sufficiently small positive s. So I is not semi-definite, and hence σ is not a minimum for L. □

8.5.1 Non-positive sectional curvature means no conjugate points.

Lemma 8. *If $\langle R(v, \sigma')v, \sigma' \rangle \geq 0$ for every v perpendicular to σ then there are no conjugate points along σ.*

Proof. Let $J \neq 0$ be a perpendicular Jacobi field along σ with $J(0) = 0$. Set $h(s) := \langle J(s), J(s) \rangle$ so $h(0) = 0$, $h'(s) = 2\langle J'(s), J(s) \rangle$ and

$$\frac{1}{2}h'' = \langle J', J' \rangle + \langle J'', J \rangle = \langle J', J' \rangle + \langle R(\sigma', J)\sigma', J \rangle$$

by Jacobi. By hypothesis, the right hand side is non-negative, and $\langle J'(0), J'(0) \rangle > 0$. Hence $h(s) > 0$ for all $s \neq 0$. □

In particular, if
$$\sec(v, w) = \frac{\langle R(w, v)v, w \rangle}{Q(v, w)}$$

is everywhere non-positive then there are no conjugate points on any geodesic.

8.6 Synge's theorem.

Theorem 17. [Synge, 1936.] *Let M be an even dimensional orientable Riemannian manifold whose sectional curvature is everywhere positive. Then every (non-constant) periodic geodesic can be shortened by a variation.*

Proof. Let $\gamma : [0, L] \to M$ be the unit speed geodesic parametrized by arc length, and we are assuming that $\gamma(L) = \gamma(0)$. Parallel translation of the tangent space always locally preserves orientation, and since we are assuming that M is orientable, it globally preserves orientation. So parallel translation all around γ is an orientation preserving map of $T_{\gamma(0)}M$ into itself. Since this map carries, $\gamma'(0)$ into itself, and is an isometry, it is an orientation preserving map, call it P, of $V := \gamma'^\perp$ into itself. So

$P : V \to V$, P has determinant one, $PP^\dagger = 1$, and $\dim V$ is odd.

According to a theorem of Euler (proved by him when $\dim V = 3$), this implies that there is a non-zero vector $y \in V$ such that $Py = y$. I recall the proof: We want to show that $\det(I - P) = 0$. Now

$$\begin{aligned}\det(I - P) &= \det(I - P^\dagger) \\ &= \det P \det(I - P^\dagger) \\ &= \det(P - PP^\dagger) \\ &= \det(P - I) \\ &= \det(-I)\det(I - P) \\ &= -\det(I - P)\end{aligned}$$

since the dimension of V is odd.

Let $y \in V$ be a vector of length one with $Py = y$, and let Y be the vector field along γ obtained by parallel translating y around γ. The condition $Py = y$ guarantees that Y is a smooth vector field along γ.

Now apply Synge's formula for a variation whose variational vector field is Y. (We can always find such a variation, for example setting $\mathbf{x}(u, v) = \exp_{\gamma(u)} vY(u)$.) In Synge's formula for the second variation, the boundary terms cancel because of the periodicity, and the term involving Y' vanishes because $Y' = 0$, as we chose Y to be parallel. So we are left with

$$L''(0) = \int_\gamma -\sec(Y, \gamma') < 0.$$

$\square$

As a corollary we obtain the following famous theorem of Synge:

Theorem 18. *Let M be an even dimensional compact orientable Riemannian manifold whose sectional curvature is everywhere positive. Then M is simply connected.*

Proof. For the proof I use the following theorem of Cartan which is intuitively obvious and whose proof I will give in the next section: In any free homotopy class of closed curves that cannot be shrunk to a point, there will be one of shortest length, and it will be a closed geodesic. (Idea: Just keep tightening any closed curve until you reach minimal length.) But by the previous theorem, there can not be any such non-constant closed curve. □

8.7 Cartan's theorem on the existence of closed geodesics.

Recall that two continuous maps $f, g : S^1 \to M$ are said to be freely homotopic if there exists a continuous map $F : S^1 \times I \to M$ (where $I := [0,1]$) such that $F(\cdot, 0) = f$ and $F(\cdot, 1) = g$. This is an equivalence relation and the corresponding equivalence class is called a free homotopy class. The trivial free homotopy class consists of those f which are freely homotopic to a point.

Theorem 19. [Cartan.] *If M is compact then any non-trivial free homotopy class contains a piecewise differentiable curve of shortest length among piecewise differentiable elements of that class and it is a geodesic.*

That every continuous path is homotopic to a piecewise differentiable one is standard (and in each local coordinate system follows from the Weierstrass approximation theorem).

Let d denote the infimum of the lengths of piecewise differentiable loops in the free homotopy class. Since this class is non-trivial, we know that $d > 0$. Let C_j be a sequence of loops in the free homotopy class whose lengths approach d. If we cover each C_j by finitely many convex neighborhood, we can shorten C_j by replacing each sub-path by a geodesic segment, so we may assume that C_j is a broken geodesic parametrized proportionally to arc length. Let $L = \sup \ell(C_j)$. Then for $0 \le s \le t \le 1$

$$d(C_j(s), C_j(t)) \le \int_s^t \|C_j'(r)\| dr \le L(t-s)$$

so the C_j are equicontinuous. Since M is compact, we can choose subsequence (which we will relabel as C_j) which converges to some continuous curve, C.

$$\ell(C_j) \to d, \quad C_j \to C.$$

Choose $0 = t_0 < t_1 \cdots < t_k = 1$ such that $C([t_i, t_{i+1}])$ lies in a convex neighborhood and let γ be the broken geodesic whose restriction to $[t_i, t_{i+1}]$ is the geodesic γ_i joining $C(t_i)$ to $C(t_j)$ and hence is the shortest curve joining these two points.

The plan is to show that $\ell(\gamma) = d$ and that γ is a smooth geodesic. In fact, the second assertion follows from the first, since if γ had any corners, we could

8.7. CARTAN ON THE EXISTENCE OF CLOSED GEODESICS.

shorten it and so get a loop in the free homotopy class whose length is $< d$. So we must show that $\ell(\gamma) = d$.

So suppose that $\ell(\gamma) > d$ and set

$$\epsilon := \frac{\ell(\gamma) - d}{2k + 1}$$

where k is the number of subintervals into which we have divided $[0, 1]$.

For large enough j we have $\quad \ell(C_j) - d < \epsilon \quad$ and

$$d(C_j(t), C(t)) < \epsilon \quad \forall t \in [0, 1].$$

Let C_j^i denote the restriction of C_j to the interval $[t_{i-1}, t_i]$. Then

$$\sum_i (\ell(C_j^i) + 2\epsilon) = \ell(C_j) + 2k\epsilon < d + (2k+1)\epsilon$$

$$= \ell(\gamma) = \sum \ell(\gamma_i).$$

Therefore there is some i such that $\ell(C_j^i) + 2\epsilon < \ell(\gamma_i)$. For large enough j we get a path joining the end points of γ_i of shorter length, contradicting the choice of γ_i. □

Chapter 9

The Hopf-Rinow theorem.

This chapter is devoted to a theorem of Hopf and Rinow which relates the various concepts of "completeness" in Riemannian geometry. In this chapter M will be a connected Riemannian manifold (with the exception of the beginning of §9.4 where semi-Riemannian manifolds are allowed).

9.1 Riemannian distance.

For $p, q \in M$ the **Riemannian distance** $d(p,q)$ is defined to be the greatest lower bound of the lengths $L(\alpha)$ of piecewise smooth curves joining p to q.

If $o \in M$ and $\epsilon > 0$ the set

$$\mathcal{N}_\epsilon(o) = \{p \in M | d(o,p) < \epsilon\}$$

is called the ϵ-**neighborhood** of o.

Recall that using Gauss's lemma we proved

1. *Let $\mathcal{U}$ be a normal neighborhood of a point o in a Riemannian manifold M. Let $\mathcal{W} \subset \mathcal{U}$ be the image of the ball of radius $\epsilon > 0$ under $\exp_o$. Then for sufficiently small ϵ, if $p \in \mathcal{W}$ then the radial geodesic segment*

$$\sigma : [0,1] \to \mathcal{W}$$

from o to p is the unique (up to reparametrization) shortest curve in M from o to p.

We can strengthen the preceding result as follows:

Proposition 24. *For $o \in M$*

1. *For $\epsilon > 0$ sufficiently small $\mathcal{N}_\epsilon(o)$ is normal.*

2. *For a normal ϵ-neighborhood the radial geodesic from o to $p \in \mathcal{N}_\epsilon(o)$ is the unique shortest curve in M from o to p and hence*

3.
$$d(o,p) = L(\sigma) = r(p).$$

Let $\mathcal{U} = \exp_o(\tilde{\mathcal{U}})$ be a normal neighborhood of o. For $\epsilon > 0$ sufficiently small, $\tilde{U}$ contains the open ball of radius ϵ about 0 in T_oM and every ball is starshaped. The image of this ball under $\exp_o$ is thus a normal neighborhood. By the preceding Proposition, if p belongs to this normal neighborhood, the radial geodesic σ from o to p is the unique shortest curve in this normal neighborhood joining o to p, and $L(\sigma) = r(p) = \|v\|$ where $v \in T_oM$ is given as $v = \exp_o^{-1}(p)$ so $L(\sigma) < \epsilon$.

We will prove the following claim:

2. *If α is a curve starting at o and leaving this normal neighborhood then $L(\alpha) \geq \epsilon$.*

Once we prove the claim, then we will know that the radial geodesic σ from o to p is the unique shortest curve in all of M from o to p and therefore $L(\sigma) = r(p) = d(o,p)$, and hence $d(o,p) < \epsilon$. On the other hand, if $q \in M$ does *not* belong to our normal neighborhood, then the claim shows that $d(o,q) \geq \epsilon$. Thus our normal neighborhood is exactly $\mathcal{N}_\epsilon(o)$.

So a proof of the claim will complete the proof of the proposition.

Proof. If α is a curve which leaves our normal neighborhood, then it meets every sphere $S(a) := \{p | r(p) = a\}$ for $a < \epsilon$. So let p be the first point of intersection of α with this sphere, and α_a the portion of α going from o to p. We know from the preceding proposition that the length of α_a is $\geq a$. Hence $L(\alpha) \geq a$ for every $a < \epsilon$ so $L(\alpha) \geq \epsilon$. □

Here is another important consequence of the claim:

Proposition 25. $d(p,q) = 0 \Rightarrow p = q$.

Proof. If $p \neq q$ then since M is Hausdorff (part of the definition of a manifold) then there will be a normal neighborhood of p which does not contain q, and hence an ϵ normal neighborhood which does not contain q and so by the claim (which we now have proved) $d(p,q) \geq \epsilon$. □

It is clear that $d(p,q) = d(q,p)$ and that the triangle inequality

$$d(p,q) + d(q,r) \geq d(p,r)$$

holds, and that $d(p,q) \geq 0$. So we have proved that d defines a metric on M. Since every neighborhood of a point of M contains an ϵ-neighborhood, and since (by the claim) the ϵ-neighborhoods are open sets, we see that **the topology induced by the distance function d is the same as the manifold topology.**

9.1.1 Some history.

In 1927, Élie Cartan published his famous text book on Riemannian geometry, and this book was updated by a much enlarged second edition in 1946. Up until the 1960's this was the only text-book on Riemannian geometry, although there were many books on general relativity. Even in the second edition of his book (page 56) Cartan writes:

> La notion générale de variété est assez difficile à définir avec précision.
>
> *The general notion of a manifold is quite difficult to define precisely.*

So even the great geometer did not have a clear definition of a manifold. The topologists did. For example Whitney proved in 1936 that any manifold of dimension d can be embedded in $\mathbb{R}^{2d+1}$.

So when I lectured at Harvard in 1960-61 on differential geometry, I could not assume that anyone knew what a differentiable manifold was, nor could I assume the exterior differential calculus. My book "Lectures on differential geometry" written in 1962 reflects this. In fact, my book, and Lang's "Introduction to differentiable manifolds" were probably the first geometry books in which a differentiable manifold is clearly defined.

9.1.2 Minimizing curves.

By the definition of d, a curve segment σ from p to q is a shortest curve segment from p to q if and only if $L(\sigma) = d(p,q)$. There may be many such shortest curves joining p to q (think of the circles of longitude from the north to the south poles on the sphere) or there may not be any (think of the plane with a point removed).

If σ is a shortest curve, we say that σ is **minimizing**. Clearly any subsegment of σ is then also minimizing. Sometimes the word **segment** is used to describe a minimizing curve segment σ which is parametrized proportionally to arc length, i.e. such that $\|\dot\sigma\|$ is constant.

We know that

3. *Any minimizing curve segment is a monotone reparametrization of an unbroken geodesic.*

9.2 Completeness and the Hopf-Rinow theorem.

9.2.1 The key proposition - de Rham's proof.

Remember that we are assuming that M is connected.

Proposition 26. *Let p be a point of M. If $\exp_p$ is defined on all of T_pM then for any $q \in M$ there is a minimizing geodesic from p to q.*

Let $\mathcal{U}$ be a normal ϵ-neighborhood of p. If $q \in \mathcal{U}$, then the result follows from what we already know. So suppose that $q \notin \mathcal{U}$.

The following clever argument is due to de Rham. The idea is to choose a geodesic which "aims at q" and is pointed in the right direction. For any $0 < r < \epsilon$ the image of the sphere of radius r in $T_p M$ maps under the exponential map at p onto a compact submanifold $S \subset \mathcal{U}$. The function $s \mapsto d(s, q)$ is continuous on S and hence achieves a minimum at some point $m \in S$. We are going to show that the geodesic segment from p to m extends to a minimizing geodesic from p to q.

As a first step we will prove:

Claim 1. $d(p, m) + d(m, q) = d(p, q)$.

Proof. Let $\alpha : [0, b]$ be any curve from p to q. It must cross S at some parameter value $0 < a < b$, i.e $\alpha(a) \in S$. Let α_1 and α_2 denote the restrictions of α to $[0, a]$ and $[a, b]$ respectively. So

$$L(\alpha) = L(\alpha_1) + L(\alpha_2).$$

But we know that $L(\alpha_1) \geq r = d(p, m)$ and by definition, $d(m, q) \leq L(\alpha_2)$. So

$$L(\alpha) \geq d(p, m) + d(m, q).$$

Hence, minimizing over all α tells us that

$$d(p, q) \geq d(p, m) + d(m, q).$$

The triangle inequality gives the reverse inequality, so we must have equality. $\square$

Let $\gamma : [0, \infty) \to M$ be the unit speed geodesic whose initial segment runs radially from p through m. Our hypothesis guarantees that γ is indeed defined for all $t > 0$. Our claim, which we just proved, asserts that for sufficiently small $t > 0$ we have

$$t + d(\gamma(t), q) = d(p, q). \tag{9.1}$$

Let T denote the set of all t such that (9.1) holds. If we could show that $s = d(p, q) \in T$, then (9.1) implies that $d(\gamma(s), q) = 0$, so that $\gamma(s) = q$ and since the length of γ restricted to $[0, s]$ is $s = d(p, q)$ we see that this restriction is a minimizing geodesic segment from p to q.

So our task is to prove that

Claim 2. $d(p, q) \in T$.

Proof. T is closed and non-empty. So it has a largest element $t_0 \leq d(p, q)$. We must show that $t_0 = d(p, q)$. Suppose not. Choose a normal neighborhood

9.2. COMPLETENESS AND THE HOPF-RINOW THEOREM.

$\mathcal{U}'$ of $\gamma(t_0)$ and a sphere in $T_{\gamma(t_0)}$ of small enough radius δ' which maps to S' diffeomorphically under $\exp_{\gamma(t_0)}$ as before and hence a point $m' \in S'$ with

$$\delta' + d(m', q) = d(\gamma(t_0), q)$$

by applying our (proven) claim to $\gamma(t_0)$ and q. Since $t_0 \in T$, we have $t_0 + d(\gamma(t_0), q) = d(p, q)$ and hence

$$t_0 + \delta' + d(m', q) = d(p, q).$$

But

$$d(p, q) \leq d(p, m') + d(m', q)$$

so

$$t_0 + \delta' = d(p, \gamma(t_0)) + d(\gamma(t_0), m') \leq d(p, m')$$

and the triangle inequality gives the reverse inequality so we have

$$d(p, \gamma(t_0)) + d(\gamma(t_0), m') = d(p, m').$$

So the broken geodesic consisting of γ from 0 to t_0 and the radial geodesic σ from $\gamma(t_0)$ to m' is a minimizing curve. But this means that the following figure is impossible:

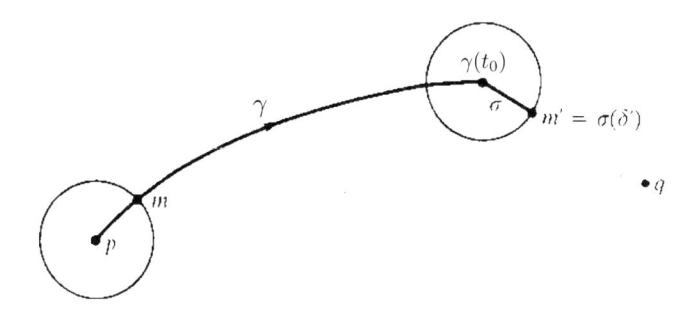

In other words γ extends past t_0 as an unbroken geodesic with $t \in T, t > t_0$, contradicting the definition of t_0. □

9.2.2 Geodesically complete manifolds.

A Riemannian manifold is called **geodesically complete** if $\exp_o$ is defined on all of $T_o M$ for all $o \in M$. It then follows from the preceding proposition that:

Proposition 27. *If M is geodesically complete then any two of its points can be joined by a minimizing geodesic.*

9.2.3 The Hopf-Rinow theorem.

Theorem 20. [Hopf-Rinow.] *The following are equivalent:*

1. *M is complete as a metric space under d, i.e. every Cauchy sequence converges.*

2. *There exists a point $p \in M$ such that $\exp_p$ is defined on all of T_pM,*

3. *M is geodesically complete.*

4. *Every closed bounded subset of M is compact.*

Notice that $3 \Rightarrow 2$ is trivial and $4 \Rightarrow 1$ since a Cauchy sequence is bounded, hence if 4 holds, its closure is compact, so the Cauchy sequence contains a convergent subsequence and hence the original sequence converges. So we must prove $1 \Rightarrow 3$ and $2 \Rightarrow 4$.

$$1 \Rightarrow 3.$$

Proof. Let $\gamma : [0, b)$ be a unit speed geodesic. We must show that γ may be extended past b if b is finite. Let $t_i \to b$. Then $d(\gamma(t_i), \gamma(t_j)) \leq |t_i - t_j|$ so $\gamma(t_i)$ form a Cauchy sequence, hence converge to some point $q \in M$ as we are assuming that M is complete under d. Then $d(\gamma(s), q) \leq d(\gamma(s), \gamma(t_i)) + d(\gamma(t_i), q)$ so the curve $\gamma(s)$ approaches q as $s \to b$. Let $\mathcal{C}$ be a convex neighborhood of q. For some t close enough to b the point $o := \gamma(t)$ lies in $\mathcal{C}$ and hence γ restricted to $[t, b)$ is a radial geodesic emanating from o. But it then can be extended as a radial geodesic until it hits the boundary of $\mathcal{C}$, which means that γ can be extended as a geodesic past b. □

$$2 \Rightarrow 4.$$

Proof. Let p be a point for which $\exp_p$ is defined on all of T_pM. So we know that for any $q \in M$, there is a minimizing geodesic $\sigma_q : [0, 1] \to M$ joining p to q. So if q ranges over a bounded closed subset A, the set $d(p, q) = \|\sigma_q'(0)\|$ is bounded, and hence the set of such $\sigma_q'(0)$ lies in some closed ball B in T_pM, which is compact. Hence $\exp_p(B)$ is compact. But $A \subset \exp_p(B)$ and a closed subset of a compact set is compact. □

This completes the proof of the Hopf-Rinow theorem.

So if M is complete, then any two points can be joined by a geodesic, in fact, by a minimizing geodesic.

The paper in which the Hopf-Rinow theorem appeared is:

Heinz Hopf und Willi Rinow *Über den begriff der vollstandigen differentialgeometrische Fläschen,* Comment. Math. Helv. **3** (1931) 209-225.

Heinz Hopf

**Born: 19 Nov. 1894 in Grabschen, Germany
(now Wroclaw, Poland)
Died: 3 June 1971 in Zollikon, Switzerland**

Willi Rinow

(1907-1973)

9.3 Hadamard's theorem.

We have proved that if the sectional curvature of a Riemannian manifold is non-positive at all points, then there are no conjugate points along any geodesic. This implies that for any $p \in M$, the exponential map $\exp_p : T_pM \to M$ is a local diffeomorphism. So we can locally pullback the metric on M to T_pM, and since the radial straight lines through the origin are geodesics, we know that T_pM is complete under the pull-back metric and, by definition the map $\exp_p$ is a local isometry of the pulled-back metric on T_p with M. We shall shortly see that this is enough to imply that $\exp_p$ is a covering map, since T_pM is simply connected. (See Theorem 7.28 in O'Neill for the elementary topological arguments involving covering spaces. I will repeat his arguments shortly.) So we get:

Theorem 21. *If M is a complete, simply connected Riemannian manifold whose sectional curvature is everywhere non-positive, then for every $p \in M$ the exponential map $\exp_p : T_pM \to M$ is a diffeomorphism.*

I am not an expert on the history of this theorem. As far as I can tell, this was first proved by von-Mangoldt in 1881 - J. Reine Angew. Math. 91 - for surfaces, then Hadamard gave two more proofs for surfaces in 1889 - J. Math. Pures Appl. - with some hint on how to do it in three dimensions. Cartan proved the full theorem in n-dimensions in 1922. So naturally it is called Hadamard's theorem since he was neither the first to prove it for surfaces, nor did he prove it in generality.

9.4 Locally isometric coverings.

Let $\phi : M \to N$ be a local isometry (with M and N connected semi-Riemannian manifolds.) Suppose that ϕ has the property that given any geodesic $\sigma : [0, 1] \to N$ with $\sigma(0) = q$, and a point $p \in M$ with $\phi(p) = q$, there is a lift $\tilde{\sigma} : [0, 1] \to M$ with $\tilde{\sigma}(0) = p$ and $\phi \circ \tilde{\sigma} = \sigma$.

Since ϕ is a local isometry, we know that $\tilde{\sigma}$ must be a geodesic. Since N is connected, every pair of points of N can be connected by a sequence of broken geodesics. Indeed, cover any curve joining the two points by a finite number of convex neighborhoods. So we know that ϕ is surjective. We want to prove

Theorem 22. *ϕ is a covering map.*

We must prove that every point $q \in N$ is *evenly covered* which means that q has neighborhood U with the property that $\phi^{-1}(U)$ is a disjoint union of open sets with the property that the restriction of ϕ to each of these open sets is a homeomorphism onto U. The idea is to use Cartan's polar map construction.

Start with a normal neighborhood U of q, and let $\tilde{U} \subset T_qN$ be the star-shaped region mapped onto U by the exponential map at q. For any $p \in \phi^{-1}(q)$ define
$$\tilde{U}(p) := d\phi_p^{-1}(\tilde{U}).$$

9.4. LOCALLY ISOMETRIC COVERINGS.

Since $d\phi_p$ is an isometry, it is a linear isomorphism, so $\tilde{U}(p)$ is a star-shaped open set in T_p.

Lemma 9. Step 1. $\exp_p$ *is defined on* $\tilde{U}(p)$ *and*

$$\phi(\exp_p(v)) = \exp_p(d\phi_p(v)) \quad \text{for } v \in \tilde{U}(p).$$

Proof. If $v \in \tilde{U}(p)$, let σ be the radial geodesic starting at q with initial velocity vector $d\phi_p(v)$. By hypothesis, this lifts (to a geodesic) $\tilde{\sigma}$ starting at p and $d\phi_p(\tilde{\sigma}'(0)) = d\phi_p(v)$ so $v = \tilde{\sigma}'(0)$. So $\exp_p(v) = \tilde{\sigma}(1)$ and the displayed equality above holds. $\square$

Define $U(p) = \exp_p(\tilde{U}(p))$. We know that ϕ maps $U(p)$ onto U. Restrict ϕ to the union of all the $U(p)$, $p \in \phi^{-1}(q)$.

Lemma 10. Step 2. *For each* $p \in \phi^{-1}(q)$, $U(p)$ *is a normal neighborhood of* p *and* ϕ *maps* $U(p)$ *diffeomorphically onto* U.

Proof. Since $d\phi_p$ and $\exp_q$ are diffeomorphisms, we know from step 1 that $\phi \circ \exp_p$ is a diffeomorphism (on $\tilde{U}(p)$). Thus $\exp_p$ is injective, and $d\phi_{\exp_p(v)} \circ d(\exp_p)_v$ is a linear isomorphism. So $d(\exp_p)_v$ is a linear isomorphism, hence the image of $\exp_p$ is an open set. Hence $U(p)$ is a normal neighborhood of p. So ϕ on $U(p)$ is the polar map

$$\phi = \exp_q \circ d\phi_p \circ (\exp_p)^{-1}$$

and so is a diffeomorphism of $U(p)$ onto the normal neighborhood U of $q \in N$. $\square$

Lemma 11. Step 3. *If* $p_1 \neq p_2$ *are elements of* $\phi^{-1}(p)$ *then* $U(p_1) \cap U(p_2) = \emptyset$.

Proof. We want to show that if $m \in U(p_1) \cap U(p_2)$ then $p_1 = p_2$. Let σ_1 and σ_2 be the radial geodesics from p_i to m, $i = 1, 2$, but with thier orientations reversed. Then $\phi \circ \sigma_i$ are reversed radial geodesics from $\phi(m)$ to q in U, and so are equal. This means that σ_1 and σ_2 are lifts of this same geodesic starting at the same point m and so are equal. So the opposite end points, i.e. p_1 and p_2 are equal. $\square$

We know that the union of the $U(p)$ as p ranges over $\phi^{-1}(p)$ is contained in $\phi^{-1}(U)$. We want to prove that

Lemma 12. Step 4.

$$\phi^{-1}(U) = \bigcup U(p).$$

Proof. Suppose that $m \in \phi^{-1}(U)$. Let σ be the reversed radial geodesic joining $\phi(m)$ to q. Its lift starting at m is a reversed radial geodesic ending at some $p \in \phi^{-1}(q)$. But then the inverse image of the reverse of this geodesic under $\exp_p$ is a ray in T_pM lying in $\tilde{U}(p)$ so $m \in U(p)$. $\square$

We now know that ϕ is a covering map. The theorem we just proved has the following important consequence in the **Riemannian case**:

Theorem 23. *Let $\phi : M \to N$ be local isometry between Riemannian manifolds with N connected. Then M is complete if and only if N is complete and ϕ is a Riemannian covering map.*

Proof. If M is complete and $\sigma : [0,1] \to N$ is a geodesic and $\phi(p) = \sigma(0) = q$, choose an initial tangent vector $v \in T_p M$ with $d\phi_p(v) = \sigma'(0)$. The geodesic γ_v starting at p with $\gamma'(0) = v$ exists for all time, in particular the image of this geodesic restricted to $[0,1]$ covers σ. So the preceding theorem shows that ϕ is a covering map. But then $\phi \circ \gamma_v$ extends σ for all time and so by Hopf-Rinow N is complete.

Conversely, if $v \in T_p M$ the geodesic in N starting at $\phi(p)$ and with initial tangent vector $d\phi_p(v)$ exists for all time since we are assuming that N is complete. Its cover exists for all time and is a geodesic. So all geodesics in M exist for all time. □

Getting back to complete manifolds with non-positive sectional curvature: If M is such a manifold and we give $T_p M$ the pulled back metric from M under $\exp_p$, the rays through the origin are geodesics, and they exist for all time. So M is complete, hence $\exp_p : T_p M \to M$ is a covering map. Hence if M is simply connected the map $\exp_p$ is a diffeomorphism, and hence M is diffeomorphic to $\mathbb{R}^n$ where $n = \dim M$.

9.5 Symmetric spaces.

Let us now use Hopf-Rinow together with Proposition 15 and Theorem 8:

Definition 3. *A semi-Riemannian manifold M is called a **symmetric space** if for every $p \in M$ there is an isometry $\zeta_p : M \to M$ whose differential at p is $-I$.*

In other words the polar map at p extends to a global isometry.

Theorem 24. *Every Riemannian symmetric space is complete.*

Proof. We show that every geodesic $\sigma : [0,b] \to$ is extendible past b. For this choose c close enough to b so that $\zeta_{\sigma(c)}$ extends σ (after a reparametrization) past b. □

If M is simply connected, the converse is true. Namely:

Theorem 25. *A complete, simply connected, locally symmetric semi-Riemannian manifold is symmetric.*

9.5. SYMMETRIC SPACES.

For the proof, see O'Neil, pages 224-226.

Suppose that M is a Riemannian symmetric space, and p and q are two points of M. Let $\sigma : [0,1] \to M$ be a geodesic with $\sigma(0) = p$ and $\sigma(1) = q$. Then the symmetry about the point $\sigma\left(\frac{1}{2}\right)$ maps p to q. So the group of isometries of M acts transitively on M. Thus Cartan reduced the study of symmetric spaces to group theory, and proceeded to classify all symmetric spaces!

The standard book on symmetric spaces is *Differential Geometry, Lie Groups, and Symmetric Spaces* by Sigurdur Helgason (2001) (AMS edition).

Chapter 10

Curvature, distance, and volume.

The purpose of this chapter to give some geometric manifestations in Riemannian geometry of the sectional curvature and the Ricci curvature. We show how the sectional curvature in a Riemmanian manifold determines the rate at which two geodesics emanating from the same point spread apart or come together. Later in the chapter we show how the Ricci curvature affects the rate of growth of volume under the exponential map. In between we prove, among other things, Myer's theorem relating curvature bounds to the global topology.

10.1 The local relation between sectional curvature and distance.

Let M be a Riemannian manifold, $p \in M$, $\mathcal{O}$ a convex neighborhood of p, with $v, w \in T_p M$. Let

$$c_0(\epsilon) := \exp_p(\epsilon v), \quad c_1(\epsilon) = \exp_p(\epsilon w)$$

and choose ϵ so small that the geodesics $s \mapsto \exp sv$ and $s \mapsto \exp sw$ lie in $\mathcal{O}$ for $s \in [0, \epsilon]$. So there is a unique geodesic joining $c_0(\epsilon)$ to $c_1(\epsilon)$ whose length $L(\epsilon)$ is the distance from $c_0(\epsilon)$ to $c_1(\epsilon)$. This geodesic is the curve given, for each fixed ϵ as $a_\epsilon : t \mapsto V(\epsilon, t)$ where

$$V(\epsilon, t) = \exp_{c_0(\epsilon)}(t \exp_{c_0(\epsilon)}^{-1}(c_1(\epsilon))).$$

We wish to prove:

Theorem 26. *If v and w are linearly independent unit vectors then*

$$L(\epsilon) = \epsilon \|v - w\| \left(1 - \frac{1}{12} \sec(v, w)(1 + \langle v, w \rangle)\epsilon^2\right) + O(\epsilon^4).$$

Thus $L(\epsilon)$ (for small values of ϵ) grows faster than the corresponding Euclidean growth $\epsilon\|v-w\|$ if the sectional curvature $K(u,v) = \sec(v,w)$ is negative and slower if $K(u,v) > 0$.

Here is a diagram illustrating the possibilities. Much of the following discussion is taken from W. Meyer, *Toponogov's theorem and applications* available on the web.

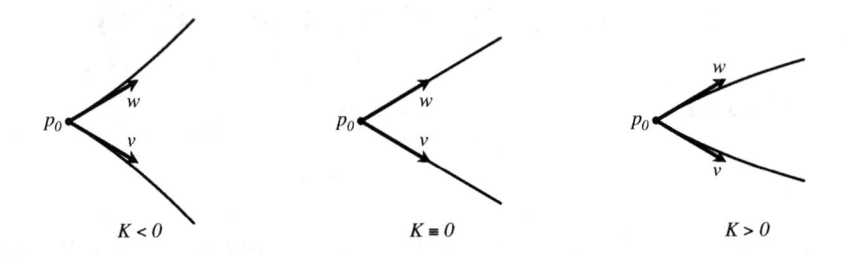

Recall that
$$V(\epsilon, t) = \exp_{c_0(\epsilon)}(t \exp_{c_0(\epsilon)}^{-1}(c_1(\epsilon))).$$

Each $a_\epsilon(\cdot) = V(\epsilon, \cdot)$ is a geodesic so we have a one parameter family of geodesics. Let
$$T := V_t \quad \text{and} \quad E = V_\epsilon$$

So $T(\epsilon, t)$ is the tangent vector to the geodesic a_ϵ and E is the Jacobi field along a_ϵ corresponding to the variation. Here is a picture describing the situation:

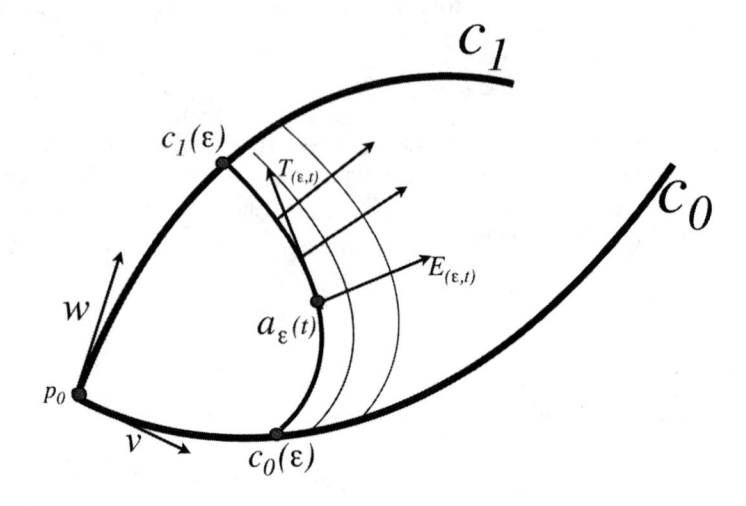

The length $L(\epsilon)$ of the geodesic a_ϵ is the same as the length of its tangent vector since the geodesic is parametrized proportional to arc length and goes from 0

10.1. SECTIONAL CURVATURE AND DISTANCE, LOCALLY.

to 1, so
$$L(\epsilon) = \|T(\epsilon, t)\|.$$
Set
$$H := L^2.$$
The plan is to compute the first four derivatives of H at 0 so as to obtain the Taylor series with remainder for H up to $O(\epsilon^5)$. Then to use this to prove the theorem.

By definition, we have $H = \langle T, T \rangle$ so
$$H'(\epsilon) = \frac{\partial}{\partial \epsilon} \langle T, T \rangle = 2 \langle T_\epsilon, T \rangle$$
where T_ϵ means covariant derivative of the vector field T along the curves $\epsilon \mapsto V(\epsilon, t)$ for each fixed t. Continuing in this way we have:

$$\begin{aligned}
H'(\epsilon) &= 2\langle T_\epsilon, T \rangle \\
H''(\epsilon) &= 2\langle T_{\epsilon\epsilon}, T \rangle + 2\langle T_\epsilon, T_\epsilon \rangle \\
H'''(\epsilon) &= 2\langle T_{\epsilon\epsilon\epsilon}, T \rangle + 6\langle T_{\epsilon\epsilon}, T_\epsilon \rangle \\
H^{iv}(\epsilon) &= 2\langle T_{\epsilon\epsilon\epsilon\epsilon}, T \rangle + 8\langle T_{\epsilon\epsilon\epsilon}, T_\epsilon \rangle + 6\langle T_{\epsilon\epsilon}, T_{\epsilon\epsilon} \rangle.
\end{aligned}$$

Since $\epsilon \mapsto V(\epsilon, t)$ is a geodesic for $t = 0$ and $t = 1$ we have
$$E_\epsilon(\epsilon, 0) = 0 \quad \text{and} \quad E_\epsilon(\epsilon, 1) = 0.$$
Since $V(0, t) \equiv p$ we have
$$T(0, t) \equiv 0.$$
We also have equality of cross first derivatives $V_{\epsilon t} = V_{t\epsilon}$ so
$$T_\epsilon = E_t.$$
Along each geodesic $t \mapsto V(\epsilon, t)$ (for fixed ϵ) we have the Jacobi equation
$$E_{tt} = R(T, E)T$$
so since $T(0, t) \equiv 0$ we have $E_{tt}(0, t) \equiv 0$. So $E(0, t)$, which is a curve in the tangent space $T_p M$, is a linear function of t. We know that $E(0, 0) = v$ and $E(0, 1) = w$ so
$$E(0, t) = v + t(w - v).$$
So
$$H'(0) = 2\langle T_\epsilon(0, t), T(0, t) \rangle = 0$$
and
$$\begin{aligned}
H''(0) &= 2\langle T_{\epsilon\epsilon}(0, t), T(0, t) \rangle + 2\langle T_\epsilon(0, t), T_\epsilon(0, t) \rangle \\
&= 2\langle T_\epsilon(0, t), T_\epsilon(0, t) \rangle \\
&= 2\langle E_t(0, t), E_t(0, t) \rangle \\
&= 2\|w - v\|^2.
\end{aligned}$$

We have computed the first two derivatives of H at 0. We will now show that $H'''(0) = 0$. We know that

$$H'''(0) = 6\langle T_{\epsilon\epsilon}, T_\epsilon\rangle(0,t) = 6\langle E_{t\epsilon}, T_\epsilon\rangle(0,t)$$

since $T_\epsilon = E_t$. We also know that

$$E_{t\epsilon} = E_{\epsilon t} + R(E,T)E$$

from our general formula for two parameter families. So

$$H'''(0) = 6\langle R(E,T)E, T_\epsilon\rangle(0,t) + 6\langle E_{\epsilon t}, T_\epsilon\rangle(0,t)$$

and the first term vanishes since $T(0,t) \equiv 0$. So we will have proved that $H'''(0) = 0$ if we can show that

$$E_\epsilon(0,t) \equiv 0. \qquad (*)$$

Proof of (*). We know that $E_\epsilon(0,0) = 0$ and $E_\epsilon(0,1) = 0$. So it is enough to prove that $E_\epsilon(0,t)$ (which is a curve in T_pM) is a linear function of t. We have

$$E_{\epsilon tt} = (R(T,E)E)_t + E_{t\epsilon t}$$

$$= (R(T,E)E)_t + T_{\epsilon\epsilon t} = (R(T,E)E)_t + R(T,E)T_\epsilon + T_{\epsilon t\epsilon}$$

$$= (R(T,E)E)_t + R(T,E)T_\epsilon + (R(T,E)T)_\epsilon$$

since $T_t \equiv 0$. We know that $T(0,t) \equiv 0$ so that the second term on the right vanishes when $\epsilon = 0$ and so does the third term since covariant derivative is a tensor derivation. As to the first term, the tensor derivation property leaves only $R(T_t, E)E$ and this vanishes at 0 since $T_t(0,t) \equiv 0$.

Computation of $H^{iv}(0)$. From $E_\epsilon(0,t) \equiv 0$ it follows that

$$E_{\epsilon t}(0,t) \equiv 0$$

Since

$$E_{\epsilon t} = E_{t\epsilon} + R(E,T)E_{t\epsilon}$$

and $T(0,t) \equiv 0$ we know that $E_{t\epsilon}(0,t) \equiv 0$. But $E_t = T_\epsilon$ so we conclude that

$$T_{\epsilon\epsilon}(0,t) \equiv 0.$$

Plugging this into the equation

$$H^{iv}(\epsilon) = 2\langle T_{\epsilon\epsilon\epsilon\epsilon}, T\rangle + 8\langle T_{\epsilon\epsilon\epsilon}, T_\epsilon\rangle + 6\langle T_{\epsilon\epsilon}, T_{\epsilon\epsilon}\rangle.$$

and evaluating at $\epsilon = 0$ gives

$$H^{iv}(0) = 8\langle T_{\epsilon\epsilon\epsilon}, T_\epsilon\rangle(0,t).$$

10.1. SECTIONAL CURVATURE AND DISTANCE, LOCALLY.

We will now massage the equation
$$H^{iv}(0) = 8\langle T_{\epsilon\epsilon\epsilon}, T_\epsilon\rangle(0,t).$$

Since $T_\epsilon = E_t$ we have
$$T_{\epsilon\epsilon\epsilon} = E_{t\epsilon\epsilon} = (R(E,T)E)_\epsilon + E_{\epsilon t\epsilon}.$$

Evaluating at $\epsilon = 0$ the first term becomes
$$8(R(E,T_\epsilon)E)(0,t) = 8R(E(0,t), E_t(0,t))E(0,t)$$
$$= 8R(v + t(w-v), w-v)(v + t(w-v))$$
$$= 8R(v,w)[v + t(w-v)].$$

Taking the scalar product with $T_\epsilon(0,t) = E_t(0,t) = (w-v)$ gives $8\langle R(v,w)v, w\rangle$. We have shown that
$$H^{iv}(0) = 8\langle R(v,w)v, w\rangle + 8\langle E_{\epsilon t \epsilon}, T_\epsilon\rangle(0,t).$$

We will now show that the second term (which must be a constant since the other two terms in the equation are) vanishes. We know that at $\epsilon = 0$ we have
$$T_{\epsilon t} = E_{tt} = 0, \; E_{\epsilon t} = 0, \; \text{and } T_{\epsilon\epsilon} = 0.$$

Now $\langle E_\epsilon, T_\epsilon\rangle_t = \langle E_{\epsilon t}, T_\epsilon\rangle + \langle E_\epsilon, T_{\epsilon t}\rangle$ so $\langle E_\epsilon, T_\epsilon\rangle_{t\epsilon} =$
$$\langle E_{\epsilon t \epsilon}, T_\epsilon\rangle + \langle E_{\epsilon t}, T_{\epsilon\epsilon}\rangle + \langle E_{\epsilon\epsilon}, T_{\epsilon t}\rangle + \langle E_\epsilon, T_{\epsilon t \epsilon}\rangle.$$

Evaluating at $\epsilon = 0$ the second and third term vanish, and so does the fourth since $T_{\epsilon t \epsilon} = T_{\epsilon \epsilon t} + R(E,T)T_\epsilon$. We have shown that at $\epsilon = 0$ we have
$$\langle E_{\epsilon t \epsilon}, T_\epsilon\rangle = \langle E_\epsilon, T_\epsilon\rangle_{t\epsilon} = \langle E_\epsilon, T_\epsilon\rangle_{\epsilon t}$$

by the equality of cross derivatives of functions. As we know that $\langle E_\epsilon, T_\epsilon\rangle_{\epsilon t}$ is a constant at $\epsilon = 0$, we conclude that $\langle E_\epsilon, T_\epsilon\rangle_\epsilon$ is a linear function of t at $\epsilon = 0$. Now
$$\langle E_\epsilon, T_\epsilon\rangle_\epsilon = \langle E_{\epsilon\epsilon}, T_\epsilon\rangle + \langle E_\epsilon, T_{\epsilon\epsilon}\rangle.$$

The second term vanishes since $T_{\epsilon\epsilon} = 0$ at $\epsilon = 0$. To show that the first term vanishes, it is enough to show that it vanishes at $t = 0$ and $t = 1$ since we know that it is a linear function of t. But at these values of t we have $E_\epsilon = 0$ since the curves $\epsilon \mapsto V(\epsilon, 0)$ and $\epsilon \mapsto V(\epsilon, 1)$ are geodesics.

The first four derivatives of H at 0:

So we have computed
$$\begin{aligned} H(0) &= 0 \\ H'(0) &= 0 \\ H''(0) &= 2\|v-w\|^2 \\ H'''(0) &= 0 \\ H^{iv}(0) &= 8\langle R(v,w)v, w\rangle = -8\langle R(w,v)v, w\rangle. \end{aligned}$$

Taylor's formula with remainder gives (since $H = L^2$)

$$L^2(\epsilon) = \epsilon^2 \|v - w\|^2 - \frac{1}{3}\epsilon^4 \langle R(w,v)v, w\rangle + O(\epsilon^5).$$

If $v \neq w$ we may take the square root of this equation to obtain

$$L(\epsilon) = \epsilon \|v - w\| - \frac{1}{6\|v - w\|}\langle R(w,v)v, w\rangle \epsilon^3 + O(\epsilon^4).$$

$$= \epsilon \|v - w\| \left(1 - \frac{1}{6}\frac{1}{\|v-w\|^2}\langle R(w,v)v, w\rangle \epsilon^2\right) + O(\epsilon^4).$$

Now suppose that $\|v\| = \|w\| = 1$ so

$$\|v - w\|^2 = 2(1 - \langle v, w\rangle) = \frac{2(1 - \langle v, w\rangle^2)}{1 + \langle v, w\rangle} = \frac{2Q(v,w)}{1 + \langle v, w\rangle}.$$

Thus we can write the above equation as

$$L(\epsilon) = \epsilon \|v - w\| \left(1 - \frac{1}{12}\sec(v,w)(1 + \langle v,w\rangle)\epsilon^2\right) + O(\epsilon^4),$$

as promised.

It would appear from the local pictures that, for negative curvature, geodesics keep spreading apart and for positive curvature they have to meet after a finite time, i.e. that there are no conjugate points in the negative sectional curvature case and that there is an upper bound to the distance between any two points in the positive curvature case.

We gave a proof of the first assertion in our discussion of Hadamard's theorem. We next turn to the second assertion. The key idea goes back to Bonnet in the case of surfaces:

The Jacobi equation $Y'' + R(Y, \gamma')\gamma' = 0$ looks something like the ordinary differential equation $y'' + ky = 0$ in one variable. Now if $k > 0$ is a constant, a solution with $y(0) = 0$ of this one variable equation is $t \mapsto \sin\sqrt{k}t$ which has a zero at $t = \pi/\sqrt{k}$. So a comparison theorem in ordinary differential equations should work.

10.2 Myer's theorem.

In fact, in the positive curvature case, we do not have to assume that *all* but only that an "average" of the sectional curvatures is positive. I will explain what I mean by "average".

10.2.1 Back to the Ricci tensor.

Recall the directional curvature operator $R_v(w) = R(v,w)v$ (so that the Jacobi equation is $Y'' = R_{\gamma'}(Y)$). Since

$$\begin{aligned}\langle R_v(w), z \rangle &= \langle R(v,w)v, z \rangle \\ &= \langle R(v,z)v, w \rangle \\ &= \langle R_v(z), w \rangle\end{aligned}$$

we see that R_v is a symmetric linear transformation on each tangent space. Also $R(v,v) = 0$ so $R_v(v) = 0$ and hence R_v preserves the orthocomplement of v. Recall that the **Ricci operator** Ric (also known as the Ricci tensor) is the operator

$$v \mapsto \text{Ric}(v) = \sum_i R_{e_i} v = \sum_i R(v, e_i) e_i$$

where the sum is over any orthonormal basis. (It is easy to check that that this is independent of the basis.)

The operator Ric is symmetric, so if $e_2, \ldots e_n$ are orthogonal to v and $\|v\| = 1$ then

$$\langle \text{Ric}(v), v \rangle = \sum_{i=2}^{n} \langle R(v, e_i) e_i, v \rangle = \sum_{i=2}^{n} \sec(v, e_i)$$

So

$$\frac{1}{n-1} \langle \text{Ric}(v), v \rangle$$

is the average of the sectional curvatures over the e_i orthogonal to v.

In general, if A is a self adjoint operator on a scalar product space then $\langle Av, v \rangle$ where v is a unit vector is a sort of average value of A (in the direction v). In quantum mechanics this is called the expectation of the observable A in the state v.

If the eigenvalues of A are $\lambda_1 \leq \cdots \leq \lambda_n$ then $\lambda_1 \leq \langle Av, v \rangle \leq \lambda_n$ for any unit vector v. So we will say that $\lambda_1 \leq A \leq \lambda_n$

Proposition 28. *Suppose that σ is a unit speed geodesic for which*

$$\frac{1}{n-1} \langle \text{Ric}(\sigma'), \sigma' \rangle > C > 0.$$

Then if σ has length $\geq \pi/\sqrt{C}$ then there are conjugate points along σ.

Proof. Let $\sigma', E_2, \ldots E_n$ be a parallel orthonormal frame field on σ. Recall the definition of I_σ: Let $\sigma : [0, b] \to M$ be a geodesic. Then

$$I_\sigma(W, W) := \int_0^b \left\{ \langle W^{\perp'}, W^{\perp'} \rangle + \langle R(W, \sigma')W, \sigma' \rangle \right\} du.$$

Take $W = fE_i$ for some function f vanishing at the end points. We get

$$\int_0^b \left[f'^2 + f^2 \langle R(E_i, \sigma')E_i, \sigma'\rangle\right] dt.$$

Summing over i gives

$$\int_0^b \left[((n-1)f'^2 - f^2 \langle \text{Ric}(\sigma'), \sigma'\rangle\right] \leq (n-1)\int_0^b \left[f'^2 - Cf^2\right] dt.$$

Taking $f = \sin(C^{\frac{1}{2}}t)$ and $b = \pi/\sqrt{C}$ the above integral vanishes. So $I_\sigma(fE_i, fE_i) \leq 0$ for at least one i. This can not happen unless there is a conjugate point on σ, as will follow from the following theorem: □

Let $I_\sigma^\perp$ be the index form I restricted to piecewise smooth vector fields along σ which are orthogonal to σ.

Theorem 27. *If there are no conjugate points along σ then $I_\sigma^\perp$ is positive definite.*

For the proof of the theorem we will use two lemmas:

Lemma 13. *If V and W are Jacobi fields along a geodesic then $\langle V', W\rangle - \langle V, W'\rangle$ is constant.*

Proof. $\langle V', W\rangle' = \langle V'', W\rangle + \langle V', W'\rangle = \langle R(V, \sigma')W, \sigma'\rangle + \langle V', W'\rangle$. This expression is symmetric in V and W. □

Lemma 14. *Let $Y_1, \ldots, Y_k$ be Jacobi fields along σ such that $\langle Y_i', Y_j\rangle = \langle Y_i, Y_j'\rangle$ for all i and j. Let $V = \sum f_i Y_i$ and set $A := \sum f_i' Y_i$, $B := \sum f_i Y_i'$. Then*

$$\langle V', V'\rangle - \langle R(\sigma', V)V, \sigma'\rangle = \langle A, A\rangle + \langle V, B\rangle'.$$

Proof. Since $V' = A + B$ we have $\langle V, B\rangle' = \langle V', B\rangle + \langle V, B'\rangle$

$$= \langle A, B\rangle + \langle B, B\rangle + \langle V, \sum f_i' Y_i'\rangle + \langle V, \sum f_i Y_i''\rangle.$$

The Jacobi equation converts the last term to $-\langle R(\sigma', V)V, \sigma'\rangle$. The third term is $\sum_{ij} f_j f_i' \langle Y_j, Y_i'\rangle = \sum_{ij} f_j f_i' \langle Y_j', Y_i\rangle = \langle A, B\rangle$. □

Proof of the theorem.

Proof. Let $Y_2, \ldots, Y_n$ be Jacobi fields vanishing at 0 and such that $Y_2'(0), \ldots, Y_n'(0)$ are a basis of $\sigma'(0)^\perp$. Since there are no conjugate points, the $Y_2, \ldots, Y_n$ form a basis of $\sigma'(u)^\perp$ for all $u \neq 0$. So if V is a piecewise smooth perpendicular vector field which vanishes at 0 and b, there are piecewise smooth functions such that $V = \sum_2^n f_i Y_i$ on $(0, b]$ and since V vanishes at 0 these have continuous extensions to $[0, b]$. Since the Y_i all vanish at 0, the hypothesis of the preceding lemma holds, by the lemma before that. So

$$\langle V', V'\rangle - \langle R(\sigma', V)V, \sigma'\rangle = \langle A, A\rangle + \langle V, B\rangle'$$

as above. Integrating gives

$$I_\sigma^\perp(V, V) = \int_0^b \langle A, A\rangle du + \langle V, B\rangle|_0^b.$$

The second term vanishes since V vanishes at 0 and b. So $I_\sigma^\perp(V, V) \geq 0$. If $I_\sigma^\perp(V, V) = 0$, then $\langle A, A\rangle \equiv 0$ so the f_i are constant and hence $= 0$ so $V = 0$. □

10.2.2 Myer's theorem.

Theorem 28. *Let M be a complete connected Riemannian manifold with $\langle \mathrm{Ric}(v), v\rangle$ $(n-1)C > 0$ for all unit vectors at all points. Then*

- *M is compact.*
- *The diameter of M is $\leq \pi/\sqrt{C}$.*
- *The fundamental group $\pi_1(M)$ is finite.*

Proof. That the maximal distance between any two points is $\leq \pi/\sqrt{C}$ follows from the preceding proposition and theorem since a geodesic can not minimize past a conjugate point. That M is compact then follows from the Hopf-Rinow theorem. The universal covering space of M is a Riemannian manifold satisfying the same condition, hence is compact. Hence the fundamental group of M is finite. □

As we have seen, the theorem is an immediate consequence of Synge's formula and Synge was aware of its consequences for the existence of conjugate points. What was not yet available at the time of Synge's paper was the Hopf-Rinow theorem. (In fact, Cartan in his famous 1922 book did not have a good definition of completeness - i.e. also did not have the Hopf-Rinow theorem at his disposal.) Synge was duly outraged that the theorem was not attributed to him.

10.3 Length variation of a Jacobi vector field.

I am now going to repeat some of the arguments of the beginning of this chapter, but at the infinitesimal level:

Let $\gamma : [0, 1] \to M$ be a geodesic with $\gamma(0) = p \in M$ and $\gamma'(0) = w \in T_pM$. Let V be a Jacobi vector field along γ with

$$V(0) = 0, \quad V'(0) = v \in T_pM.$$

We wish to find the first four terms in the Taylor expansion of $\langle V(t), V(t) \rangle$. For this we compute the first four derivatives of $\langle V(t), V(t) \rangle$ and then set $t = 0$: We have:

$$\begin{aligned}
\langle V, V \rangle' &= 2\langle V', V \rangle \\
\langle V, V \rangle'' &= 2\langle V'', V \rangle + 2\langle V', V' \rangle \\
&= 2\langle R(\gamma', V)\gamma', V \rangle + 2\langle V', V' \rangle \\
\langle V, V \rangle''' &= 2\langle R(\gamma', V)'\gamma', V \rangle + 2\langle R(\gamma, V)\gamma', V' \rangle + 4\langle V'', V' \rangle \\
&= 2\langle R(\gamma', V)'\gamma', V \rangle + 6\langle R(\gamma, V)\gamma', V' \rangle \\
\langle V, V \rangle^{(iv)} &= 8\langle R(\gamma', V)'\gamma', V' \rangle + 2\langle R(\gamma', V)''\gamma', V \rangle \\
&\quad + 6\langle R(\gamma', V)\gamma', V'' \rangle.
\end{aligned}$$

Setting $t = 0$ we see that the first and third derivatives vanish, and

$$\langle V, V \rangle''(0) = 2\langle V', V' \rangle(0) = 2\langle v, v \rangle.$$

In the fourth derivative, the second and third terms above vanish, and in computing $R(\gamma', V)'$ the only term that survives evaluation at 0 is $R(\gamma', V')$. So $\langle V, V \rangle^{(iv)}(0) = 8\langle R(\gamma', V')'\gamma', V' \rangle(0) = 8\langle R(w, v)w, v \rangle$.

We have shown that the first and third derivatives of $\langle V, V \rangle$ vanish at $t = 0$ and the second and fourth derivatives at $t = 0$ are given by

$$\langle V, V \rangle''(0) = 2\langle v, v \rangle, \qquad \langle V, V \rangle^{(iv)}(0) = 8\langle R(w, v)w, v \rangle.$$

So we get

$$\langle V, V \rangle(t) = \langle v, v \rangle t^2 + \frac{1}{3}\langle R(w, v)w, v \rangle t^4 + O(t^5).$$

The coefficients of t^2 and t^4 in the right hand side of the above expression are quadratic in v. So we can apply polarization to conclude that

Proposition 29. *Let U and V be Jacobi fields along a geodesic γ with $U(0) = V(0) = 0$, $U'(0) = u$, $V'(0) = v$. Then*

$$\langle U(t), V(t) \rangle = \langle u, v \rangle t^2 + \frac{1}{3}\langle R(w, u)w, v \rangle t^4 + O(t^5) \tag{10.1}$$

where $w = \gamma'(0)$.

10.3.1 Riemann's formula for the metric in a normal neighborhood.

I want to use the preceding discussion to get an expression for the metric in a normal neighborhood up to third order. This formula goes back to Riemann. Recall that in normal coordinates, the first derivatives of the metric tensor vanish at the origin. In other words,

$$g_{ij}(x) = g_{ij}(0) + O(\|x\|^2).$$

10.3. LENGTH VARIATION OF A JACOBI VECTOR FIELD.

Riemann's formula gives the second order terms. More precisely,

$$g_{jk}(x) = \delta_{jk} + \frac{1}{3}\sum_{i\ell} R_{ijk\ell} x^i x^\ell + O(\|x\|^3). \tag{10.2}$$

For this recall the following theorem:

Theorem 29. *Let $o \in M$, $x \in T_o(M)$ and $v \in T_o(M)$ considered as an element of $T_x(T_o(M))$. Then*

$$d(\exp_o)_x(v) = \dot{V}(1)$$

where V is the unique Jacobi field along the geodesic γ_x such that $V(0) = 0$ and $V'(0) = v$.

Let us reprove this theorem giving a parametrized version: So let $w \in T_o(M)$ and consider the geodesic $\gamma(t) = \exp_o(tw)$. Let c be a curve in T_oM with $c(0) = w$, $\dot{c}(0) = v$, for example, $c(s) = w + sv$. Then

$$\mathbf{x}(t,s) := \exp_o(tc(s))$$

is a geodesic variation of γ with $V(t) = \mathbf{x}_s(t,0)$ given by

$$V(t) = d(\exp_o)_{tw}(tv) = t\, d\exp_o(tw)(v).$$

Clearly $V(0) = 0$ and $V'(0) = \mathbf{x}_{st}(0,0) = \mathbf{x}_{ts}(0,0)$. But $\mathbf{x}_s(t,0) = \exp_o(tv)$ and differentiating with respect to t at $t = 0$ gives v. So we have proved:

$$d(\exp_o(tw))(v) = t^{-1}V(t) \tag{10.3}$$

where $V(t)$ is the Jacobi field along $\gamma(t) = \exp_o(tw)$ with $V(0) = 0$ and $V'(0) = v$.

Now let v range over an orthonormal basis $e_1, \ldots e_n$ of T_oM and let $Y_1, \ldots, Y_n$ the corresponding Jacobi fields. By the definition of normal coordinates, we have

$$\partial_j(\exp_o(tw)) = d(\exp_o)_{tw}e_j$$

and this equals $t^{-1}Y_j(t)$ by (10.3). So

$$\partial_j(\exp_o(tw)) = t^{-1}Y_j(t).$$

Hence $g_{jk}(\exp_o(tw)) = \langle \partial_j(\exp_o(tw)), \partial_k(\exp_o(tw))\rangle$ is given by

$$t^{-2}\langle Y_j, Y_k\rangle(t) = t^{-2}\left(t^2\langle e_j, e_k\rangle + \frac{t^4}{3}\langle R(w,e_j)w, e_k\rangle + O(t^5)\right)$$

$$= \delta_{jk} + \frac{t^2}{3}\langle R(w,e_j)w, e_k\rangle + O(t^3). \tag{10.4}$$

Substituting $w = t^{-1}x$ into this last expression gives Riemann's formula (10.2). In view of the symmetry-antisymmetry properties of the Riemann curvature tensor

$$R_{ijk\ell} = R(e_i, e_j, e_k, e_\ell)$$

formula (10.2) is frequently written as

$$ds^2 = dx_1^2 + \cdots + dx_n^2$$

$$+ \frac{1}{12} \sum_{ijk\ell} R_{ijk\ell}(x_i dx_j - x_j dx_i)(x_k dx_\ell - x_\ell dx_k) + O(\|x\|^3). \tag{10.5}$$

10.4 The Ricci tensor and volume growth.

The last item in this chapter is an estimate on the Riemannian volume of the exponential map of a small ball in the tangent space $T_m M$. Let V_n denote the volume of the unit ball in Euclidean space, so that the volume of the Euclidean ball of radius r is $V_n r^n$.

For sufficiently small r, the exponential map is a diffeomorphism, so we can ask about the Riemannian volume of its image. I want to prove that this volume is

$$V(n)r^n - \frac{r^{n+2}}{6(n+2)} \int_S \text{Ric}(u,u) du + O(r^{n+3}) \tag{10.6}$$

where the integration is over the unit sphere, S in $T_m M$ (in its Euclidean metric) and du is the spherical measure.

The proof will be yet another variant on the Cartan philosophy.

Let $x \in T_m M$ have length 1 and let γ_x be the geodesic with $\gamma_x(0) = m$ and $\gamma'(0) = x$. Suppose that r_0 is such that the geodesic exists for $0 \le r \le r_0$. Let $y \in T_m M$. Then we know that if we think of y as an element of $T_{rx}(T_m M))$ then

$$d(\exp_m)_{rx}(y) = Y(r)$$

where Y is the Jacobi field along γ_x with the initial conditions

$$Y(0) = 0, \quad Y'(0) = \frac{y}{r}.$$

So it is clear that

$$Y(r) = \frac{r_0}{r} Y(r_0).$$

Let $\tilde{Y}(r)$ be the element of $T_m M$ obtained by taking $Y(r)$ and parallel translating it back along γ_x to m.

So let Y_r be the Jacobi field along γ_x with the initial conditions $Y(0) = 0, Y'(0) = \frac{y}{r}$, and let $\tilde{Y}_r(s)$ be the curve in $T_m M$ obtained by parallel transport of $Y_r(s)$ back along γ_x to m. So

$$\tilde{Y}(r) = \frac{r_0}{r} \tilde{Y}_{r_0}(r)$$

10.4. THE RICCI TENSOR AND VOLUME GROWTH.

and therefore

$$\tilde{Y}(r) = \frac{r_0}{r}\left(\tilde{Y}_{r_0}(0) + r\tilde{Y}'_{r_0} + \frac{r^2}{2!}\tilde{Y}''_{r_0} + \cdots\right)$$

$$= r_0\left(\tilde{Y}'_{r_0}(0) + \frac{r}{2}\tilde{Y}''_{r_0}(0) + \frac{r^2}{3!}\tilde{Y}'''_{r_0}(0) + \cdots\right)$$

since $\tilde{Y}_{r_0}(0) = 0$.

The definition of covariant derivative tells us that $\tilde{Y}^{(k)}_{r_0}(0) = Y^{(k)}_{r_0}(0)$. Now $Y_{r_0}(0) = 0$ and the Jacobi equation tells us that $Y''_{r_0}(0) = 0$ as well. To compute $Y'''_{r_0}(0)$ we differentiate the Jacobi equation

$$Y''_{r_0} = R(\gamma'_x, Y_{r_0}))\gamma'_x$$

and set $r = 0$. The only term that survives is the value at 0 of

$$R(\gamma'_x(0), Y'_{r_0}(0))\gamma'_x(0) = \frac{1}{r_0}R(x,y)x.$$

So we have established that the Taylor expansion of $\tilde{Y}(r)$ at $r = 0$ is

$$y + \frac{r^2}{6}R(x,y)x + \cdots.$$

The formula for the volume of a region U in a coordinate system in a Riemannian manifold is

$$\int_U |\det(g_{ij})|^{\frac{1}{2}} dx_1 \ldots dx_n$$

where $g_{ij} = \langle \partial_i, \partial_j \rangle$. More generally, we can let $F_1, \ldots, F_n$ be any frame field on the coordinate neighborhood which is orthonormal with respect to the Euclidean metric given by the coordinates and take $g_{ij} = \langle F_i, F_j \rangle$.

For the normal coordinates we will proceed as follows: for the unit vector x take $y_2, \ldots, y_n$ to complete x to an orthonormal basis and use this frame all along the ray determined by x.

By Gauss's lemma, the vector γ'_x is a unit vector and is orthogonal to the image of the $y_2, \ldots, y_n$ under the exponential map, and parallel translation preserves scalar product, so the determinant we have to compute at the point rx is the same as the determinant of the $(n-1) \times (n-1)$ matrix whose entries are

$$\langle \tilde{Y}_i(r), \tilde{Y}_j(r) \rangle = \langle y_i + \frac{r^2}{6}R(x,y_i)x + \cdots, y_j + \frac{r^2}{6}R(x,y_j)x + \cdots \rangle$$

$$= \delta_{ij} + \frac{r^2}{3}\langle R(x,y_i)x, y_j \rangle + O(r^3).$$

(We have already encountered this formula in the preceding section, see (10.4).) Now the determinant of a matrix of the form $I + tA + O(t^2)$ is

$$1 + t\operatorname{tr} A + O(t^2)$$

and hence the square root of this determinant (for small values of t) is

$$1 + \frac{1}{2}t \operatorname{tr} A + O(t^2).$$

Thus we must integrate the function

$$1 + \frac{s^2}{6} \sum \langle R(x, y_i)x, y_i \rangle + O(s^3)$$

over the ball $\|s\| \leq r$ to get the volume of its image under the exponential map. But

$$-\sum \langle R(x, y_i)x, y_i \rangle = \operatorname{Ric}(x, x)$$

and the Euclidean measure in spherical coordinates is $s^{n-1}dsdu$ where du is the measure on the unit sphere.

The integral of the first term above over the ball of radius r is just the Euclidean volume of the ball. The integral of the second term is

$$-\frac{1}{6}\int_0^r s^{n+1}ds \int_S \operatorname{Ric}(u,u)du = -\frac{r^{n+2}}{6(n+2)}\int_S \operatorname{Ric}(u,u)du. \quad \square$$

Chapter 11

Review of special relativity.

In this chapter we give a rapid review of some aspects of special relativity. It is not meant as a substitute for a full course on the subject.

11.1 Two dimensional Lorentz transformations.

11.1.1 Two dimensional Minkowski spaces.

We study a two dimensional vector space with (pseudo-)scalar product $\langle \, , \, \rangle$ of signature $+\,-$. A Lorentz transformation is a linear transformation which preserves the scalar product. In particular it preserves

$$||\mathbf{u}||^2 := \langle \mathbf{u}, \mathbf{u} \rangle$$

(where with the usual abuse of notation this expression can be positive negative or zero). In particular, every such transformation must preserve the "light cone" consisting of all $\mathbf{u}$ with $||\mathbf{u}||^2 = 0$.

Two dimensional Lorentz transformations in "light cone coordinates'.

All such two dimensional spaces are isomorphic. In particular, we can choose our vector space to be $\mathbb{R}^2$ with (pseudo-)metric given by

$$\left\| \begin{pmatrix} u \\ v \end{pmatrix} \right\|^2 = uv.$$

The light cone consists of the coordinate axes, so every Lorentz transformation must carry the axes into themselves or interchange the axes. A transformation which preserves the axes is just a diagonal matrix. Hence the (connected component of) the Lorentz group consists of all matrices of the form

$$\begin{pmatrix} r & 0 \\ 0 & r^{-1} \end{pmatrix}, \quad r > 0.$$

231

Two dimensional Lorentz transformations in t, x coordinates.

We introduce (t, x) coordinates by
$$u = t + x$$
$$v = t - x$$
or
$$\begin{pmatrix} u \\ v \end{pmatrix} = \begin{pmatrix} 1 & 1 \\ 1 & -1 \end{pmatrix} \begin{pmatrix} t \\ x \end{pmatrix}$$
so $\left\| \begin{pmatrix} t \\ x \end{pmatrix} \right\|^2 = t^2 - x^2$.

Notice that $\begin{pmatrix} 1 & 1 \\ 1 & -1 \end{pmatrix}^2 = 2 \begin{pmatrix} 1 & 0 \\ 0 & 1 \end{pmatrix}$, so if

$$\begin{pmatrix} u \\ v \end{pmatrix} = \begin{pmatrix} 1 & 1 \\ 1 & -1 \end{pmatrix} \begin{pmatrix} t \\ x \end{pmatrix}$$

$$\begin{pmatrix} u' \\ v' \end{pmatrix} = \begin{pmatrix} r & 0 \\ 0 & r^{-1} \end{pmatrix} \begin{pmatrix} u \\ v \end{pmatrix}$$

$$\begin{pmatrix} u' \\ v' \end{pmatrix} = \begin{pmatrix} 1 & 1 \\ 1 & -1 \end{pmatrix} \begin{pmatrix} t' \\ x' \end{pmatrix}$$

then $\begin{pmatrix} t' \\ x' \end{pmatrix} = \frac{1}{2} \begin{pmatrix} 1 & 1 \\ 1 & -1 \end{pmatrix} \begin{pmatrix} r & 0 \\ 0 & r^{-1} \end{pmatrix} \begin{pmatrix} 1 & 1 \\ 1 & -1 \end{pmatrix} \begin{pmatrix} t \\ x \end{pmatrix}$.

Relativistic "velocity".

$$\begin{pmatrix} t' \\ x' \end{pmatrix} = \frac{1}{2} \begin{pmatrix} 1 & 1 \\ 1 & -1 \end{pmatrix} \begin{pmatrix} r & 0 \\ 0 & r^{-1} \end{pmatrix} \begin{pmatrix} 1 & 1 \\ 1 & -1 \end{pmatrix} \begin{pmatrix} t \\ x \end{pmatrix}.$$

Multiplying out the matrices gives
$$\begin{pmatrix} t' \\ x' \end{pmatrix} = \gamma \begin{pmatrix} 1 & w \\ w & 1 \end{pmatrix} \begin{pmatrix} t \\ x \end{pmatrix} \tag{11.1}$$

where
$$\gamma := \frac{r + r^{-1}}{2} \tag{11.2}$$

$$w := \frac{r - r^{-1}}{r + r^{-1}}. \tag{11.3}$$

The parameter w is called the "velocity" and is, of course, restricted by
$$|w| < 1. \tag{11.4}$$

11.1. TWO DIMENSIONAL LORENTZ TRANSFORMATIONS.

w determines γ.

We have

$$1 - w^2 = \frac{r^2 + 2 + r^{-2} - r^2 + 2 - r^{-2}}{(r + r^{-1})^2}$$

$$= \frac{4}{(r + r^{-1})^2}$$

so

$$\gamma = \frac{1}{\sqrt{1 - w^2}}. \tag{11.5}$$

Thus w determines γ.

w determines r.

Similarly, we can recover r from w:

$$r = \sqrt{\frac{1 + w}{1 - w}}.$$

So we can use w to parameterize the Lorentz transformations. We write

$$L_w := \gamma \begin{pmatrix} 1 & w \\ w & 1 \end{pmatrix}.$$

11.1.2 Addition law for velocities.

It is useful to express the multiplication law in terms of the velocity parameter. If

$$w_1 = \frac{r - r^{-1}}{r + r^{-1}}$$

$$w_2 = \frac{s - s^{-1}}{s + s^{-1}}$$

then

$$\frac{rs - (rs)^{-1}}{rs + (rs)^{-1}} = \frac{\frac{r-r^{-1}}{r+r^{-1}} + \frac{s-s^{-1}}{s+s^{-1}}}{1 + \frac{s-s^{-1}}{s+s^{-1}} \cdot \frac{r-r^{-1}}{r+r^{-1}}}$$

so we obtain

$$L_{w_1} \circ L_{w_2} = L_w \quad \text{where } w = \frac{w_1 + w_2}{1 + w_1 w_2}. \tag{11.6}$$

This is know as the "addition law for velocities".

11.1.3 Hyperbolic angle aka "rapidity".

One also introduces the "hyperbolic angle", actually a real number, ϕ by

$$r = e^\phi$$

so

$$\gamma = \cosh \phi = \frac{1}{\sqrt{1-w^2}}$$

and

$$L_w = \begin{pmatrix} \cosh \phi & \sinh \phi \\ \sinh \phi & \cosh \phi \end{pmatrix}.$$

Here

$$w = \tanh \phi.$$

Notice that the hyperbolic angle is additive: If we write $L(\phi) = L_w, w = \tanh \phi$ then $L(\phi_1) \circ L(\phi_2) = L(\phi_1 + \phi_2)$.

For any $\begin{pmatrix} t \\ x \end{pmatrix}$ with $t > 0$ and $t^2 - x^2 = 1$, we must have $t > x$ and $t - x = (t+x)^{-1}$ so

$$\begin{pmatrix} t+x \\ t-x \end{pmatrix} = \begin{pmatrix} r & 0 \\ 0 & r^{-1} \end{pmatrix} \begin{pmatrix} 1 \\ 1 \end{pmatrix} \qquad r = t+x.$$

This shows that the group of all (one dimensional proper) Lorentz transformations, $\{L_w\}$, acts simply transitively on the hyperbola

$$\left\| \begin{pmatrix} t \\ x \end{pmatrix} \right\|^2 = 1, \quad t > 0.$$

This means that if $\begin{pmatrix} t \\ x \end{pmatrix}$ and $\begin{pmatrix} t' \\ x' \end{pmatrix}$ are two points on this hyperbola, there is a unique L_w with

$$L_w \begin{pmatrix} t \\ x \end{pmatrix} = \begin{pmatrix} t' \\ x' \end{pmatrix}.$$

If

$$\begin{pmatrix} t \\ x \end{pmatrix} = L_z \begin{pmatrix} 1 \\ 0 \end{pmatrix}$$

this means that

$$\begin{pmatrix} t' \\ x' \end{pmatrix} = L_w L_z \begin{pmatrix} 1 \\ 0 \end{pmatrix} = L_z L_w \begin{pmatrix} 1 \\ 0 \end{pmatrix}$$

and so

$$\left\langle \begin{pmatrix} t \\ x \end{pmatrix}, \begin{pmatrix} t' \\ x' \end{pmatrix} \right\rangle = tt' - xx' = \left\langle \begin{pmatrix} 1 \\ 0 \end{pmatrix}, L_w \begin{pmatrix} 1 \\ 0 \end{pmatrix} \right\rangle.$$

11.1. TWO DIMENSIONAL LORENTZ TRANSFORMATIONS.

Writing $w = \tanh \phi$ as above we have

$$\langle \mathbf{u}, \mathbf{u}' \rangle = \cosh \phi, \quad \mathbf{u} = \begin{pmatrix} t \\ x \end{pmatrix} \mathbf{u}' = \begin{pmatrix} t' \\ x' \end{pmatrix},$$

and ϕ is called the **hyperbolic angle** between $\mathbf{u}$ and $\mathbf{u}'$.

More generally, if we don't require $||\mathbf{u}|| = ||\mathbf{u}'|| = 1$ but merely $||\mathbf{u}|| > 0$, $||\mathbf{u}'|| > 0, t > 0, t' > 0$ we define the hyperbolic angle between them to be the hyperbolic angle between the corresponding unit vectors so

$$\langle \mathbf{u}, \mathbf{u}' \rangle = ||\mathbf{u}|| \, ||\mathbf{u}'|| \cosh \phi.$$

11.1.4 Proper time.

Material particles and proper time.

A **material particle** is a curve $\alpha : \tau \mapsto \alpha(\tau)$ whose tangent vector $\alpha'(\tau)$ has positive t coordinate everywhere and satisfies

$$||\alpha'(\tau)|| \equiv 1.$$

Of course, this fixes the parameter τ up to an additive constant. τ is called the **proper time** of the material particle. It is to be thought of as as the "internal clock" of the material particle. For an unstable particle, for example, it is this internal clock which tells the particle that its time is up. Let ∂_0 denote unit vector in the t direction,

$$\partial_0 := \begin{pmatrix} 1 \\ 0 \end{pmatrix}.$$

Let us write $t(\tau)$ for the t coordinate of $\alpha(\tau)$ and $x(\tau)$ for its x coordinate so that

11.1.5 Time dilation.

$$\alpha(\tau) = \begin{pmatrix} t(\tau) \\ x(\tau) \end{pmatrix} \quad \alpha' := \frac{d\alpha}{d\tau} = \begin{pmatrix} dt/d\tau \\ dx/d\tau \end{pmatrix}, \quad \text{so}$$

$$\begin{aligned} \frac{dt}{d\tau} &= \langle \partial_0, \alpha' \rangle, \\ &= \cosh \phi \\ &= \frac{1}{\sqrt{1-w^2}} \geq 1, \end{aligned}$$

where

$$w := \frac{dx}{dt} = \frac{dx/d\tau}{dt/d\tau} \tag{11.7}$$

is the "velocity" of the particle measured in the t, x coordinate system.

Time dilation in elementary particle physics.

Thus the internal clock of a moving particle appears to run slow in any coordinate system where it is not at rest. This phenomenon, known as "time dilatation" is observed all the time in elementary particle physics. For example, fast moving muons make it from the upper atmosphere to the ground before decaying due to this effect.

11.1.6 The Lorentz-Fitzgerald contraction.

Simultaneity in a frame.

Let α and β be material particles whose trajectories are parallel straight lines. Once we have chosen a Minkowski basis, we have a notion of "simultaneity" relative to that basis, meaning that we can adjust the arbitrary additive constant in the definition of the proper time of each particle so that the two parallel straight lines are given by

$$\tau \mapsto \begin{pmatrix} a\tau \\ b\tau + c \end{pmatrix}, \text{ and } \tau \mapsto \begin{pmatrix} a\tau \\ b\tau + c + \ell \end{pmatrix}.$$

We can then think of the configuration as the motion of the end points of a "rigid rod" of length ℓ. The length ℓ depends on our notion of simultaneity.

The Lorentz-Fitzgerald "contraction".

For example, suppose we apply a Lorentz transformation L_w to obtain $a = 1, b = 0$ (and readjust the additive constants in the clocks to achieve simultaneity). The corresponding frame is called the rest frame of the rod and the its length, ℓ_{rest}, called the rest length of the rod is related to our "laboratory frame" by

$$\ell_{\text{rest}} = (\cosh \phi) \, \ell_{\text{lab}}$$

or

$$\ell_{\text{lab}} = \sqrt{1 - w^2} \ell_{\text{rest}}, \tag{11.8}$$

a moving object "contracts" in the direction of its motion.

This is the Lorentz-Fitzgerald contraction which was discovered before special relativity in the context of electromagnetic theory, and can be considered as a forerunner of special relativity. As an effect in the laboratory, it is not nearly as important as time dilatation.

11.1.7 The reverse triangle inequality.

Consider any interval, say $[0, T]$, on the t axis, and let $0 < s < T$. The curve $t^2 - x^2 = s^2$ bends away from the origin. In other words, all other vectors with t coordinate equal to s have smaller Minkowski length:

$$\left\| \begin{pmatrix} s \\ x \end{pmatrix} \right\|^2 < s^2, \quad x \neq 0.$$

11.1. TWO DIMENSIONAL LORENTZ TRANSFORMATIONS.

The length of any timelike vector $\mathbf{u} := \begin{pmatrix} s \\ x \end{pmatrix}$ is $< s$ if $x \neq 0$. Similarly, the Minkowski length of the (timelike) vector, $\mathbf{v}$, joining $\begin{pmatrix} s \\ x \end{pmatrix}$ to $\begin{pmatrix} T \\ 0 \end{pmatrix}$ is $< T - s$. We conclude that

$$||\mathbf{u} + \mathbf{v}|| \geq ||\mathbf{u}|| + ||\mathbf{v}|| \qquad (11.9)$$

with equality holding only if $\mathbf{u}$ and $\mathbf{v}$ actually lie on the t axis. There is nothing special in this argument about the t axis, or the fact that we are in two dimensions. It holds for any pair of forward timelike vectors, with equality holding if and only if the vectors are collinear. Inequality (11.9) is known as the **reverse triangle inequality**.

The classical way of putting this is to say that the time measured by a clock moving along a (timelike) straight line path joining the events P and Q is *longer* than the time measured along any (timelike forward) broken path joining P to Q. It is also called the "twin effect". The twin moving along the broken path (if he survives the bumps) will be younger than the twin who moves along the uniform path. This was known as the *twin paradox*. It is no paradox, just an immediate corollary of the reverse triangle inequality.

11.1.8 Physical significance of the Minkowski distance.

We wish to give an interpretation of the Minkowski square length (due originally to Robb (1936)) in terms of signals and clocks. Consider points $\begin{pmatrix} t_1 \\ 0 \end{pmatrix}$ and $\begin{pmatrix} t_2 \\ 0 \end{pmatrix}$ on the t axis which are joined to the point $\begin{pmatrix} t \\ x \end{pmatrix}$ by light rays (lines parallel to $t = x$ or $t = -x$). Then (assuming $t_2 > t > t_1$)

$$\begin{aligned} t - t_1 &= x \quad \text{so} \\ t_1 &= t - x \quad \text{and} \\ t_2 - t &= x \quad \text{so} \\ t_2 &= t + x \end{aligned}$$

hence

$$t_1 t_2 = t^2 - x^2. \qquad (11.10)$$

This equation has the following significance: Point $P = \begin{pmatrix} 0 \\ 0 \end{pmatrix}$ at rest or in uniform motion wishes to communicate with point $Q = \begin{pmatrix} t \\ x \end{pmatrix}$. It records the time, t_1 on its clock when a light signal was sent to Q and the time t_2 when the answer was received (assuming an instantaneous response.) Even though the individual times depend on the coordinates, their product, $t_1 t_2$ gives the square of the Minkowski norm of the vector joining P to Q.

Robb's book *Geometry of Time and Space*, Cambridge University Press (1936) is available online.

11.1.9 Energy-momentum

In classical mechanics, a momentum vector is usually considered to be an element of the cotangent space, i.e the dual space to the tangent space. Thus in our situation, where we identify all tangent spaces with the Minkowski plane itself, a "momentum" vector will be a row vector of the form $\mu = (E, p)$. For a material particle the associated momentum vector, called the "energy momentum vector" in special relativity, is a row vector with the property that the evaluation map

$$v \mapsto \mu(v)$$

for any vector v is a positive multiple of the scalar product evaluation

$$v \mapsto \langle v, \alpha'(\tau) \rangle.$$

Rest mass.

In other words, evaluation under μ is the same as scalar product with $m\alpha'$ where m, is an invariant of the material particle known as the **rest mass**. The rest mass is an invariant of the particle in question, constant throughout its motion. So in the rest frame of the particle, where $\alpha' = \partial_0$, the energy momentum vector has the form $(m, 0)$. Here m is identified (up to a choice of units, and we will have more to say about units later) with the usual notion of mass, as determined by collision experiments, for example. In a general frame we will have

$$\mu = (E, p), \quad E^2 - p^2 = m^2. \tag{11.11}$$

In this frame we have

$$p/E = w \tag{11.12}$$

where w is the velocity as defined in (11.7). We can solve equations (11.11) and (11.12) to obtain

$$E = \frac{m}{\sqrt{1 - w^2}} \tag{11.13}$$

$$p = \frac{mw}{\sqrt{1 - w^2}}. \tag{11.14}$$

For small values of w we have the Taylor expansion

$$\frac{1}{\sqrt{1 - w^2}} = 1 + \frac{1}{2}w^2 + \cdots$$

and so we have

$$E \doteq m + \frac{1}{2}mw^2 + \cdots \tag{11.15}$$

$$p \doteq mw + \frac{1}{2}mw^3 + \cdots. \tag{11.16}$$

11.1. TWO DIMENSIONAL LORENTZ TRANSFORMATIONS.

The first term in (11.16) looks like the classical expression $p = mw$ for the momentum in terms of the velocity if we think of m as the classical mass, and the second term in (11.15) looks like the classical expression for the kinetic energy.

We are thus led to the following modification of the classical definitions of energy and momentum. Associated to any object there is a definite value of m called its rest mass. If the object is at rest in a given frame, its rest mass coincides with the classical notion of mass; when it is in motion relative to a given frame, its energy momentum vector is of the form (E, p) where E and p are determined by equations (11.13) and (11.14). We have been implicitly assuming that $m > 0$ which implies that $|w| < 1$.

Conservation of energy.

We can supplement these particles by particles of rest mass 0 whose energy momentum vector satisfy (11.11), so have the form $(E, \pm E)$. These correspond to particles which move along light rays $x = \pm t$. The law of conservation of energy momentum says that in any collision the total energy momentum vector is conserved.

11.1.10 Psychological units.

Our description of two dimensional Minkowski geometry has been in terms of "natural units" where the speed of light is one. Points in our two dimensional space time are called *events*. They record when and where something happens. If we record the total events of a single human consciousness (say roughly 70 years measured in seconds) and several thousand meters measured in seconds, we get a set of events which is enormously stretched out in one particular time direction compared to space direction, by a factor of something like 10^{18}. Being very skinny in the space direction as opposed to the time direction we tend to have a preferred splitting of spacetime with space and time directions picked out, and to measure distances in space with much smaller units, such as meters, than the units we use (such as seconds) to measure time.

Of course, if we use a small unit, the corresponding numerical value of the measurement will be large; in terms of human or "ordinary units" space distances will be greatly magnified in comparison with time differences. This suggests that we consider variables T and X related to the natural units t and x by $T = c^{-1}t$, $X = x$ or

$$\begin{pmatrix} T \\ X \end{pmatrix} = \begin{pmatrix} c^{-1} & 0 \\ 0 & 1 \end{pmatrix} \begin{pmatrix} t \\ x \end{pmatrix}.$$

The light cone $|t| = |x|$ goes over to $|X| = c|T|$ and we say that the "speed of light is c in ordinary units". Similarly, the time-like hyperbolas $t^2 - x^2 = k > 0$ become very flattened out and are almost the vertical lines $T =$ const., lines of "simultaneity".

Lorentz transformations in psychological units.

To find the expression for the Lorentz transformations in ordinary units, we must conjugate the Lorentz transformation, L, by the matrix $\begin{pmatrix} c^{-1} & 0 \\ 0 & 1 \end{pmatrix}$ so

$$\begin{aligned} M &= \begin{pmatrix} c^{-1} & 0 \\ 0 & 1 \end{pmatrix} L \begin{pmatrix} c & 0 \\ 0 & 1 \end{pmatrix} \\ &= \begin{pmatrix} \cosh\phi & c^{-1}\sinh\phi \\ c\sinh\phi & \cosh\phi \end{pmatrix} \\ &= \gamma \begin{pmatrix} 1 & c^{-1}w \\ cw & 1 \end{pmatrix}, \end{aligned}$$

where $L = L_w$. Of course w is a pure number in natural units. In psychological units we must write $w = v/c$, the ratio of a velocity (in units like meters per second) to the speed of light.

Then

$$M = M_v = \gamma \begin{pmatrix} 1 & \frac{v}{c^2} \\ v & 1 \end{pmatrix}, \quad \gamma = \frac{1}{(1 - \frac{v^2}{c^2})^{1/2}}. \qquad (11.17)$$

Since we have passed to new coordinates in which

$$\left\| \begin{pmatrix} T \\ X \end{pmatrix} \right\|^2 = c^2 T^2 - X^2,$$

the corresponding metric in the dual space will have the energy component divided by c. As we have used cap for energy and lower case for momentum, we shall continue to denote the energy momentum vector in psychological units by (E, p) and we have

$$\|(E,p)\|^2 = \frac{E^2}{c^2} - p^2.$$

We still must see how these units relate to our conventional units of mass.

For this, observe that we want the second term in

$$E \doteq m + \frac{1}{2}mw^2 + \cdots \qquad (11.15)$$

to look like kinetic energy when E is replaced by E/c, so we must rescale by $m \mapsto mc$. Thus we get

$$\|(E,p)\|^2 = \frac{E^2}{c^2} - p^2 = m^2 c^2. \qquad (11.18)$$

11.2. MINKOWSKI SPACE.

$E = mc^2$.

So in psychological coordinates we rewrite (11.11)-(11.15) as (11.18) together with

$$\frac{p}{E} = \frac{v}{c^2} \tag{11.19}$$

$$E = \frac{mc^2}{(1 - v^2/c^2)^{1/2}} \tag{11.20}$$

$$p = \frac{mv}{(1 - v^2/c^2)^{1/2}} \tag{11.21}$$

$$E \doteq mc^2 + \frac{1}{2}mv^2 + \cdots \tag{11.22}$$

$$p \doteq mv + \frac{1}{2}m\frac{v^3}{c^2} + \cdots. \tag{11.23}$$

Of course at velocity zero we get the famous Einstein formula $E = mc^2$.

11.1.11 The Galilean limit.

In "the limit" $c \to \infty$ the transformations M_v become

$$G_v = \begin{pmatrix} 1 & 0 \\ v & 1 \end{pmatrix}$$

which preserve T and send $X \mapsto X + vT$. These are known as Galilean transformations. They satisfy the more familiar addition rule for velocities:

$$G_{v_1} \circ G_{v_2} = G_{v_1+v_2}.$$

11.2 Minkowski space.

Since our everyday space is three dimensional, the correct space for special relativity is a four dimensional Lorentzian vector space. This key idea is due to Minkowski. In a famous lecture at Cologne in September 1908 he says

> Henceforth space by itself, and time by itself are doomed to fade away into mere shadows, and only a kind of union of the two will preserve an independent reality.

Hermann Minkowski

**Born: 22 June 1864 in Alexotas, Russian Empire
(now Kaunas, Lithuania)
Died: 12 Jan. 1909 in Gottingen, Germany**

Much of what we did in the two dimensional case goes over unchanged to four dimensions. Of course, velocity, w or v, become vectors, $\mathbf{w}$ and $\mathbf{v}$ as does momentum, $\mathbf{p}$ instead of p. So in any expression a term such as v^2 must be replaced by $||\mathbf{v}||^2$, the three dimensional norm squared, etc.. With this modification the key formulas of the preceding section go through. We will not rewrite them. The reverse triangle inequality and so the twin effect go through unchanged.

Of course there are important differences: the light cone is really a cone, and not two light rays, the space-like vectors form a connected set, the Lorentz group is six dimensional instead of one dimensional. We will study the Lorentz group in four dimensions in a later section. In this section we will concentrate on two-particle collisions, where the relative angle between the momenta gives an additional ingredient in four dimensions.

11.2.1 The Compton effect.

We consider a photon (a "particle" of mass zero) impinging on a massive particle (say an electron) at rest. After the collision the the photon moves at an angle,

11.2. MINKOWSKI SPACE.

θ, to its original path. The frequency of the light is changed as a function of the angle: If λ is the incoming wave length and λ' the wave length of the scattered light then

$$\lambda' = \lambda + \frac{h}{mc}(1 - \cos\theta), \tag{11.24}$$

where h is Planck's constant and m is the mass of the target particle. The expression

$$\frac{h}{mc}$$

is known as the **Compton wave length** of a particle of mass m.

Compton's derivation of the Compton effect.

Compton derived (11.24) from the conservation of energy momentum as follows: We will work in natural units where $c = 1$. Assume Einstein's formula

$$E_{\text{photon}} = h\nu \tag{11.25}$$

for the energy of the photon, where ν is the frequency, or equivalently,

$$E_{\text{photon}} = \frac{h}{\lambda} \tag{11.26}$$

where λ is the wave length. Work in the rest frame of the target particle, so its energy momentum vector is $(m, 0, 0, 0)$.

Take the x-axis to be the direction of the incoming photon, so its energy momentum vector is $(\frac{h}{\lambda}, \frac{h}{\lambda}, 0, 0)$. Assume that the collision is *elastic* so that the outgoing photon still has mass zero and the recoiling particle still has mass m. Choose the y-axis so that the outgoing photon and the recoiling particle move in the x, y plane. Then the outgoing photon has energy momentum $(\frac{h}{\lambda'}, \frac{h}{\lambda'}\cos\theta, \frac{h}{\lambda'}\sin\theta, 0)$ while the recoiling particle has energy momentum $(E, p_x, p_y, 0)$ and conservation of energy momentum together with the assumed elasticity of the collision yield

$$\frac{h}{\lambda} + m = \frac{h}{\lambda'} + E$$
$$\frac{h}{\lambda} = \frac{h}{\lambda'}\cos\theta + p_x$$
$$0 = \frac{h}{\lambda'}\sin\theta + p_y$$
$$m^2 = E^2 - p_x^2 - p_y^2.$$

Substituting the second and third equations into the last gives

$$E^2 = m^2 + \frac{h^2}{\lambda^2} + \frac{h^2}{\lambda'^2} - 2\frac{h^2}{\lambda\lambda'}\cos\theta$$

while the first equation yields

$$E^2 = m^2 + \frac{h^2}{\lambda^2} + \frac{h^2}{\lambda'^2} + 2\left[m\frac{h}{\lambda} - m\frac{h}{\lambda'} - \frac{h^2}{\lambda\lambda'}\right].$$

Comparing these two equations gives Compton's formula, (11.24).

Arthur Holly Compton

Time magazine cover - January 13, 1936

Born: September 10, 1892 in Wooster, Ohio, USA
Died: March 15, 1962 in Berkeley, California

Three startling predictions from Compton's formula.

Notice that Compton's formula makes three startling predictions: that the shift in wavelength is independent of the wavelength of the incoming radiation, the explicit nature of the dependence of this shift on the scattering angle, and an experimental determination of h/mc, in particular, if h and c are known, of the mass, m, of the scattering particle. These were the results of Compton's formula and experiments.

Some history.

It is worth recalling the historical importance of Compton's experiments (1923). At the end of the nineteenth century, statistical mechanics, which had been

11.2. MINKOWSKI SPACE.

enormously successful in explaining many aspects of thermodynamics, yielded wrong, and even non-sensical, predictions when it came to the study of the electromagnetic radiation emitted by a hot body - the study of "blackbody radiation". In 1900 Planck showed that the paradoxes could be resolved and a an excellent fit to the experimental data achieved if one assumed that the electromagnetic radiation is emitted in packets of energy given by (11.25) where h is a constant, now called Planck's constant, with value

$$h = 6.26 \times 10^{-27} \text{erg s}.$$

For Planck, this quantization of the energy of radiation was a property of the emission process in blackbody radiation. In 1905 Einstein proposed the radical view that (11.25) was a property of the electromagnetic field itself, and not of any particular emission process. Light, according to Einstein, is quantized according to (11.25). He used this to explain the *photoelectric effect*: When light strikes a metallic surface, electrons are emitted. According to Einstein, an incoming light quantum of energy $h\nu$ strikes an electron in the metal, giving up all its energy to the electron, which then uses up a certain amount of energy, w, to escape from the surface. The electron may also use up some energy to reach the surface. In any event, the escaping electron has energy

$$E \leq h\nu - w$$

where w is an empirical property of the material, called the work function.

The startling consequence here is that the maximum energy of the emitted electron depends only on the frequency of the radiation, but not on the intensity of the light beam. Increasing the intensity will increase the number of electrons emitted, but not their maximum energy. Einstein's theory was rejected by the entire physics community. With the temporary exception of Stark (who later became a vicious nazi and attacked the theory of relativity as a Jewish plot) physicists could not accept the idea of a corpuscular nature to light, for this seemed to contradict the well established interference phenomena which implied a wave theory, and also contradicted Maxwell's equations, which were the cornerstone of all of theoretical physics. For a typical view, let us quote at length from Millikan (of oil drop fame) whose experimental result gave the best confirmation of Einstein's predictions for the photoelectric effect. In his Nobel lecture (1924) he writes

> After ten years of testing and changing and learning and sometimes blundering, all efforts being directed from the first toward the accurate experimental measurement of the energies of emission of photoelectrons, now as a function of temperature, now of wavelength,now of material (contact e.m.f. relations), this work resulted, contrary to my own expectation, in the first direct experimental proof in 1914 of the exact validity, within narrow limits of experimental error, of the Einstein equation, and the first direct photoelectric determination of Planck's h.

But despite Millikan's own experimental verification of Einstein's formula for the photoelectric effect, he did not regard this as confirmation of Einstein's theory of quantized radiation. On the contrary, in his paper, "A direct Photoelectric Determination of Planck's h" *Phy. Rev. 7 (1916)*355-388 where he presents his experimental results he writes:

> ... the semi-corpuscular theory by which Einstein arrived at his equation seems at present to wholly untenable....[Einstein's] bold, not to say reckless [hypothesis] seems a violation of the very conception of electromagnetic disturbance...[it] flies in the face of the thoroughly established facts of interference.... Despite... the apparently complete success of the Einstein equation, the physical theory of which it was designed to be the symbolic expression is found so untenable that Einstein himself, I believe, no longer holds to it, and we are in the position of having built a perfect structure and then knocked out entirely the underpinning without causing the building to fall. It stands complete and apparently well tested, but without any visible means of support. These supports must obviously exist, and the most fascinating problem of modern physics is to find them.
>
> Experiment has outrun theory, or, better, guided by an erroneous theory, it has discovered relationships which seem to be of the greatest interest and importance, but the reasons for them are as yet not at all understood.

Of course, Millikan was mistaken when he wrote that Einstein himself had abandoned his own theory. In fact, Einstein extended his theory in 1916 to include the quantization of the momentum of the photon. But for Millikan, as for most physicists, Einstein's hypothesis of the light quantum was clearly "an erroneous theory".

By the way, it is amusing to compare Millikan's actual state of mind in 1916 (which was the accepted view of the entire physics community outside of Einstein) with his fallacious account of it in his autobiography (1950) pp. 100-101, where he writes about his experimental verification of Einstein's equation for the photoelectric effect:

> This seemed to me, as it did to many others, a matter of very great importance, for it rendered what I will call Planck's 1912 explosive or trigger approach to the problem of quanta completely untenable and proved simply and irrefutedly, I thought, *that the emitted electron that escapes with the energy $h\nu$ gets that energy by the direct transfer of $h\nu$ units of energy from the light to the electron* and hence scarcely permits of any interpretation than that which Einstein had originally suggested, namely that of the semi-corpuscular or photon theory of light itself.

11.2. MINKOWSKI SPACE.

Self-delusion or outright mendacity? In general I have found that one can not trust the accounts given by scientists of their own thought processes, especially those given many years after the events.

In any event, it was only with the Compton experiment, that Einstein's formula, (11.25) was accepted as a property of light itself.

For a detailed history see the book *The Compton Effect* by Roger H. Stuewer, Science History Publications, New York 1975, from which I have taken the above quotes.

11.2.2 Natural Units.

In this section I will make the paradoxical argument that Planck's constant and (11.25) have a purely classical interpretation: Like c, Planck's constant, h, may be viewed as a conversion factor from natural units to conventional units.

For this I will again briefly call on a higher theory, symplectic geometry. In that theory, conserved quantities are associated to continuous symmetries. More precisely, if G is a Lie group of symmetries with Lie algebra g, the moment map, Φ for a Hamiltonian action takes values in g^*, the *dual space* of the Lie algebra. A basis of g determines a dual basis of g^*. In the case at hand, the Lie algebra in question is the algebra of translations, and the moment map yields the (total) energy-momentum vector. Hence if we measure translations in units of length, then the corresponding units for energy momentum should be inverse length. In this sense the role of Planck's constant in (11.26) is a conversion factor from natural units of inverse length to the conventional units of energy. So we interpret $h = 6.626 \times 10^{-27}$ erg s as the conversion factor from the natural units of inverse seconds to the conventional units of ergs.

In order to emphasize this point, let us engage in some historical science fiction: Suppose that mechanics had developed before the invention of clocks. So we could observe trajectories of particles, their collisions and deflections, but not their velocities. For instance, we might be able to observe tracks in a bubble chamber or on a photographic plate. If our theory is invariant under the group of translations in space, then linear momentum would be an invariant of the particle; if our theory is invariant under the group of three dimensional Euclidean motions, the symplectic geometry tells that $||\mathbf{p}||$, the length of the linear momentum is an invariant of the particle. In the absence of a notion of velocity, we might not be able to distinguish between a heavy particle moving slowly or a light particle moving fast.

Without some way of relating momentum to length, we would introduce "independent units" of momentum, perhaps by combining particles in various ways and by performing collision experiments. But symplectic geometry tells us that the "natural" units of momentum should be inverse length, and that de Broglie's equation

$$||\mathbf{p}|| = \frac{h}{\lambda} \qquad (11.27)$$

gives Planck's constant as a conversion factor from natural units to conventional units. In fact, the crucial experiment was the photo-electric effect, carried out

CHAPTER 11. REVIEW OF SPECIAL RELATIVITY.

in detail by Millikan.

The above discussion does not diminish, even in retrospect, from the radical character of Einstein's 1905 proposal. Even in terms of "natural units" the startling proposal is that it is a single particle, the photon, which interacts with a single particle, the electron to produce the photoelectric effect. It is this "corpuscular" picture which was so difficult to accept. Furthermore, it is a bold hypothesis to identify the "natural units" of the photon momentum with the inverse wave length.

For reasons of convenience physicists frequently prefer to use $\hbar := h/2\pi$ as the conversion factor.

One way of choosing natural units is to pick some particular particle and use its rest mass as the mass unit. Suppose we pick the proton. Then m_P, the mass of the proton is the basic unit of mass, and ℓ_P, the Compton wave length of the proton is the basic unit of length. Also t_P, the time it takes for light to travel the distance of one Compton wave length, is the basic unit of time. These are known as the Planck mass, Planck length and Planck time.

The conversion factors to the cgs system (using $\hbar$) are:

$$\begin{aligned} m_P &= 1.672 \times 10^{-24} g \\ \ell_P &= .211 \times 10^{-13} cm \\ t_P &= 0.07 \times 10^{-23} sec. \end{aligned}$$

We will oscillate between using natural units and familiar units. Usually, we will derive the formulas we want in natural units, where the computations are cleaner and then state the results in conventional units which are used in the laboratory.

Chapter 12

The star operator and electromagnetism.

12.1 Definition of the star operator.

12.1.1 The induced scalar product on exterior powers.

We start with a finite dimensional vector space V over the real numbers which carries two additional pieces of structure: an orientation and a non-degenerate scalar product. The scalar product, $\langle\ ,\ \rangle$ determines a scalar product on each of the spaces $\wedge^k V$ which is fixed by the requirement that it take on the values

$$\langle x_1 \wedge \cdots \wedge x_k, y_1 \wedge \cdots \wedge y_k \rangle = \det\left(\langle x_i, y_j \rangle\right)$$

on decomposable elements. This scalar product is non-degenerate. Indeed, starting from an "orthonormal" basis $e_1, \ldots, e_n$ of V, the basis $e_{i_1} \wedge \cdots \wedge e_{i_k}$, $i_1 < \cdots < i_k$ is an "orthonormal" basis of $\wedge^k$ where

$$\langle e_{i_1} \wedge \cdots \wedge e_{i_k}, e_{i_1} \wedge \cdots \wedge e_{i_k} \rangle = (-1)^r$$

where r is the number of the i_j with $\langle e_{i_j}, e_{i_j} \rangle = -1$. In particular,

If the scalar product on V is positive definite, then so is the induced metric on each of the $\wedge^k V$.

The induced scalar product on $\wedge^2$ of Minkowski space has signature (3,3).

Suppose, like in our universe, V is four dimensional with signature $(1,3)$. So we have an "orthonormal" basis e_0, e_1, e_2, e_3 with $\langle e_0, e_0 \rangle = 1$ and $\langle e_1, e_1 \rangle = \langle e_2, e_2 \rangle = \langle e_3, e_3 \rangle = -1$. On $\wedge^2(V)$ we have the basis consisting of $e_0 \wedge e_i, i = 1, 2, 3$ and $e_i \wedge e_j$ with $1 \leq i < j \leq 3$. The first three satisfy

$$\langle e_0 \wedge e_i, e_0 \wedge e_i \rangle = -1$$

249

and the last three satisfy

$$\langle e_i \wedge e_j, e_i \wedge e_j \rangle = 1.$$

So the induced scalar product on the six dimensional space $\wedge^2(V)$ has signature (3,3).

The star operator

Back to the general case: There are exactly two elements in the one dimensional space $\wedge^n V$, $n = \dim V$ which satisfy

$$\langle v, v \rangle = \pm 1.$$

Here the ± 1 is determined by the signature (p, q) (p pluses and q minuses) of the scalar product:

$$\langle v, v \rangle = (-1)^q.$$

An orientation of a vector space amounts to choosing one of the two half lines (rays) of non-zero elements in $\wedge^n V$. Hence for an oriented vector space with non-degenerate scalar product there is a well defined unique basis element

$$v \in \wedge^n V \quad \langle v, v \rangle = (-1)^q.$$

Wedge product always gives a bilinear map from $\wedge^k V \times \wedge^{n-k} V \to \wedge^n V$ But now we have a distinguished basis element for the one dimensional space, $\wedge^n V$. The wedge product allows us to assign to each element of $\lambda \in \wedge^k V$ the linear function, ℓ_λ on $\wedge^{n-k} V$ given by

$$\lambda \wedge \omega = \ell_\lambda(\omega) v \quad \forall \, \omega \in \wedge^{n-k} V.$$

But since the induced scalar product on $\wedge^{n-k} V$ is non-degenerate, any linear function ℓ is given as $\ell(\omega) = \langle \tau, \omega \rangle$ for a unique $\tau = \tau(\ell)$.

So there is a unique element

$$\star \lambda \in \wedge^{n-k} V$$

determined by

$$\lambda \wedge \omega = \langle \star \lambda, \omega \rangle v \quad \forall \, \omega \in \wedge^{n-k} V. \tag{12.1}$$

This is our convention with regard to the star operator. In short, we have defined a linear map

$$\star : \wedge^k V \to \wedge^{n-k} V$$

for each $0 \leq k \leq n$ which is determined by (12.1).

Let us choose an "orthonormal" basis of V as above, but being sure to choose our "orthonormal" basis to be oriented, which means that

$$v = e_1 \wedge \cdots e_n.$$

12.1. DEFINITION OF THE STAR OPERATOR.

Let $I = (i_1, \ldots, i_k)$ be a $k-$ subset of $\{1, \ldots, n\}$ with its elements arranged in order, $i_1 < \cdots < i_k$ so that the

$$e_I := e_{i_1} \wedge \cdots \wedge e_{i_k}$$

form an "orthonormal" basis of $\wedge^k V$. Let I^c denote the complementary set of $I \subset \{1, \ldots, n\}$ with its elements arranged in increasing order. Thus e_{I^c} is one of the basis elements, $\{e_J\}$ where J ranges over all $(n-k)$ subsets of $\{1, \ldots, n\}$.

We have

$$e_I \wedge e_J = 0 \quad \text{if } J \neq I^c$$

while

$$e_I \wedge e_{I^c} = (-1)^\pi v$$

where $(-1)^\pi$ is the sign of the permutation required to bring the entries in $e_I \wedge e_{I^c}$ back to increasing order. Thus

$$\star e_I = (-1)^{\pi + r(I^c)} e_{I^c} \tag{12.2}$$

where $(-1)^{\pi+r} := (-1)^\pi (-1)^r$ and

$$r(J) \text{ is the number of } j \in J \text{ with } \langle e_j, e_j \rangle = -1,$$

i.e.

$$(-1)^{r(J)} = \langle e_J, e_J \rangle. \tag{12.3}$$

We should make explicit the general definition of the star operator for the extreme cases $k = 0$ and $k = n$. We have $\wedge^0 V = \mathbb{R}$ for any vector space V, and the scalar product on $\mathbb{R}$ is the standard one assigning to each real number its square. Taking the number 1 as a basis for $\mathbb{R}$ thought of as a one dimensional vector space over itself, this means that $\langle 1, 1 \rangle = 1$. Wedge product by an element of $\wedge^0 V = \mathbb{R}$ is just ordinary multiplication by of a vector by a real number.

So,

$$v \wedge 1 = 1 \wedge v = v$$

and the definition

$$v \wedge 1 = \langle \star v, 1 \rangle v$$

requires that

$$\star v = 1 \tag{12.4}$$

no matter what the signature of the scalar product on V is.

On the other hand, $\star 1 = \pm v$. We determine the sign from

$$1 \wedge v = v = \langle \star 1, v \rangle v$$

so

$$\star 1 = \langle v, v \rangle v = (-1)^q v \tag{12.5}$$

in accordance with our general rule.

Applying $\star$ twice gives a linear map of $\wedge^k V$ into itself for each k. We claim that

$$\star^2 = (-1)^{k(n-k)+q} \text{ id}. \tag{12.6}$$

Proof. Since both sides are linear operators it suffices to verify this equation on basis elements, e.g. on elements of the form e_I, and by relabeling if necessary we may assume, without loss of generality, that $I = \{1, \ldots, k\}$. Then

$$\star(e_1 \wedge \cdots \wedge e_k) = (-1)^{r(I^c)} e_{k+1} \wedge \cdots \wedge e_n,$$

while

$$\star(e_{k+1} \wedge \cdots \wedge e_n) = (-1)^{k(n-k)+r(I)} e_1 \wedge \cdots \wedge e_k$$

since there are $n - k$ transpositions needed to bring each of the e_i, $i \leq k$, past $e_{k+1} \wedge \cdots \wedge e_n$. Since $r(I) + r(I^c) = q$, (12.6) follows. □

12.2 Does $\star : \wedge^k V \to \wedge^{n-k} V$ determine the metric?

The star operator depends on the metric and on the orientation. Clearly, changing the orientation changes the sign of the star operator.

Let us discuss the question of when the star operator determines the scalar product. We claim, as a preliminary, that it follows from the definition that

$$\lambda \wedge \star\omega = (-1)^q \langle \lambda, \omega \rangle v \quad \forall \, \lambda, \omega \in \wedge^k \tag{12.7}$$

for any $0 \leq k \leq n$.

Proof. We have really already verified this formula for the case $k = 0$ or $k = n$.

For any intermediate k, we observe that both sides are bilinear in λ and ω, so it suffices to verify this equation on basis elements, i.e when $\lambda = e_I$ and $\omega = e_K$ where I and K are k–subsets of $\{1, \ldots, n\}$. If $K \neq I$ then $\langle e_I, e_K \rangle = 0$, while K^c and I have at least one element in common, so $e_I \wedge \star e_K = 0$. Hence both sides equal zero. So we must only check the equation for $I = K$, and without loss of generality we may assume (by relabeling the indices) that $I = \{1, 2, \ldots, k\}$. Then the left hand side of (12.7) is

$$(-1)^{r(I^c)} v$$

while the right hand side is $(-1)^{q+r(I)} v$ by (12.2). Since $q = r(I) + r(I^c)$ the result follows. □

By the way, a useful consequence of (12.7) is

$$\lambda \wedge \star\omega = \omega \wedge \star\lambda, \tag{12.8}$$

since the right hand side of (12.7) is symmetric in λ and ω.

One might think that (12.7) implies that $\star$ acting on $\wedge^k V$, $k \neq 0, n$ determines the scalar product, but this is not quite true. Here is the simplest (and very important) counterexample. Take $V = \mathbb{R}^2$ with the standard positive definite scalar product and $k = 1$. So $\star : \wedge^1 V = V \to V$. In terms of an oriented

12.2. DOES $\star : \wedge^K V \to \wedge^{N-K} V$ DETERMINE THE METRIC?

orthonormal basis we have $\star e_1 = e_2, \star e_2 = -e_1$, thus $\star$ is (counterclockwise) rotation through ninety degrees. Any (non-zero) multiple of the standard scalar product will determine the same notion of angle, and hence the same $\star$ operator. Thus, in two dimensions, the $\star$ operator only determines the metric up to scale.

The reason for the breakdown in the argument is that the v occurring on the right hand side of (12.7) depends on the choice of metric. It is clear from (12.7) that the star operator acting on $\wedge^k V$ determines the induced scalar product on $\wedge^k V$ up to scale. Indeed, let $\langle\ ,\ \rangle'$ denote a second scalar product on V. Let v' denote the element of $\wedge^n V$ determined by the scalar product $\langle\ ,\ \rangle'$, so

$$v' = av$$

for some non-zero constant, $a > 0$. Finally, for purposes of the present argument, let us use more precise notation and denote the scalar products induced on $\wedge^k V$ by $\langle\ ,\ \rangle_k$ and $\langle\ ,\ \rangle'_k$. Then (12.7) implies that

$$\langle\ ,\ \rangle'_k = \frac{1}{a}\langle\ ,\ \rangle_k. \tag{12.9}$$

For example, suppose that we know that the original scalar products on V differ by a positive scalar factor, say

$$\langle\ ,\ \rangle' = c\langle\ ,\ \rangle, \quad c > 0.$$

Then

$$\langle\ ,\ \rangle'_k = c^k \langle\ ,\ \rangle$$

while

$$v' = \frac{1}{c^{n/2}} v$$

since $\langle v, v \rangle'_n = c^n \langle v, v \rangle$. Hence the fact that the star operators are the same on $\wedge^k V$ implies that $c = 1$ for any k other than $k = \frac{n}{2}$. This was exactly the point of breakdown in our two dimensional example where $n = 2, k = 1$.

In general, if $\langle\ ,\ \rangle$ is positive definite, and $\langle\ ,\ \rangle'$ is any other non-degenerate scalar product, then the principal axis theorem (the diagonalization theorem for symmetric matrices) from linear algebra says that we can find a basis $e_1, \ldots, e_n$ which is orthonormal for $\langle\ ,\ \rangle$ and orthogonal with respect to $\langle\ ,\ \rangle'$ with

$$\langle e_i, e_i \rangle' = s_i, \quad s_i \neq 0.$$

Then

$$\langle e_I, e_I \rangle' = s_{i_1} \cdots s_{i_k} \langle e_I, e_I \rangle, \quad I = \{i_1, \ldots, i_k\}.$$

The only way that (12.9) can hold for a given $0 < k < n$ is for all the s_i to be equal. Let s denote this common value of the s_i. Then $a = |s|^{-n/2}$ and we can conclude that $s = \pm 1$ if $k \neq n/2$ and in fact that $s = 1$ if, in addition, k is odd.

I don't know how to deal the case of a general (non-definite) scalar product in so straightforward a manner. Perhaps you can work this out. But let me deal with the case of importance to us, a Lorentzian metric on a four dimensional space, so a metric of signature $(1,3)$ or $(3,1)$. For $k = 1$, we know from the above discussion that the star operator determines the metric completely. The case $k = 3$ reduces to the case $k = 1$ since $\star^2 = (-1)^{3 \cdot 1 + 1}\text{id} = \text{id}$ in this degree. The only remaining case is $k = 2$, where we know that $\star$ only determines $\langle\,,\,\rangle_2$ up to a scalar. So the best we can hope for is that $\star : \wedge^2 V \to \wedge^2 V$ determines $\langle\,,\,\rangle$ up to a scalar multiple. This is true, but I won't go into the details of the proof. But here is a useful fact:

Proposition 30. *Let V be an oriented four dimensional vector space with a non-degenerate bilinear form of signature $(1,3)$. Then $\star : \wedge^2 V \to \wedge^2 V$ is self adjoint relative to $\langle\,,\,\rangle_2$, i.e.*

$$\langle \star\lambda, \omega \rangle = \langle \lambda, \star\omega \rangle \quad \forall\, \lambda, \omega \in \wedge^2 V. \quad (*)$$

Proof. Take $\lambda = e_0 \wedge e_1$ and $\omega = e_2 \wedge e_3$, so that

$$\begin{aligned}\star(e_0 \wedge e_1) &= e_2 \wedge e_3 \\ \star(e_2 \wedge e_3) &= -e_0 \wedge e_1 \quad \text{so}\end{aligned}$$

$$\langle \star\lambda, \omega \rangle = 1 = \langle \lambda, \star\omega \rangle \quad \forall\, \lambda, \omega \in \wedge^2 V,$$

for this choice of λ and ω. A moment's reflection will show that this is the typical case where the scalar products in $(*)$ applied to "orthonormal" basis elements are not zero. $\square$

12.3 The star operator on forms.

If M is an oriented semi-Riemannian manifold, we can consider the star operator associated to each cotangent space. Thus, operating pointwise, we get a star operator mapping k–forms into $(n-k)$forms, where $n = \dim M$:

$$\star : \Omega^k(M) \to \Omega^{n-k}(M).$$

Many of the important equations of physics have simple expressions in terms of the star operator on forms.

We now describe some of them. In fact, all of the equations we shall write down will be for various star operators of flat space of two, three and four dimensions. But the general formulation goes over unchanged for curved spaces or spacetimes.

12.3.1 Some equations of mathematical physics.

The star operator on $\mathbb{R}^2$.

We take as our orthonormal frame of forms to be dx, dy and the orientation two form to be $v := dx \wedge dy$. Then

$$\star dx = dy, \qquad \star dy = -dx$$

12.3. THE STAR OPERATOR ON FORMS.

as we have already seen. . .

For any pair of smooth real valued functions u and v, let

$$\omega := udx - vdy.$$

The Cauchy-Riemann equations.

Then $d\omega = -\left(\frac{\partial v}{\partial x} + \frac{\partial u}{\partial y}\right) dx \wedge dy$ and $\star\omega = udy + vdx$ so $d\star\omega = \left(\frac{\partial u}{\partial x} - \frac{\partial v}{\partial y}\right) dx \wedge dy$ so the equations

$$d \star \omega = 0, \quad d\omega = 0 \tag{12.10}$$

are the famous Cauchy-Riemann equations:

$$\frac{\partial u}{\partial x} = \frac{\partial v}{\partial y}$$

$$\frac{\partial v}{\partial x} = -\frac{\partial v}{\partial y}.$$

These equations say that the Jacobian matrix

$$\begin{pmatrix} \frac{\partial u}{\partial x} & \frac{\partial u}{\partial y} \\ \frac{\partial v}{\partial x} & \frac{\partial v}{\partial y} \end{pmatrix}$$

has the form

$$\begin{pmatrix} a & b \\ -b & a \end{pmatrix}.$$

If $r^2 := a^2 + b^2 \neq 0$ we can write $a = r\cos\phi$, $b = r\sin\phi$ for a suitable angle ϕ and we see that a matrix of the above form represents a rotation through angle ϕ followed by a dilatation by the factor r. In other words, at points where the Jacobian matrix does not vanish, the map $(x, y) \mapsto (u, v)$ is conformal.

Relation to Maxwell's equations.

We will find later on that Maxwell's equations in the absence of sources has exactly the same form as (12.10), except that for Maxwell's equations ω is a two form on Minkowski space instead of being a one form on the plane.

Relation to complex analysis.

If we allow complex valued forms, write $f = u + iv$ and $dz = dx + idy$ then the above pair of equations can be written as

$$d[fdz] = 0.$$

It then follows from Stokes' theorem that the integral of fdz around the boundary of any region where f is defined (and smooth) must be zero. This is known as the Cauchy integral theorem.

CHAPTER 12. THE STAR OPERATOR AND ELECTROMAGNETISM.

For $\mathbb{R}^3$.

We have the orthonormal coframe field dx, dy, dz, with $v = dx \wedge dy \wedge dz$, so $\star 1 = v$,

$$\begin{align} \star dx &= dy \wedge dz \\ \star dy &= -dx \wedge dz \\ \star dz &= dx \wedge dy \end{align}$$

with

$$\star^2 = 1$$

in all degrees. Let

$$\Delta := \frac{\partial^2}{\partial x^2} + \frac{\partial^2}{\partial y^2} + \frac{\partial^2}{\partial z^2}.$$

Let

$$\begin{align} \theta &= a\,dx + b\,dy + c\,dz \\ \Omega &= A\,dx \wedge dy + B\,dx \wedge dz + C\,dy \wedge dz. \end{align}$$

I will use subscripts to denote partial derivatives so $a_x = \frac{\partial a}{\partial x}$ and $b_{xz} = \frac{\partial^2 b}{\partial x \partial z}$ etc. We have

$$\begin{align}
d\theta &= (b_x - a_y)dx \wedge dy + (c_x - a_z)dx \wedge dz + (c_y - b_z)dy \wedge dz \\
\star d\theta &= (b_x - a_y)dz - (c_x - a_z)dy + (c_y - b_z)dx \\
d \star d\theta &= (b_{xx} - a_{xy} - c_{yz} + b_{zz})dx \wedge dz \\
&\quad + (-c_{xx} + a_{xz} - c_{yy} + b_{yz})dx \wedge dy \\
&\quad + (b_{xy} - a_{yy} + c_{xz} - a_{zz})dy \wedge dz \\
\star d \star d\theta &= (b_{xy} - a_{yy} + c_{xz} - a_{zz})dx - (b_{xx} - a_{xy} - c_{yz} + b_{zz})dy \\
&\quad + (-c_{xx} + a_{xz} - c_{yy} + b_{yz})dz \\
\star\theta &= a\,dy \wedge dz - b\,dx \wedge dz + c\,dx \wedge dy \\
d \star \theta &= (a_x + b_y + c_z)dx \wedge dy \wedge dz \\
\star d \star \theta &= a_x + b_y + c_z \\
d \star d \star \theta &= (a_{xx} + b_{xy} + c_{xz})dx + (a_{xy} + b_{yy} + c_{yz})dy \\
&\quad (a_{xz} + b_{yz} + c_{zz})dz.
\end{align}$$

We have shown that

$$\begin{align}
\star d \star d\theta &= (b_{xy} - a_{yy} + c_{xz} - a_{zz})dx - (b_{xx} - a_{xy} - c_{yz} + b_{zz})dy \\
&\quad + (-c_{xx} + a_{xz} - c_{yy} + b_{yz})dz
\end{align}$$

and

$$\begin{align}
d \star d \star \theta &= (a_{xx} + b_{xy} + c_{xz})dx + (a_{xy} + b_{yy} + c_{yz})dy \\
&\quad (a_{xz} + b_{yz} + c_{zz})dz.
\end{align}$$

12.3. THE STAR OPERATOR ON FORMS.

Subtracting the second equation from the first gives

$$\star d \star d\theta - d \star d \star \theta = -(\Delta a)dx - (\Delta b)dy - (\Delta c)dz \qquad (12.11)$$

From $\star^2 = \text{id}$ it then follows that we have a similar formula for Ω namely

$$-\star d \star d\Omega + d \star d \star \Omega = -(\Delta A)dx \wedge dy - (\Delta B)dx \wedge dz - (\Delta C)dy \wedge dz. \qquad (12.12)$$

For $\mathbb{R}^{1,3}$.

We will choose the metric to be of type $(1,3)$ so that we have the "orthonormal" coframe field cdt, dx, dy, dz with

$$\langle cdt, cdt \rangle = 1$$

and

$$\langle dx, dx \rangle = \langle dy, dy \rangle = \langle dz, dz \rangle = -1.$$

We will choose

$$v = cdt \wedge dx \wedge dy \wedge dz.$$

This fixes the star operator. But I am faced with an awkward notational problem in the next section when we will discuss the Maxwell equations and the relativistic London equations: We will want to deal with the star operator on $\mathbb{R}^3$ and $\mathbb{R}^{1,3}$ simultaneously, in fact in the same equation. So I will use $\star_3$ to denote the three dimensional star operator and $\star_4$ to denote the four dimensional star operator.

So

$$\star_4(cdt \wedge dx \wedge dy \wedge dz) = 1, \quad \star_4 1 = -cdt \wedge dx \wedge dy \wedge dz$$

while

$$\begin{aligned}
\star_4 cdt &= -dx \wedge dy \wedge dz \\
\star_4 dx &= -cdt \wedge dy \wedge dz \\
\star_4 dy &= cdt \wedge dx \wedge dz \\
\star_4 dz &= -cdt \wedge dx \wedge dy
\end{aligned}$$

which we can summarize as

$$\begin{aligned}
\star_4 cdt &= -\star_3 1 \\
\star_4 \theta &= -cdt \wedge \star_3 \theta \quad \text{for} \\
\theta &= adx + bdy + cdz
\end{aligned}$$

258 CHAPTER 12. THE STAR OPERATOR AND ELECTROMAGNETISM.

and
$$\star_4(cdt \wedge dx) = dy \wedge dz$$
$$\star_4(cdt \wedge dy) = -dx \wedge dz$$
$$\star_4(cdt \wedge dz) = dx \wedge dy$$
$$\star_4(dx \wedge dy) = -cdt \wedge dz$$
$$\star_4(dx \wedge dz) = cdt \wedge dy$$
$$\star_4(dy \wedge dz) = -cdt \wedge dx.$$

Notice that the last three equations follow from the preceding three because $\star_4^2 = -\text{id}$ as a map on two forms in $\mathbb{R}^{1,3}$.

We can summarize these last six equations as

$$\star_4(cdt \wedge \theta) = \star_3\theta, \quad \star_4\Omega = -cdt \wedge \star_3\Omega,$$

for $\Omega = A dx \wedge dy + B dx \wedge dz + C dy \wedge dz$.

I want to make it clear that in these equations $\theta = a\,dx + b\,dy + c\,dz$ where the functions $a, b,$ and c can depend on all four variables, t, x, y and z. Similarly Ω is a linear combination of $dx \wedge dy, dx \wedge dz$ and $dx \wedge dz$ whose coefficients can depend on all four variables. So we may think of θ and Ω as forms on three space which depend on time.

We have $\star_4^2 = \text{id}$ on one forms and on three forms which checks with

$$\star_4(cdt \wedge dx \wedge dy) = -dz$$

or, more generally,

$$\star_4(cdt \wedge \Omega) = -\star_3 \Omega.$$

12.4 Electromagnetism.

12.4.1 Two non-relativistic regimes.

We begin with two regimes in which we solely use the star operator on $\mathbb{R}^3$. Then we will pass to the full relativistic theory.

Electrostatics.

The objects of the theory are:

- A linear differential form, E, called the **electric field strength**. A point charge e experiences the force eE. The integral of E along any path gives the voltage drop along that path. The units of E are

$$\frac{\text{voltage}}{\text{length}} = \frac{\text{energy}}{\text{charge} \cdot \text{length}}.$$

12.4. ELECTROMAGNETISM.

- The **dielectric displacement**, D, which is a two form.
  In principle, we could measure $D(v_1, v_2)$ where $v_1, v_2 \in T\mathbb{R}^3_x \sim \mathbb{R}^3$ are a pair of vectors as follows: construct a parallel-plate capacitor whose plates are metal parallelograms determined by hv_1, hv_2 where h is a small positive number. Place these plates with the corner at x touch them together, then separate them. They acquire charges $\pm Q$. The orientation of $\mathbb{R}^3$ picks out one of these two plates which we call the top plate. Then

$$D(v_1, v_2) = \lim_{h \to 0} \frac{\text{charge on top plate}}{h^2}.$$

The units of D are

$$\frac{\text{charge}}{\text{area}}.$$

- The *charge density* which is a three form, ρ. (We identify densities with three forms since we have an orientation.)

The key equations in the theory are:

$$dE = 0$$

which, in a simply connected region implies that that $E = -du$ for some function, u called the potential.

The integral of D over the boundary surface of some three dimensional region is the total charge in the region. This is *Gauss' law*:

$$\int_{\partial U} D = \int_U \rho$$

which, by Stokes, can be written differentially as

$$dD = \rho.$$

(I will use units which absorb the traditional 4π into ρ.)

Finally there is a *constitutive equation* relating E and D. In an isotropic medium it is given by

$$D = \epsilon \star E$$

where ϵ is called the dielectric factor. (Here, and in the next section, I am writing $\star$ istead of $\star_3$ since all the action is taking place in three dimensions.)

In a homogeneous medium ϵ it is a constant, called the dielectric constant. In particular, the dielectric constant of the vacuum is denoted by ϵ_0. The units of ϵ_0 are

$$\frac{\text{charge}}{\text{area}} \times \frac{\text{charge} \cdot \text{length}}{\text{energy}} = \frac{(\text{charge})^2}{\text{energy} \cdot \text{length}}.$$

The laws of electrostatics, since they involve the star operator, determine the three dimensional Euclidean geometry of space.

260 CHAPTER 12. THE STAR OPERATOR AND ELECTROMAGNETISM.

Magnetoquasistatics.

In this regime, it is assumed that there are no static charges, so $\rho = 0$, and that Maxwell's term $\partial D/\partial t$ can be ignored; energy is stored in the magnetic field rather than in capacitors.

The fundamental objects are:

- A one form E giving the electric force field. The force on a charge e is eE, as before.

- A two form B giving the magnetic induction or the magnetic flux density. The force on a current element I (which is a vector) is $i(I)B$ where i denotes interior product.

- The current flux, J which is a two form [measured in (amps)/(area)].

- A one form, H called the magnetic excitation or the magnetic field.

The integral of H over the boundary, C of a surface S is equal to the flux of current through the surface. This is *Ampère's law.*

$$\int_C H = \int_S J \tag{12.13}$$

Faraday's law of induction says that

$$-\frac{d}{dt}\int_S B = \int_C E. \tag{12.14}$$

By Stokes' theorem, the differential form of Ampere's law is

$$dH = J, \tag{12.15}$$

and of Faraday's law is

$$\frac{\partial B}{\partial t} = -dE. \tag{12.16}$$

Faraday's law implies that the time derivative of dB vanishes. But in fact we have the stronger assertion (Hertz's law)

$$dB = 0. \tag{12.17}$$

Equations (12.15, (12.16), and (12.17) are the structural laws of electrodynamics in the magnetoquasistatic approximation. We must supplement them by constituitive equations:

One of these is

$$B = \mu \star H, \tag{12.18}$$

where $\star$ denotes the star operator in three dimensions.

According to ère's law, H has units

$$\frac{\text{charge}}{\text{time} \cdot \text{length}}$$

12.4. ELECTROMAGNETISM.

while according to Faraday's law B has units

$$\frac{\text{energy} \cdot \text{time}}{\text{charge} \cdot (\text{length})^2}$$

so that μ has units

$$\frac{\text{energy} \cdot (\text{time})^2}{(\text{charge})^2 \cdot \text{length}}.$$

Thus $\epsilon \cdot \mu$ has units

$$\frac{(\text{time})^2}{(\text{length})^2} = (\text{velocity})^{-2}$$

and it was Maxwell's great discovery, the foundation stone of all that has happened since in physics, that

$$\frac{1}{\epsilon_0 \mu_0} = c^2$$

where c is the speed of light. (This discussion is a bit premature in our present regime of quasimagnetostatics where D plays no role.)

We need one more constituitive equation, to relate the current to the electromagnetic field. In ordinary conductivity, one mimics the equation

$$V = RI$$

for a resistor in a network by *Ohm's law*:

$$J = \sigma \star E. \tag{12.19}$$

According to the Drude theory (as modified by Sommerfeld) the charge carriers are free electrons and σ can be determined semi-empirically from a model involving the mean free time between collisions as a parameter. Notice that in ordinary conductivity the charge carrier is something external to the electromagnetic field, and σ is not regarded as a fundamental constant of nature (like c, say) but is an empirical parameter to be derived from another theory, say statistical mechanics. In fact, Drude proposed the theory of the free electron gas in 1900, some three years after the discovery of the electron, by J.J. Thompson, and it had a major success in explaining the law of Wiedemann and Franz, relating thermal conductivity to electrical conductivity.

However, if you look at the lengthy article on conductivity in the 1911 edition of the Encyclopedia Britannica, written by J.J. Thompson himself, you will find no mention of electrons in the section on conductivity in solids. The reason is that Drude's theory gave absolutely the wrong answer for the specific heat of metals, and this was only rectified in 1925 in the brilliant paper by Sommerfeld where he replaces Maxwell Boltzmann statistics by the Fermi-Dirac statistics. All this is explained in a solid state physics course. I repeat my main point - σ is not a fundamental constant and the source of J is external to the electromagentic fields.

12.4.2 Maxwell's equations.

The laws of quasi-magnetostatics take on a very suggestive form when written in four dimensions rather than three, and when an important modification is made to Ampère's law. This modification was introduced by Maxwell: We can combine the laws

$$dB = 0 \quad \text{Hertz}$$
$$\frac{\partial B}{\partial t} = -dE \quad \text{Faraday's law of induction}$$

into the single law

$$dF = 0 \tag{12.20}$$

if we set

$$F := B + E \wedge dt. \tag{12.21}$$

Here, the d occurring on the left in Hertz's law, and on the right in Faraday's law is the three dimensional d-operator, d_3. So the full four dimensional d operator, when applied to B gives $dB = d_3 B + \frac{\partial B}{\partial t} \wedge dt$. That is why $dF = 0$ is equivalent to the two laws above.

In electrostatics the assumption is that there is no current, i.e. $J = 0$, and that the charge density does not depend on t.

In quasimagnetostatics the charge density was ignored but a current was present. For the full equations of electromagnetism, one assumes that there is a charge density and a current, so one considers the three form

$$j := \rho\, dx \wedge dy \wedge dz - J \wedge dt.$$

"Conservation of charge" then demands that

$$dj = 0. \tag{12.22}$$

Locally, this says that there is a two form G such that

$$dG = 4\pi j. \tag{12.23}$$

If we write

$$G = D - H \wedge dt \tag{12.24}$$

then the dt component of the right hand side of (12.23) is

$$d_3 H = \frac{\partial D}{\partial t} + 4\pi J.$$

This is Ampère's law, with the modification that the displacement current $\frac{\partial D}{\partial t}$ is added to the right hand side of Ampère's original law. This brilliant idea of modifying Ampère's law by adding the displacement current is due to Maxwell. The "space component" of (12.23) is $d_3 D = 4\pi \rho\, dx \wedge dy \wedge dz$ as in electrostatics.

12.4.3 Natural units and Maxwell's equations.

According to the Planck-Einstein equation

$$E = h\nu$$

where E is the energy and ν is the frequency. We may think of this equation as expressing as a conversion factor between inverse time (the unit of frequency) and energy. So we may choose units in which $h = 1$ so that the units of energy are inverse time. If we choose units so that $c = 1$, then time and length have the same units, so the units of energy are inverse time. In terms of these units, the units of the permeability become

$$\frac{\text{energy} \cdot (\text{time})^2}{(\text{charge})^2 \cdot \text{length}} = \text{charge}^{-2}.$$

The units of the permittivity become

$$\frac{(\text{charge})^2}{\text{energy} \cdot \text{length}} = \text{charge}^2.$$

So these are inverses of one another as expected.

In our natural units, B has units $\text{charge}^{-1} \cdot \text{area}^{-1}$ so the integral of B over any two dimensional surface in four space has units of charge^{-1}. The integral of E over any curve has units of $(\text{energy})/(\text{charge})$ so the integral of $E \wedge dt$ over any surface in space-time has units

$$\frac{(\text{energy}) \times (\text{time})}{(\text{charge})} = (\text{charge})^{-1}.$$

In short, the integral of F over any two dimensional surface in space time has units $(\text{charge})^{-1}$.

From its definition, the integral of D over any surface has units of charge. From Ampère's law, and Stokes's theorem the integral of H over a curve has the same units as as the flux of a current through a surface, which has units of $(\text{charge})/(\text{time})$. So the integral of $H \wedge dt$ over a surface in space-time also has units of charge. In short, the integral of G over a surface in space-time has units of charge.

From the preceding it follows that the integral of $F \wedge G$ over a four dimensional region in space time is a scalar.

In our natural units, $\mu_0 = \epsilon_0^{-1}$, so we can write $B = \mu_0 \star_3 H$ as

$$H = \epsilon_0 \star_3 B,$$

or

$$H \wedge dt = \epsilon_0 \star_4 B.$$

Also $D = \epsilon_0 \star_3 E$ translates into
$$D = -\epsilon_0 \star_4 (E \wedge dt)$$
in four dimensions. So we can combine these into
$$G = -\epsilon_0 \star_4 F. \tag{12.25}$$
This is equivalent to
$$F = \mu_0 \star_4 G. \tag{12.26}$$

12.4.4 The Maxwell equations with a source term.

Allowing for the presence of a j, but assuming that the medium has the same permeability and permittivity as the vacuum, the full Maxwell equations become
$$dF = 0, \quad d \star_4 F = -4\pi\mu_0 j. \tag{12.27}$$
The equation $dF = 0$ implies that (locally) $F = dA$ for some one form A. We may try to solve the full Maxwell equations by looking for an A which satisfies the "Lorentz gauge condition"
$$d \star_4 A = 0.$$
Then (12.27) becomes the equation
$$(d \star_4 d \star_4 + \star_4 d \star_4 d) A = -4\pi\mu_0 \star_4 j. \tag{12.28}$$
The operator $d \star_4 d \star_4 + \star_4 d \star_4 d$ sends one forms into one forms.

Problem. Let $A = adx + bdy + cdz + fdt$. What are the coefficients of dx, dy, dz and dt in
$$(d \star_4 d \star_4 + \star_4 d \star_4 d) A \quad ?$$

Maxwell's equations as a source equation.

Let us show how (12.27) fits into general philosophy of "source equations of physics" that we shall expound in Chapter 14 : Consider the **Poincaré** density
$$\mathfrak{P}(A) := -\frac{1}{2} dA \wedge \star_4 dA \tag{12.29}$$
with the idea that we want to consider behavior of the "function" given by the integral of this density with respect to variations of compact support. Notice that this density really depends only on $F = dA$.

Suppose that B is a one form of compact support. We have
$$\frac{d}{ds}(d(A + sB) \wedge \star_4(d(A + sB)))\bigg|_{s=0} = dB \wedge \star_4 dA + dA \wedge \star_4 dB$$
$$d(B \wedge \star_4 dA) = dB \wedge \star_4 dA - B \wedge d \star_4 dA \quad \text{and from (12.8)}$$
$$dB \wedge \star_4 dA = dA \wedge \star_4 dB \quad \text{so}$$
$$\frac{d}{ds} \int_R \mathfrak{P}(A + sB)\bigg|_{s=0} = \int_R B \wedge d \star_4 dA$$

12.5. THE LONDON EQUATIONS.   265

where $R \subset \mathbb{R}^4$ is any compact region containing the support of B in its interior (by Stokes.) If we want this to equal $-4\pi\mu_0 \int_{\mathbb{R}^4} B \wedge j$ for all B of compact support we obtain (12.27).

12.5 The London equations.

In the superconducting domain, it is natural to mimic a network inductor which satisfies the equation
$$V = L\frac{dI}{dt}.$$
So the London brothers (1933) introduced the equation
$$E = \Lambda \star_3 \frac{\partial J}{\partial t}, \tag{12.30}$$
where Λ is an empirical parameter similar to the conductance, but the analogue of inductance of a circuit element. Equation (12.30) is known as the **first London equation**.

If we assume that Λ is a constant, we have
$$\frac{\partial}{\partial t} \star_3 dH = \star_3 \frac{\partial J}{\partial t} = \frac{1}{\Lambda} E.$$
Setting $H = \mu^{-1} \star_3 B$, applying d, and using (12.16) we get
$$\frac{\partial}{\partial t}\left(d \star d \star_3 B + \frac{\mu}{\Lambda} B\right) = 0. \tag{12.31}$$
From this one can deduce that an applied external field will not penetrate, but not the full Meissner effect expelling all magnetic fields in any superconducting region.

Here is a sample argument about the non-penetration of imposed magnetic fields into a superconducting domain: Since $dB = 0$ we can write
$$d \star_3 d \star_3 B = (d \star_3 d \star_3 + \star_3 d \star_3 d)B = -\triangle B,$$
where $\triangle$ is the usual three dimensional Laplacian applied to the coefficients of B (using(12.12)). Suppose we have a situation which is invariant under translation in the x and z direction. For example an infinite slab of width $2a$ with sides at $y = \pm a$ parallel to the $y = 0$ plane. Then assuming the solution also invariant, (12.31) becomes
$$\left(\frac{\mu}{\Lambda} - \frac{\partial^2}{\partial y^2}\right)\frac{\partial B}{\partial t} = 0.$$
If we assume symmetry with respect to $y = 0$ in the problem, we get
$$\frac{\partial B}{\partial t} = C(t) \cosh\frac{y}{\lambda},$$

266 CHAPTER 12. THE STAR OPERATOR AND ELECTROMAGNETISM.

where

$$\lambda = \sqrt{\frac{\Lambda}{\mu}}$$

is called the *penetration depth* of the superconducting material. It is typically of order $.1\mu m$. Suppose we impose some time dependent external field which takes on the the value

$$b(t)dx \wedge dy,$$

for example, on the surface of the slab. Continuity then gives

$$\frac{\partial B}{\partial t} = b'(t)\frac{\cosh y/\lambda}{\cosh a/\lambda}.$$

The quotient on the right decays exponentially with penetration y/λ. So externally applied magnetic fields do not penetrate, in the sense that the time derivative of the magnetic flux vanishes exponentially within a few multiples of the penetration depth. But the full Meissner effect says that all magnetic fields in the interior are expelled.

So the Londons proposed strengthening (12.31) by requiring that the expression in parenthesis in (12.31) be actually zero, instead of merely assuming that it is a constant. Since $d \star B = \mu dH = \mu J$ (assuming that μ is a constant) we get

$$d \star_3 J = -\frac{1}{\Lambda} B. \tag{12.32}$$

Equation (12.32) is known as **the second London equation**.

12.5. THE LONDON EQUATIONS.

Heinz and Fritz London, 1953 Cambridge, photo: K. Mendelssohn

Fritz London (1900-1954)
Heinz London (1907-1970)

12.5.1 The London equations in relativistic form.

We can write the two London equations in relativistic form, by letting

$$j = -J \wedge dt$$

be the three form representing the current in space time. In general, we write

$$j := \rho dx \wedge dy \wedge dz - J \wedge dt \qquad (12.33)$$

as the three form in space time giving the relativistic "current", but in the quasistatic regime $\rho = 0$.

We have

$$\star_4 j = \frac{1}{c} \star_3 J,$$

a one form on space time with no dt component (under our assumption of the absence of static charge in our space time splitting). So

$$cd(\star_4 j) = d_{space} \star_3 J - \frac{\partial \star_3 J}{\partial t} \wedge dt,$$

where the d on the left is the full d operator on space time. (From now on, until the end of this section, we will be in space-time, and so use d to denote the full

268 CHAPTER 12. THE STAR OPERATOR AND ELECTROMAGNETISM.

d operator in four dimensions, and use d_{space} to denote the three dimensional d operator.)

We recall that in the relativistic treatment of Maxwell's equations, the electric field and the magnetic induction are combined to give the electromagnetic field

$$F = B + E \wedge dt$$

so that Faraday's law, (12.16), and Hertz's law, (12.17) are combined into the single equation,

$$dF = 0, \qquad (12.34)$$

known as the *first Maxwell equation*. We see that the two London equations can also be combined to give

$$d \star_4 c \Lambda j = -F, \qquad (12.35)$$

which implies (12.34). This suggests that superconductivity involves modifying Maxwell's equations, in contrast to ordinary conductivity which is supplementary to Maxwell's equations.

Maxwell's equations again.

To see the nature of this modification, we recall the second Maxwell equation which involves the two form

$$G = D - H \wedge dt$$

where D is the "dielectric displacement", as above. Recall that

$$d_{space} D$$

gives the density of charge according to Gauss' law. The *second Maxwell equation* combines Gauss' law and Maxwell's modification of Ampere's law into the single equation

$$dG = j, \qquad (12.36)$$

where the three current, j is given by (12.33). The product $(\epsilon \mu)^{-1/2}$ has the units of velocity, as we have seen, and let us us assume that we are in the vacuum or in a medium for which this velocity is c, the same value as the vacuum.

So using the corresponding Lorentz metric on space time to define our $\star_4$ operator the combined constituitive relations can be written as

$$G = -\frac{1}{c\mu} \star_4 F,$$

or using units where $c = 1$ more simply as

$$G = -\frac{1}{\mu} \star_4 F. \qquad (12.37)$$

From now on, we will use "natural" units in which $c = 1$ and in which energy and mass have units $(\text{length})^{-1}$.

12.5. THE LONDON EQUATIONS.

12.5.2 Comparing Maxwell and London.

The material in this subsection, especially the comments at the end, might be acceptable in the mathematics department. You should be warned that they do not reflect the currently accepted physical theories of superconductivity, and hence might encounter some trouble in the physics department.

In classical electromagnetic theory, j is regarded as a source term in the sense that one introduces a one form, A, the four potential, with

$$F = -dA$$

and Maxwell's equations become the variational equations for the Lagrangian with Lagrange density

$$\mathcal{L}_M(A,j) = \frac{1}{2} dA \wedge \star_4 dA - \mu A \wedge j. \tag{12.38}$$

This means the following: $\mathcal{L}_M(A,j)$ is a four form on $\mathbb{R}^{1,3}$ and we can imagine the "function"

$$\mathbf{L}_M(A,j) \text{"} := \text{"} \int_{\mathbb{R}^{1,3}} \mathcal{L}_M(A,j).$$

It is of course not defined because the integral need not converge. But if C is any smooth one form with compact support, the variation

$$d(\mathbf{L}_M)_{(A,j)}[C] \text{"} := \text{"} \frac{d}{ds} \mathbf{L}_M(A + sC, j)_{|s=0}$$

is well defined, as described above, i.e we replace the integral over $\mathbb{R}^{1,3}$ by the integral over any compact region R containing the support of C, and the result is independent of the choice of such R.

The condition that this variation vanish for all such C gives Maxwell's equations.

Problem. Show that these variational equations do indeed give Maxwell's equations. Use $d(C \wedge \star_4 A) = dC \wedge \star_4 dA - C \wedge d \star_4 dA$ and the fact that $\tau \wedge \star_4 \omega = \omega \wedge \star_4 \tau$ for two forms.

In particular, one has gauge invariance: A is only determined up to the addition of a closed one form, and the Maxwell equations become

$$d \star_4 dA = \mu j. \tag{12.39}$$

For the London equations, if we apply $\star_4$ to (12.35) and use (12.37) we get

$$\star_4 d \star_4 j = \frac{\mu}{\Lambda} G,$$

and so by the second Maxwell equation, (12.36) we have

$$d \star_4 d \star_4 j = \frac{1}{\lambda^2} j. \qquad (12.40)$$

We no longer restrict j by requiring the absence of stationary charge, but do observe that "conservation of charge". i.e. $dj = 0$ is a consequence of (12.40).

If we set

$$\star_4 \Lambda j = A, \qquad (12.41)$$

we see that the Maxwell Lagrange density (12.38) is modified to become the "Proca" Lagrange density

$$\mathcal{L}_L(A) = \frac{1}{2}\left(dA \wedge \star_4 dA - \frac{1}{\lambda^2} A \wedge \star_4 A\right). \qquad (12.42)$$

A number of remarks are in order:

1. The London equations have no gauge freedom.

2. The Maxwell equations in free space (that is with $j = 0$) are conformally invariant. This is a general property of the star operator on middle degrees, in our case from $\wedge^2$ to $\wedge^2$, as we have seen. But the London equations involve the star operator from $\wedge^1$ to $\wedge^3$ and hence depend on, and determine, the actual metric and not just on the conformal class. This is to be expected in that the Meissner effect involves the penetration depth, λ.

3. Since the units of λ are length, the units of $1/\lambda^2$ are (mass)^2 as is to be expected. So the London modification of Maxwell's equations can be expressed as the addition of a masslike term to the massless photons. In fact, substituting a plane wave with four momentum **k** directly into (12.40) shows that **k** must lie on the mass shell $\mathbf{k}^2 = 1/\lambda^2$. In quantum mechanical language, we would say that the Maxwell equations (with no source term) correspond to particles of mass zero and spin one, while the Proca equations correspond to massive particles of spin one.

4. Since the Maxwell equations are the mass zero limit of the Proca equations, one might say that the London equations represent the more generic situation from the mathematical point of view. Perhaps the "true world" is always superconducting and we exist in some limiting case where the photon can be considered to have mass zero.

5. On the other hand, if one starts from a firm belief in gauge theories, then one would regard the mass acquisition as the result of spontaneous symmetry breaking via the Higgs mechanism. See Chapter 17. In the standard treatment one gets the Higgs field as the spin zero field given by a Cooper pair. But since the electrons are not needed for charge transport, as no external source term occurs in (12.40), one might imagine an entirely different origin for the Higgs field. Do we need electrons for superconductivity? We don't use them to give mass to quarks or leptons in the standard model.

Chapter 13

Preliminaries to the study of the Einstein field equations.

13.1 Preliminaries to the preliminaries.

13.1.1 Densities and n-forms.

If we regard $\mathbb{R}^n$ as a differentiable manifold, the law for the change of variables for an integral involves the absolute value of the Jacobian determinant. This is different from the law of change of variables of a function (which is just substitution).[But it is close to the transition law for an n-form which involves the Jacobian determinant (not its absolute value).] For this reason we can not expect to integrate functions on a manifold. The objects that we *can* integrate are known as **densities**. We briefly recall two equivalent ways of defining these objects:

1. **Coordinate chart description.** A density ρ is a rule which assigns to each coordinate chart (U, α) on M (where U is an open subset of M and $\alpha : U \to \mathbb{R}^n$) a function ρ_α defined on $\alpha(U)$ subject to the following transition law: If (W, β) is a second chart then

$$\rho_\alpha(v) = \rho_\beta(\beta \circ \alpha^{-1}(v)) \cdot |\det J_{\beta \circ \alpha^{-1}}|(v) \quad \text{for} \ \ v \in \alpha(U \cap W) \quad (13.1)$$

where $J_{\beta \circ \alpha^{-1}}$ denotes the Jacobian matrix of the diffeomorphism

$$\beta \circ \alpha^{-1} : \ \beta(U \cap V) \to \alpha(U \cap V).$$

Of course (13.1) is just the change of variables formula for an integrand in $\mathbb{R}^n$.

2. **Tangent space description.** If V is an n-dimensional vector space, let $|\wedge V^*|$ denote the space of (real or complex valued) functions of n-tuplets of vectors which satisfy

$$\sigma(Av_1, \ldots, Av_n) = |\det A|\sigma(v_1, \ldots, v_n) \qquad (13.2)$$

for any linear transformation A of V. The space $|\wedge V^*|$ is clearly a one-dimensional vector space. A density ρ is then a rule which assigns to each $x \in M$ an element of $|\wedge TM_x^*|$. Naturally, we demand that these vary smoothly with x.

The relation between these two descriptions is the following: Let ρ be a density according to the tangent space description. Thus $\rho_x \in |\wedge TM_x^*|$ for every $x \in M$. Let (U, α) be a coordinate chart with coordinates $x^1, \ldots, x^n$. Then on U we have the vector fields

$$\frac{\partial}{\partial x^1}, \ldots, \frac{\partial}{\partial x^n}.$$

We can then evaluate ρ_x on the values of these vector fields at any $x \in U$, and so define

$$\rho_\alpha(\alpha(x)) = \rho_x\left(\left(\frac{\partial}{\partial x^1}\right)_x, \ldots, \left(\frac{\partial}{\partial x^n}\right)_x\right).$$

If (W, β) is a second coordinate chart with coordinates $y^1, \ldots, y^n$ then on $U \cap W$ we have

$$\frac{\partial}{\partial x^j} = \sum \frac{\partial y^i}{\partial x^j} \frac{\partial}{\partial y^i}$$

and

$$J_{\beta \circ \alpha^{-1}} = \left(\frac{\partial y^i}{\partial x^j}\right)$$

so (13.1) follows from (13.2).

Notation.

If (U, α) is a coordinate chart with coordinates $x^1, \ldots, x^n$ then the density defined on U by $\rho_\alpha \equiv 1$, that is by

$$\rho_x\left(\left(\frac{\partial}{\partial x^1}\right)_x, \ldots, \left(\frac{\partial}{\partial x^n}\right)_x\right) = 1 \ \forall \, x \in U$$

is denoted by $|dx|$. Every other density then has the local description $G|dx|$ on U where G is a function.

I may on occasion forget to put in the vertical bars add simply write dx.

13.1. PRELIMINARIES TO THE PRELIMINARIES.

13.1.2 Densities of arbitrary order.

Let $s \in \mathbb{C}$. If V is an n-dimensional vector space we let $|\wedge V^*|^s$ denote the space of real or complex valued functions on n-tuplets of vectors which satisfy

$$\sigma(Av_1,\ldots,Av_n) = |\det A|^s \sigma(v_1,\ldots,v_n) \quad (13.2)_s,$$

and make the corresponding definition of a density of order s on a manifold. It is easy to check that the product of a density of order s_1 and a density of order s_2 is well defined and is a density of order $s_1 + s_2$.

If ρ is a density of order s, then $\overline{\rho}$ (with the obvious definition) is a density of order $\overline{s}$. So if ρ_1 and ρ_2 are densities of order $\frac{1}{2}$ and of compact support, the product $\rho_1 \cdot \overline{\rho}_2$ is a density of order one and can be integrated. This leads to a pre-Hilbert space which can then be completed to give a Hilbert space intrinsically associated to a differentiable manifold.

Notational comment: In the relativity literature, the word "weight" is used instead of our "order".

13.1.3 Pullback of a density under a diffeomorphism.

If $\phi : N \to M$ is a diffeomorphism and if ρ is a density (of order one) on M, then the pull back $\phi^*\rho$ is the density on N defined by

$$(\phi^*\rho)_z(v_1,\ldots,v_n)$$
$$:= \rho_{\phi(z)}(d\phi_z(v_1),\ldots,d\phi_z(v_n)) \quad z \in N, \ v_1,\ldots,v_n \in TN_z.$$

(It is easy to check that this is indeed a density, i.e. that (13.2) holds at each $z \in N$.)

13.1.4 The Lie derivative of a density.

In particular, if X is a vector field on M generating a one parameter group

$$t \mapsto \phi_t = \exp tX$$

of diffeomorphisms, we can form the Lie derivative

$$L_X \rho := \frac{d}{dt} \phi_t^* \rho_{|t=0}.$$

We will need a local description of this Lie derivative. We can derive such a local description from Weil's formula for the Lie derivative of a differential form by the following device: Suppose that the manifold M is orientable and that we have chosen an orientation of M. This means that we have chosen a system of coordinate charts such that all the Jacobian determinants $\det J_{\beta \circ \alpha^{-1}}$ are positive. Relative to this system of charts, we can drop the absolute value sign in (13.1) since $\det J_{\beta \circ \alpha^{-1}} > 0$. But (13.1) without the absolute value signs is just the transition law for an n-form on the n-dimensional manifold M. In other words, *once we have chosen an orientation* on an orientable manifold M we can identify densities with n-forms. A fixed chart (U, α) carries the orientation coming from $\mathbb{R}^n$ and our identification amounts to identifying the density $|dx|$ with the n-form $dx^1 \wedge \cdots \wedge dx^n$. If τ is an n-form on an n-dimensional manifold then Weil's formula

$$L_X \tau = i(X) d\tau + di(X)\tau$$

reduces to

$$L_X \tau = di(X)\tau$$

since $d\tau = 0$ as there are no non-zero $(n+1)$ forms on an n-dimensional manifold. In terms of local coordinates this gives

$$di(X)\tau = \left(\sum_{i=1}^{n} \partial_i (G X^i) \right) dx^1 \wedge \cdots \wedge dx^n \quad (13.3)$$

where

$$\partial_i := \frac{\partial}{\partial x^i} \quad \text{and} \quad \tau = G dx.$$

13.1.5 The divergence of a vector field relative to a density.

It is useful to express this formula somewhat differently. It makes no sense to talk about a numerical value of a density ρ at a point x since ρ is not a function. But it *does* make sense to say that ρ does not vanish at x, since if $\rho_\alpha(\alpha(x)) \neq 0$ then (13.1) implies that $\rho_\beta(\beta(x)) \neq 0$. Suppose that ρ is a density which does not vanish anywhere. Then any other density on M is of the form $f \cdot \rho$ where f is a function. If X is a vector field, so that $L_X \rho$ is another density, then $L_X \rho$ is of the form $f \rho$ where f is a function, called the **divergence** of the vector field X relative to the non-vanishing density ρ and denoted by $\mathrm{div}_\rho(X)$.

In symbols,

$$L_X \rho = (\mathrm{div}_\rho(X)) \cdot \rho.$$

We can then rephrase

$$di(X)\tau = \left(\sum_{i=1}^{n} \partial_i (G X^i) \right) dx^1 \wedge \cdots \wedge dx^n \quad (13.3)$$

as saying that

$$\mathrm{div}_\rho(X) = \frac{1}{G} \sum_{i=1}^{n} \partial_i (G X^i) \quad (13.4)$$

The **divergence theorem** asserts that if X has compact support, then

$$\int_M \mathrm{div}_\rho \cdot \rho = 0.$$

13.2 The divergence of a vector field on a semi-Riemannian manifold.

Suppose that **g** is a semi-Riemann metric on an n-dimensional manifold, M. Then **g** determines a density, call it g, which assigns to every n tangent vectors, $\xi_1, \ldots, \xi_n$ at a point x the "volume" of the parallelepiped that they span:

$$g: \quad \xi_1, \ldots, \xi_n \mapsto |\det(\langle \xi_i, \xi_j \rangle)|^{\frac{1}{2}}. \tag{13.5}$$

If we replace the ξ_i by $A\xi_i$ where $A: TM_x \to TM_x$ the determinant is replaced by

$$\det((A\xi_i, A\xi_j)) = \det(A(\langle \xi_i, \xi_j \rangle)A^*) = (\det A)^2 \det((\xi_i, \xi_j))$$

so we see that (13.2) is satisfied. So g is indeed a density, and since the metric is non-singular, the density g does not vanish at any point.

So if X is a vector field on M, we can consider its divergence $\operatorname{div}_g(X)$ with respect to g. Since g will be fixed for the rest of this section, we may drop the subscript g and simply write div X. So

$$\operatorname{div} X \cdot g = L_X g. \tag{13.6}$$

We now come to an important alternative formula for the divergence of a vector field on a semi-Riemannian manifold:

We we can form the covariant differential of X with respect to the connection determined by **g**,

$$\nabla X.$$

It assigns an element of $\operatorname{Hom}(TM_p, TM_p)$ to each $p \in M$ according to the rule

$$\xi \mapsto \nabla_\xi X.$$

The trace of this operator is a number, assigned to each point, p, i.e. a function known as the "contraction" of ∇X, so

$$C(\nabla X) := f, \quad f(p) := \operatorname{tr}(\xi \mapsto \nabla_\xi X).$$

We wish to prove the following very important formula

$$\operatorname{div} X = C(\nabla X). \tag{13.7}$$

We will prove this by computing both sides in a coordinate chart with coordinates, say, $x^1, \ldots, x^n$. Let $|dx| = dx^1 dx^2 \cdots dx^n$ denote the standard density (the one which assigns constant value one to the $\partial_1, \ldots, \partial_n$, $\partial_i := \partial/\partial x^i$). Then

$$g = G|dx|, \quad G = |\det(\langle \partial_i, \partial_j \rangle)|^{\frac{1}{2}} = (\epsilon \det(\langle \partial_i, \partial_j \rangle))^{\frac{1}{2}}$$

where

$$\epsilon := \operatorname{sgn} \det(\langle \partial_i, \partial_j \rangle).$$

Recall the local formula (13.4) for the divergence:
$$\mathrm{div} X = \frac{1}{G}\sum_i \partial_i(X^i G).$$

Write
$$\Delta := \det(\langle \partial_i, \partial_j \rangle)$$

so
$$\frac{1}{G}\partial_i G = \frac{1}{\sqrt{\epsilon\Delta}}\frac{1}{2\sqrt{\epsilon\Delta}}\frac{\partial(\epsilon\Delta)}{\partial x^i}$$
$$= \frac{1}{2}\frac{1}{\Delta}\frac{\partial\Delta}{\partial x^i}$$

independent of whether $\epsilon = 1$ or -1.

I will now give two proofs of (13.7), a short proof and a long proof.

The short proof. Both sides of (13.7) are invariantly defined, so it is enough to prove (13.7) at the center of a normal coordinate system. Near the center of a normal coordinate system $g_{ij} = \delta_{ij} + O(\|x\|^2)$ so both $\partial_i G = 0$ and the Γ^i_{jk} vanish at the center of the normal coordinate system and at that center we have

$$\mathrm{div}(X) = C(\nabla X) = \sum_i \frac{\partial X^i}{\partial x_i}, \tag{13.8}$$

the formula for the divergence as is taught in high school for the case of Euclidean three space.

The long proof valid in any coordinate system. To compute $\frac{1}{\Delta}\frac{\partial\Delta}{\partial x^i}$, let us use the standard notation
$$g_{ij} := \langle \partial_i, \partial_j \rangle$$

so
$$\Delta = \det(g_{ij}) = \sum_j g_{ij}\Delta^{ij}$$

where we have expanded the determinant along the i-th row and the Δ^{ij} are the corresponding cofactors. If we think of Δ as a function of the n^2 variables, g_{ij} then, since none of the Δ_{ik} (for a fixed i) involve g_{ij}, we conclude from the above cofactor expansion that

$$\frac{\partial\Delta}{\partial g_{ij}} = \Delta^{ij} \tag{13.9}$$

and hence by the chain rule that

$$\frac{\partial\Delta}{\partial x^k} = \sum_{ij}\Delta^{ij}\frac{\partial g_{ij}}{\partial x^k}.$$

13.2. DIVERGENCE ON A SEMI-RIEMANNIAN MANIFOLD.

But
$$\frac{1}{\Delta}(\Delta^{ij}) = (g_{ij})^{-1},$$

the inverse matrix of (g_{ij}), which is usually denoted by

$$(g^{kl})$$

so we have

$$\frac{\partial \Delta}{\partial x^k} = \Delta \sum_{ij} g^{ij} \frac{\partial g_{ij}}{\partial x^k}$$

or

$$\frac{1}{G}\partial_k G = \frac{1}{2} \sum_{ij} g^{ij} \frac{\partial g_{ij}}{\partial x^k}.$$

Recall that

$$\Gamma^a_{bc} := \frac{1}{2} \sum_r g^{ar} \left(\frac{\partial g_{rb}}{\partial x^c} + \frac{\partial g_{rc}}{\partial x^b} - \frac{\partial g_{bc}}{\partial x^r} \right)$$

so

$$\sum \Gamma^a_{ba} = \frac{1}{2} \sum_{ar} g^{ar} \frac{\partial g_{ar}}{\partial x^b}$$

or

$$\sum_a \Gamma^a_{ka} = \frac{1}{G} \frac{\partial G}{\partial x^k}. \tag{13.10}$$

On the other hand, we have

$$\nabla_{\partial_i} X = \sum \frac{\partial X^j}{\partial x^i} \partial_j + \sum \Gamma^j_{ik} X^k \partial_j$$

so

$$C(\nabla X) = \sum \frac{\partial X^j}{\partial x^j} + \frac{1}{G} \sum X^j \frac{\partial G}{\partial x^j},$$

proving

$$\mathrm{div} X = C(\nabla X). \tag{13.7}.$$

The reason for my extended local computation is that for later use I will need to use one step of the long proof: From

$$\frac{\partial \Delta}{\partial g_{ij}} = \Delta^{ij} \tag{13.9}$$

we can conclude, as above, that

$$\frac{\partial G}{\partial g_{ij}} = \frac{1}{2} G g^{ij}. \tag{13.11}$$

13.3 The Lie derivative of a semi-Riemannian metric.

We wish to prove that
$$L_V \mathbf{g} = \mathcal{S}\nabla(V\downarrow). \qquad (13.12)$$
The left hand side of this equation is the Lie derivative of the metric $\mathbf{g}$ with respect to the vector field V. It is a rule which assigns a symmetric bilinear form to each tangent space. By definition, it assigns to any pair of vector fields, X and Y, the value
$$(L_V \mathbf{g})(X,Y) = V\langle X,Y\rangle - \langle [V,X],Y\rangle - \langle X,[V,Y]\rangle.$$
The right hand side of (13.12) means the following: $V\downarrow$ denotes the linear differential form whose value at any vector field Y is
$$(V\downarrow)(Y) := \langle V,Y\rangle.$$
In tensor calculus terminology, $\downarrow$ is the "lowering operator", and it commutes with covariant differential. We now turn to the proof of (13.12):

Proof. Since $\downarrow$ commutes with ∇, we have
$$\nabla(V\downarrow)(X,Y) = \nabla_X(V\downarrow)(Y) = \langle \nabla_X V, Y\rangle.$$
The symbol $\mathcal{S}$ in (13.12) denotes symmetric sum, so that the right hand side of (13.12) when applied to X, Y is
$$\langle \nabla_X V, Y\rangle + \langle \nabla_Y V, X\rangle.$$
But now (13.12) follows from the identities
$$\begin{aligned} L_V\langle X,Y\rangle = V\langle X,Y\rangle &= \langle \nabla_V X, Y\rangle + \langle X, \nabla_V Y\rangle \\ \nabla_V X - [V,X] &= \nabla_X V \\ \nabla_V Y - [V,Y] &= \nabla_Y V. \end{aligned}$$

$\square$

We remind the reader that a Killing vector field was defined in Chapter 6 to be a vector field X which satisfies $L_X \mathbf{g} \equiv 0$. In view of (13.12) this is the same as $\mathcal{S}\nabla(V\downarrow) \equiv 0$.

13.4 The covariant divergence of a symmetric tensor field.

The contraction of symmetric contravariant tensor field with a linear differential form.

13.4. THE DIVERGENCE OF A SYMMETRIC TENSOR FIELD.

Let **T** be a symmetric "contravariant" tensor field (of second order), so that in any local coordinate system **T** has the expression

$$\mathbf{T} = \sum T^{ij} \partial_i \partial_j, \quad T^{ij} = T^{ji}.$$

If θ is a linear differential form, then we can "contract" **T** with θ to obtain a vector field, $\mathbf{T} \cdot \theta$: In local coordinates, if

$$\theta = \sum a_i dx^i$$

then

$$\mathbf{T} \cdot \theta = \sum T^{ij} a_j \partial_i.$$

The covariant divergence of a symmetric tensor field.

We can also form the covariant differential, $\nabla \mathbf{T}$ which then assigns to every linear differential form a linear transformation of the tangent space at each point, and then form the contraction, $C(\nabla \mathbf{T})$. (Since **T** is symmetric, we don't have to specify on "which of the upper indices" we are contracting.) Here and in what follows, the the covariant position (the"lower index") with which we are contracting is the position associated to ∇.

We make the

Definition 4.
$$\text{div } \mathbf{T} := C(\nabla \mathbf{T}),$$

called the **covariant divergence** *of* **T**.

It is a vector field.

13.4.1 The meaning of the condition div **T** = 0.

The purpose of the rest of this section and the next section is to explain the geometrical significance of the condition

$$\text{div } \mathbf{T} = 0 \qquad (13.13)$$

which occurs in general relativity. In general relativity equation (13.13 is a necessary condition for T to appear on the right hand side of the Einstein field equations, and in these equations T is regarded as as describing the distribution of matter-energy in the universe. This is explained in the next chapter.

In the standard physics literature, condition (13.13) is regarded as a conservation of energy condition - see for example Wald, pp. 69 - 70. We will find that (13.13), and its re-interpretation and generalization - see equation (13.16) below - have a "group theoretical" meaning which transcends the specific form of the Einstein field equations, but refer only to their symmetry properties.

T as a linear functional on symmetric covariant tensor fields.

If **S** is a "covariant" symmetric tensor field so that

$$\mathbf{S} = \sum S_{ij} dx^i dx^j$$

in local coordinates, let $\mathbf{S} \bullet \mathbf{T}$ denote the double contraction. It is a function, given in local coordinates by

$$\mathbf{S} \bullet \mathbf{T} = \sum S_{ij} T^{ij}.$$

Thus **T** can be regarded as a linear function on the space of all covariant symmetric tensors of compact support by the rule

$$\mathbf{S} \mapsto \int_M \mathbf{S} \bullet \mathbf{T} g,$$

where g is the volume density associated to **g**.

Let V be a vector field of compact support. Then $L_V \mathbf{g}$ is a symmetric covariant tensor field of compact support. We claim that

Proposition 31. *Equation*

$$\operatorname{div} \mathbf{T} = 0 \tag{13.13}$$

is equivalent to

$$\int_M (L_V \mathbf{g}) \bullet \mathbf{T} g = 0 \tag{13.14}$$

for all vector fields V of compact support.

Proof. Let $\theta := V \downarrow$ so $\mathbf{T} \cdot \theta$ is a vector field of compact support, and so

$$\int_M C(\nabla(\mathbf{T} \cdot \theta)) g = \int_M L_{\mathbf{T} \cdot \theta} g = 0 \tag{13.15}$$

by the divergence theorem. (Recall our notation: the symbol $\cdot$ denotes a "single" contraction, so that $\mathbf{T} \cdot \theta$ is a vector field.)

On the other hand,

$$\nabla(\mathbf{T} \cdot \theta) = (\nabla \mathbf{T}) \cdot \theta + \mathbf{T} \cdot \nabla \theta,$$

So

$$C(\nabla(\mathbf{T} \cdot \theta)) = C((\nabla \mathbf{T})) \cdot \theta + C(\mathbf{T} \cdot \nabla \theta).$$

Hence, by (13.15), the condition $\int C((\nabla \mathbf{T})) \cdot \theta g = 0$ is equivalent to the condition

$$\int C(\mathbf{T} \cdot \nabla \theta) g = 0$$

which we now examine:

13.5. ANALYZING THE CONDITION $\ell(L_V \mathbf{G}) = 0$.

We have
$$2C(\mathbf{T} \cdot \nabla \theta) = 2\mathbf{T} \bullet \nabla \theta$$
$$= \mathbf{T} \bullet L_V \mathbf{g},$$

using the fact that $\mathbf{T}$ is symmetric and
$$L_V \mathbf{g} = \mathcal{S}\nabla(V \downarrow). \tag{13.12}$$

So $\int_M \mathbf{T} \bullet L_V \mathbf{g} \, g = 0$ for all V of compact support if and only if $\int_M (\text{div } \mathbf{T} \cdot \theta) g = 0$ for all θ of compact support. If div $\mathbf{T} \not\equiv 0$, we can find a point p and a linear differential form θ such that div $\mathbf{T} \cdot \theta(p) > 0$ at some point, p. Multiplying θ by a blip function ϕ if necessary, we can arrange that θ has compact support and div$\mathbf{T} \cdot \theta \geq 0$ so that $\int_M (\text{div } \mathbf{T} \cdot \theta) g > 0$.

$\square$

13.4.2 Generalizing the condition of the vanishing of the covariant divergence.

Let us write $\ell_\mathbf{T}$ for the linear function on the space of smooth covariant tensors of compact support given by
$$\ell_\mathbf{T}(\mathbf{S}) := \int_M \mathbf{S} \bullet \mathbf{T} g.$$

We can rewrite (13.14) as
$$\ell(L_V \mathbf{g}) = 0 \quad \forall V \text{ of compact support} \tag{13.16}$$

when $\ell = \ell_\mathbf{T}$. But we can consider other types of linear functions ℓ, as we shall see.

13.5 Analyzing the condition $\ell(L_V \mathbf{g}) = 0$ for all V of compact support.

We begin by giving an interpretation of the space of degree two covariant symmetric tensor fields of compact support, in the presence of a semi-Riemannian metric:

Let $\mathcal{M}$ denote the space of all semi-Riemann metrics on a manifold, M, say all with a fixed signature. If $\mathbf{g} \in \mathcal{M}$ is a particular metric, and if $\mathbf{S}$ is a compactly supported symmetric tensor field of degree two, then
$$\mathbf{g} + t\mathbf{S}$$
is again a metric of the same signature for sufficiently small $|t|$.

So we can regard $\mathbf{S}$ as the infinitesimal variation in $\mathbf{g}$ along this "line segment" of metrics. On the other hand, if $\mathbf{g}_t$ is any curve of metrics depending

smoothly on t, and with the property that $\mathbf{g}_t = \mathbf{g}$ outside some fixed compact set, K, then
$$\mathbf{S} := \frac{d\mathbf{g}_t}{dt}_{|t=0}$$
is a symmetric tensor field of compact support.

So if we imagine that $\mathcal{M}$ is to be thought of as a "manifold", then we might imagine that $T_{\mathbf{g}}\mathcal{M}$ consists of all symmetric tensor fields or degree two, and that the "compactly supported part of this tangent space", call it $T_{\mathbf{g}}^c\mathcal{M}$ consists of all symmetric tensor fields or degree two of compact support.

Notice that these two "tangent spaces" are defined independently of $\mathbf{g}$. That is, the "tangent spaces" at every metric is identified with the fixed vector space of all symmetric tensor fields of degree two. We have "trivialized the tangent bundle to $\mathcal{M}$".

How to think of the space of all $L_V \mathbf{g}$.

If V is a smooth vector field of compact support, it generates a flow ϕ_t on M, and ϕ_t is a diffeomorphism of compact support, meaning that it is the identity outside a compact set.

We can then consider the variation $\phi_t^* \mathbf{g}$ of $\mathbf{g}$. The tangent vector to this one parameter family of variations is $L_V \mathbf{g}$.

So let $\mathcal{G}$ denote the group of all diffeomorphisms of M each of which equals the identity outside a compact set. $\mathcal{G}$ acts on the space $\mathcal{M}$ via pull-back: the action of $\phi \in \mathcal{G}$ on a metric $\mathbf{g}$ is to send it into
$$(\phi^{-1})^* \mathbf{g}.$$
So we can consider the orbit $\mathcal{O}_{\mathbf{g}} = \mathcal{G} \cdot \mathbf{g}$ of $\mathbf{g}$ under $\mathcal{G}$. The "tangent space" to this orbit at $\mathbf{g}$ is then the subspace of $T_{\mathbf{g}}^c \mathcal{M}$ consisting of all $L_V \mathbf{g}$ where V is a vector field of compact support on M.

In many situations, we would like to consider two metrics which transformed into one another by a diffeomorphism of compact support as being "the same". So the condition on ℓ given by (13.16) says that (infinitesimally) ℓ is insensitive to replacing $\mathbf{g}$ by $\phi_t^* \mathbf{g}$.

For example, suppose that F is a "function" on $\mathcal{M}$ which is invariant under the action of $\mathcal{G}$. In particular, it is constant on $\mathcal{O}_{\mathbf{g}}$. So if $\ell = dF_{\mathbf{g}}$ then ℓ will satisfy (13.16).

In many texts, the philosophy is to replace $\mathcal{M}$ by its quotient space $\mathcal{M}/\mathcal{G}$ by the $\mathcal{G}$ action. See for example the marvelous text *Einstein Manifolds* by "Arthur L. Besse".

But my philosophy is not to pass to the quotient. Rather, condition (13.16) has interesting consequences, as we shall see.

13.5. ANALYZING THE CONDITION $\ell(L_V \mathbf{G}) = 0$.

We can ask about condition (13.16) for different types of linear functions, ℓ. For example, consider a "delta tensor concentrated at a point $p \in M$", that is a linear function of the form

$$\ell(\mathbf{S}) = \mathbf{S}(p) \bullet t$$

where t is a ("contravariant") symmetric two tensor defined at the point $p \in M$. We claim that

No (non-zero) linear function of this type can satisfy

$$\ell(L_V \mathbf{g}) = 0 \quad \forall\, V \text{ of compact support.} \tag{13.16}$$

Proof. Let W be a vector field of compact support and let ϕ be a smooth function which vanishes at p. Set $V = \phi W$. Then

$$\nabla V \downarrow = d\phi \otimes W \downarrow + \phi \nabla W \downarrow$$

and the second term vanishes at p. Since

$$L_V \mathbf{g} = \mathcal{S} \nabla (V \downarrow), \tag{13.12}$$

and t is symmetric, condition (13.16) says that

$$0 = t \bullet (d\phi(p) \otimes W \downarrow (p)) = [t \cdot W \downarrow (p)] \cdot d\phi(p).$$

This says that the tangent vector $t \cdot (W \downarrow)(p)$ yields zero when applied to the function ϕ:

$$t \cdot (W \downarrow)(p)\phi = 0.$$

This is to hold for all ϕ vanishing at p, which implies that

$$t \cdot (W \downarrow)(p) = 0.$$

Now given any tangent vector, $w \in TM_p$ we can always find a vector field W of compact support such that $W(p) = w$. Hence the preceding equation implies that $t \cdot w \downarrow = 0 \ \forall w \in TM_p$ which implies that $t = 0$. $\square$

13.5.1 What does condition (13.16) say for a tensor field concentrated along a curve?

Let us turn to the next simplest case, a "delta tensor field concentrated on a curve". That is, let $\gamma : I \to M$ be a smooth curve and let τ be a continuous function which assigns to each $s \in I$ a symmetric contravariant tensor, $\tau(s)$ at the point $\gamma(s)$. Define the linear function ℓ_τ on the space of covariant symmetric tensor fields of compact support by

$$\ell_\tau(\mathbf{S}) = \int_I \mathbf{S}(\gamma(s)) \bullet \tau(s) ds.$$

284 CHAPTER 13. PRELIMINARIES TO THE EINSTEIN EQUATIONS.

Let us examine the implications of (13.16) for $\ell = \ell_\tau$.

Once again, let us choose $V = \phi W$, this time with $\phi = 0$ on γ. We then get that

$$\int_I [\tau(s) \cdot (W \downarrow)(s)] \phi(s) ds = 0$$

for all vector fields W and all functions ϕ of compact support vanishing on γ. This implies that for each s,

the tangent vector $\tau(s) \cdot (w \downarrow)$ is tangent to the curve γ for any tangent vector w at $\gamma(s)$. (∗)

(For otherwise we could find a function ϕ which vanished on γ and for which $[\tau(s) \cdot w]\phi \neq 0$. By extending w to a vector field W with $W(\gamma(s)) = w$ and modifying ϕ if necessary so as to vanish outside a small neighborhood of $\gamma(s)$ we could then arrange that the integral on the left hand side of the preceding equation would not vanish.)

The symmetry of $\tau(s)$ then implies that

$$\tau(s) = c(s) \gamma'(s) \otimes \gamma'(s)$$

for some scalar function, c. Indeed, in local coordinates suppose that $\gamma'(s) = \sum v^i \partial_{i\gamma(s)}$ and $\tau(s) = \sum t^{ij} \partial_{i\gamma(s)} \partial_{j\gamma(s)}$. Apply condition (∗) successively to the basis vectors $w = (dx^i_{\gamma(s)}) \uparrow$ so that $W \downarrow = dx^i_{\gamma(s)}$: We conclude that $t^{ij} = c^i v^j$ and hence from $t^{ij} = t^{ji}$ that $t^{ij} = cv^i v^j$. (Indeed, if some $v^i \neq 0$, say $v^1 \neq 0$ then $c^j v^1 = c^1 v^j$ implies that $c^j = c^1 v^j / v^1$ so $t^{ij} = c^j v^i = cv^i v^j$ with $c = c^1/v^1$.)

Let us assume that

$$\tau(s) \neq 0$$

anywhere so $c(s) \neq 0$ anywhere. Changing the parameterization means multiplying $\gamma'(s)$ by a scalar factor, and hence multiplying τ by a positive factor. So by reparametrizing the curve we can arrange that $\tau = \pm \gamma' \otimes \gamma'$. To avoid carrying around the $\pm$ sign, let us assume that $\tau = \gamma' \otimes \gamma'$. Since multiplying τ by -1 does not change the validity of (13.16), we may make this choice without loss of generality.

Again let us choose $V = \phi W$, but this time with no restriction on ϕ, but let us use the fact that $\tau(s) = \gamma'(s) \otimes \gamma'(s)$. We get

$$\begin{aligned} \tau \bullet \nabla V \downarrow &= \tau \bullet [d\phi \otimes W \downarrow + \phi \nabla W \downarrow] \\ &= (\gamma' \phi)\langle \gamma', W\rangle + \phi \langle \nabla_{\gamma'} W, \gamma'\rangle \\ &= \gamma' (\phi \langle \gamma', W\rangle) - \phi \langle W, \nabla_{\gamma'} \gamma'\rangle. \end{aligned}$$

The passage from the second line to the third consists of two applications of Leibnitz's rule:

$$\gamma'(\phi\langle \gamma', W\rangle) = (\gamma'\phi)\langle \gamma', W\rangle + \phi \gamma'\langle \gamma', W\rangle$$

and

$$\gamma'\langle W, \gamma'\rangle = \langle \nabla_{\gamma'} W, \gamma'\rangle + \langle W, \nabla_{\gamma'} \gamma'\rangle.$$

13.6. THREE DIFFERENT CHARACTERIZATIONS OF A GEODESIC.

We have proved that

$$\tau \bullet \nabla V \downarrow = \gamma' \left(\phi \langle \gamma', W \rangle \right) - \phi \langle W, \nabla_{\gamma'} \gamma' \rangle$$

when $V = \phi W$.

The integral of this expression must vanish for every vector field and every function ϕ of compact support. We claim that this implies that $\nabla_{\gamma'} \gamma' \equiv 0$, i.e. that γ **is a geodesic!**

Indeed, suppose that $\nabla_{\gamma'(s)} \gamma'(s) \neq 0$ for some value, s_0, of s. We could then find a tangent vector w at $\gamma(s_0)$ such that $\langle w, \nabla_{\gamma'(s_0)} \gamma'(s_0) \rangle = 1$ and then extend w to a vector field W, and so $\langle W, \nabla_{\gamma'(s)} \gamma'(s) \rangle > 0$ for all s near s_0. Now choose $\phi \geq 0$ with $\phi(s_0) = 1$ and of compact support. Indeed, choose ϕ to have support contained in a small neighborhood of $\gamma(s_0)$, so that

$$\int_I \gamma' \left(\phi \langle \gamma', W \rangle \right) ds = \left(\phi \langle \gamma', W \rangle \right)(\gamma(b)) - \left(\phi \langle \gamma', W \rangle \right)(\gamma(a)) = 0$$

where $a < s_0 < b$ are points in I with $\gamma(b)$ and $\gamma(a)$ outside the support of ϕ. We are thus left with

$$\ell_\tau(L_{\phi W}(\mathbf{g})) = - \int_a^b \phi \langle W, \nabla_{\gamma'} \gamma' \rangle ds < 0,$$

contradicting our hypothesis. Conversely, if γ is a geodesic and $\tau = \gamma' \otimes \gamma'$ then

$$\tau \bullet \nabla V \downarrow = \langle \nabla_{\gamma'} V, \gamma' \rangle = \gamma' \langle V, \gamma' \rangle - \langle V, \nabla_{\gamma'} \gamma' \rangle.$$

The second term vanishes since γ is a geodesic, and the integral of the first term vanishes so long as γ extends beyond the support of V or if γ is a closed curve. We have thus proved a remarkable theorem of Einstein, Infeld and Hoffmann:

Theorem 30. *If τ is a continuous (contravariant second order) nowhere vanishing symmetric tensor field along a curve γ whose associated linear function, ℓ_τ satisfies (13.16) then we can reparametrize γ so that it becomes a geodesic and so that $\tau = \pm \gamma' \otimes \gamma'$. Conversely, if τ is of this form and if γ is unbounded or closed then ℓ_τ satisfies (13.16).*

(Here "unbounded" means that for any compact region, K, there are real numbers a and b such that $\gamma(s) \notin K$, $\forall s > b$ or $< a$.)

The Einstein, Infeld, Hoffmann paper is "The Gravitational Equations and the Problem of Motion", *Annals of Mathematics*, **39** (1938). The treatment presented above follows J.M. Souriau "Modèle de particule à spin dans le champ électromagntique et gravitationnel" *Annales de l'institut Henri Poincaré*, **20** (1974).

13.6 Three different looking characterizations of a geodesic.

Our starting definition of a geodesic (in terms of a connection) was that it was self-parallel (i.e. had acceleration zero). This generalized the notion of a straight line in Euclidean geometry as "always pointing in the same direction".

In the case of the Levi-Civita connection associated to a semi-Riemannian metric, we saw that a geodesic could be characterized as being stationary relative to variations of arc length. (In Riemannian geometry, this generalized the statement in Euclidean geometry that "a straight line is the shortest distance between two points".)

We now have a third characterization of a geodesic as given by the theorem of Einstein, Infeld and Hoffmann.

The rest of this chapter is devoted to establishing some more formulas that I will need in discussing the Einstein field equations. I will show that the space of all connections on a manifold is an affine space. This implies that its "tangent space" at any point can be identified with a fixed vector space.

We will then think of the Levi-Civita theorem as providing a map from the space $\mathcal{M}$ of semi-Riemaniann metrics to the space of connections, and compute its differential.

We will also compute the differential of the map which assigns to each metric its Ricci curvature, and the differential of the map which assigns to each metric its scalar curvature.

13.7 The space of connections as an affine space.

Let let ∇ and ∇' be two (linear) connections on a manifold M. Then

$$\nabla_{fX}(gY) - \nabla'_{fX}(gY) = fg(\nabla_X Y - \nabla'_X Y).$$

In other words, the map

$$A : (X, Y) \to \nabla_X Y - \nabla'_X Y$$

is a tensor; its value at any point p depends only on the values of X and Y at p. So

$$A = \nabla - \nabla'$$

is a tensor field of type (1,2), i.e. of type $T^* \otimes T^* \otimes T$ (one which assigns to every tangent vector at $p \in M$ an element of $\text{Hom}(TM_p, TM_p)$).

Conversely, if A is any such tensor field and if ∇' is any connection then $\nabla = \nabla' + A$ is another connection. Thus the space of all connections is an *affine space* whose associated *linear space* is the space of all A's. We will be interested in torsion free connections, in which case the A's are restricted to being symmetric: $A_X Z = A_Z X$. (Check this as an exercise.) Let $\mathcal{A}$ denote the space of all such (smooth) symmetric A and let $\mathcal{C}$ denote the space of all torsion free connections.

13.8. THE LEVI-CIVITA MAP AND ITS DERIVATIVE.

So we can identify the "tangent space" to $\mathcal{C}$ at any connection ∇ with the space $\mathcal{A}$, because in any affine space we can identify the tangent space at any point with the associated linear space. In symbols, we may write

$$T\mathcal{C}_\nabla = \mathcal{A},$$

independent of the particular ∇. Once again we will be interested in variations of compact support in the connection, so we will want to consider the space

$$\mathcal{A}_{\text{compact}}$$

consisting of tensor fields of our given type of compact support.

13.8 The Levi-Civita map and its derivative.

We can think of Levi-Civita's theorem as defining a map (call it $L.C.$) from the space of semi-Riemannian metrics on a manifold to the space of torsion free connections:

$$L.C.: \mathcal{M} \to \mathcal{C}.$$

The value of $L.C.(\mathbf{g})$ at any point depends only on g_{ij} and its first derivatives at the point, and hence the differential of the Levi-Civita map can be considered as a linear map

$$d(L.C.)_\mathbf{g} : T\mathcal{M}_{\text{compact}} \to \mathcal{A}_{\text{compact}}.$$

(The spaces on both sides are independent of $\mathbf{g}$ but the differential definitely depends on $\mathbf{g}$.) In what follows, we will let A denote the value of this differential at a given $\mathbf{g}$ and $\mathbf{S} \in T\mathcal{M}_{\text{compact}}$:

$$A := d(L.C.)_\mathbf{g}[\mathbf{S}].$$

13.8.1 The Riemann curvature and the Ricci tensor as a maps.

The map R associates to every metric its Riemann curvature tensor. The map Ric associates to every metric its Ricci curvature. For reasons that will become apparent in the next chapter, we need to compute the differentials of these maps.

The curvature is expressed in terms of the connection by:

$$R_{XY} = [\nabla_X, \nabla_Y] - \nabla_{[X,Y]}.$$

So we may think of the right hand side of this equation as defining a map, curv, from the space of connections to the space of tensors of curvature type. Differentiating this expression using Leibniz's rule gives, for any $A \in \mathcal{A}$,

$$(d\text{curv}_\nabla[A])(X,Y) = A_X \nabla_Y + \nabla_X A_Y - A_Y \nabla_X - \nabla_Y A_X - A_{[X,Y]}. \quad (13.17)$$

We have

$$[X,Y] = \nabla_X Y - \nabla_Y X$$

288 CHAPTER 13. PRELIMINARIES TO THE EINSTEIN EQUATIONS.

so
$$A_{[X,Y]} = A_{\nabla_X Y} - A_{\nabla_Y X}.$$

On the other hand, the covariant differential, ∇A of the tensor field A with respect to the connection, ∇ is given by

$$(\nabla A)(X,Y)Z = \nabla_X(A_Y Z) - A_{\nabla_X Y} Z - A_Y \nabla_X Z$$

or, more succinctly,

$$(\nabla A)(X,Y) = \nabla_X A_Y - A_{\nabla_X Y} - A_Y \nabla_X.$$

So we have the equations

$$\begin{aligned}
(d\mathrm{curv}_\nabla[A])(X,Y) &= A_X \nabla_Y + \nabla_X A_Y - A_Y \nabla_X - \nabla_Y A_X - A_{[X,Y]}. \quad (13.17)\\
A_{[X,Y]} &= A_{\nabla_X Y} - A_{\nabla_Y X}.\\
(\nabla A)(X,Y) &= \nabla_X A_Y - A_{\nabla_X Y} - A_Y \nabla_X, \text{ so}\\
(\nabla A)(Y,X) &= \nabla_Y A_X - A_{\nabla_Y X} - A_X \nabla_Y
\end{aligned}$$

From this we see that

$$(d\mathrm{curv}_\nabla[A])(X,Y) = (\nabla A)(X,Y) - (\nabla A)(Y,X).$$

If we let $\tilde{\nabla} A$ denote the tensor obtained from ∇A by $\tilde{\nabla} A(X,Y) = \nabla A(Y,X)$ we can write this equation even more succinctly as

$$d\mathrm{curv}_\nabla[A] = \nabla A - \tilde{\nabla} A. \qquad (13.18)$$

If we substitute $A = d(L.C.)_\mathbf{g}[S]$ into this equation we get, by the chain rule, the value of $dR_\mathbf{g}[S]$. Taking the contraction, C, which yields the Ricci tensor from the Riemann tensor, we obtain

$$d\mathrm{Ric}_\mathbf{g}[S] = C(\nabla A - \tilde{\nabla} A). \qquad (13.19)$$

Let $\hat{\mathbf{g}}$ denote the contravariant symmetric tensor corresponding to $\mathbf{g}$, the scalar product induced by $\mathbf{g}$ on the cotangent space at each point. Thus, for example, the scalar curvature, S, is obtained from the Ricci curvature by contraction with $\hat{\mathbf{g}}$:

$$S = \hat{\mathbf{g}} \bullet \mathrm{Ric}.$$

13.9 An important integral identity.

Contract equation (13.19) with $\hat{\mathbf{g}}$ and use the fact that ∇ commutes with contraction with $\hat{\mathbf{g}}$ and with C. We claim that the result is

$$\hat{\mathbf{g}} \bullet d\mathrm{Ric}_\mathbf{g}[S] = C(\nabla V) \qquad (13.20)$$

where V is the vector field

$$V := C(A) \uparrow - \hat{\mathbf{g}} \bullet A. \qquad (13.21)$$

13.9. AN IMPORTANT INTEGRAL IDENTITY.

Proof. The contraction C applied to $\tilde{\nabla} A$ involves contracting the (unique) contravariant position against the position corresponding to the covariant derivative. We pass the double contraction with $\hat{\mathbf{g}}$ through these operations and so obtain $C(\nabla(\hat{\mathbf{g}} \bullet A))$. The contraction C applied to ∇A is just $\nabla(C(A))$ since covariant differential commutes with contraction. Taking the double contraction of this with $\hat{\mathbf{g}}$ yields $C(\nabla C(A) \uparrow)$. □

We have $C(\nabla V) = \text{div } V$. Also V has compact support since **S** does. Hence we obtain, from the divergence theorem, the following important result:

$$\int_M \hat{\mathbf{g}} \bullet d\text{Ric}_{\mathbf{g}}[\mathbf{S}] \, g = 0. \tag{13.22}$$

Chapter 14

Die Grundlagen der Physik.

This was the title of Hilbert's 1915 paper. It sounds a bit audacious, but let us try to put the ideas in a general context.

14.1 The structure of physical laws.

14.1.1 The Legendre transformation.

The Legendre transformation in one variable.

Let f be a function of one real variable. We can consider the map $t \mapsto f'(t)$ which is known as the Legendre transformation, or the "point slope transformation", $\mathcal{L}(f)$, associated to f. For example, if $f = \frac{1}{2}kt^2$ then the associated Legendre transformation is the linear map $t \mapsto kt$. As for any transformation, we might be interested in computing its inverse. That is, find the (or a) point t with a given value of $f'(t)$.

The Legendre transformation in two variables.

For a function, f, of two variables we can make the same definition and pose the same question: Define $\mathcal{L}(f)$ as the map

$$\begin{pmatrix} x \\ y \end{pmatrix} \mapsto (\partial f/\partial x,\ \partial f/\partial y).$$

Given (a, b) we may ask to solve the equations

$$\begin{aligned} \partial f/\partial x &= a \\ \partial f/\partial y &= b \end{aligned}$$

for $\begin{pmatrix} x \\ y \end{pmatrix}$.

The Legendre transformation in general.

The general situation is as follows: Suppose that $\mathcal{M}$ is a manifold whose tangent bundle is trivialized, i.e. that we are given a smooth identification of $T\mathcal{M}$ with $\mathcal{M} \times \mathcal{V}$; all the tangent spaces are identified with a fixed vector space, $\mathcal{V}$. Of course this also gives an identification of all the cotangent spaces with the fixed vector space $\mathcal{V}^*$. In this situation, if F is a function on $\mathcal{M}$, the associated Legendre transformation is the map

$$\mathcal{L}(F) : \mathcal{M} \to \mathcal{V}^*, \quad \mathbf{x} \mapsto dF_{\mathbf{x}}.$$

In particular, given $\ell \in \mathcal{V}^*$, we may ask to find $\mathbf{x} \in \mathcal{M}$ which solves the equation

$$dF_{\mathbf{x}} = \ell. \tag{14.1}$$

14.1.2 Inverting the Legendre transformation as the "source equation" of physics.

I will now try to explain that equation (14.1) is the "source equation" of physics, with the caveat, that the function F might not really be defined, but its variation (under a suitable class of perturbations) i.e. the left hand side of (14.1) will be defined. This best explained by example:

14.2 The Newtonian example.

In Newtonian physics, the background is the Euclidean geometry of three dimensional space (God's *sensorium* to use Newton's terminology). The objects are conservative force fields which are linear differential forms that are closed. With a mild loss of generality let us consider "potentials" instead of force fields, so the objects are functions, ϕ on Euclidean three space.

So our space $\mathcal{M}$ consists of all (smooth) functions. Since $\mathcal{M}$ is a vector space, its tangent space is automatically identified with $\mathcal{M}$ itself, so $\mathcal{V} = \mathcal{M}$. The force field associated with the potential ϕ is $-d\phi$, and its "energy density" at a point is one half the square of its Euclidean length. That is, the energy density is given by

$$\frac{1}{2}(\phi_x^2 + \phi_y^2 + \phi_z^2)$$

where subscript denotes partial derivative.

The **energy density** is given by

$$\frac{1}{2}(\phi_x^2 + \phi_y^2 + \phi_z^2).$$

We would like to define the function F to be the "total energy"

$$F(\phi) = \frac{1}{2}\int_{\mathbb{R}^3} (\phi_x^2 + \phi_y^2 + \phi_z^2) dxdydz$$

14.2. THE NEWTONIAN EXAMPLE. 293

but there is no reason to believe that this integral need converge. However, suppose that s is a smooth function of compact support, K.

Thus s vanishes outside the closed bounded set, K. For any bounded set, B, the integral
$$F^B(\phi) := \frac{1}{2}\int_B (\phi_x^2 + \phi_y^2 + \phi_z^2)dxdydz$$
converges, and the derivative
$$\frac{dF^B[\phi+ts]}{dt}\bigg|_{t=0} = \int_{\mathbb{R}^3}(\phi_x s_x + \phi_y s_y + \phi_z s_z)dxdydz$$
exists and is independent of B so long as $B \supset K$.

So it is reasonable to define the right hand side of this equation as dF_ϕ evaluated at s:
$$dF_\phi[s] := \int_{\mathbb{R}^3}(\phi_x s_x + \phi_y s_y + \phi_z s_z)dxdydz$$
even though the function F itself is not defined.

Of course to do so, we must not take $\mathcal{V} = \mathcal{M}$ but take $\mathcal{V}$ to be of the subspace consisting of smooth functions of compact support.

A linear function on $\mathcal{V}$ is just a "generalized density", but in Euclidean space, with Euclidean volume density $dxdydz$ we may identify densities with functions. Suppose that ρ is a smooth function, and we let ℓ_ρ be the corresponding element of $\mathcal{V}^*$,
$$\ell_\rho(s) = \int_{\mathbb{R}^3} s\rho dxdydz.$$
Equation (14.1) with $\ell = \ell_\rho$ becomes
$$\int_{\mathbb{R}^3}(\phi_x s_x + \phi_y s_y + \phi_z s_z)dxdydz = \int_{\mathbb{R}^3} s\rho dxdydz \quad \forall s \in \mathcal{V},$$
which is to be regarded as an equation for ϕ where ρ is given.

We have $\phi_x s_x + \phi_y s_y + \phi_z s_z = (\phi_x s)_x + (\phi_y s)_y + (\phi_z s)_z + s\Delta\phi$ where Δ is the Euclidean Laplacian,
$$\Delta\phi = -(\phi_{xx} + \phi_{yy} + \phi_{zz}).$$
Thus, since the total derivatives $(s\phi_x)_x$ etc. contribute zero to the integral, equation (14.1) becomes the Poisson equation
$$\Delta\phi = \rho.$$

As we know, a solution to this equation is given by convolution with the $1/r$ potential:
$$\phi(x,y,z) = \frac{1}{4\pi}\int \frac{\rho(\xi,\eta,\zeta)}{\sqrt{(x-\xi)^2 + (y-\eta)^2 + (z-\zeta)^2}}d\xi d\eta d\zeta$$
if ρ has compact support, for example, so that this integral converges. In this sense Euclidean geometry determines the $1/r$ potential.

14.3 The passive equations.

Symmetries of the function F may lead to constraints on the right hand side of (14.1). In our example of a function of two variables, suppose that the function f on the plane is invariant under rotations. Thus f would have to be a function of the radius, r, and hence the right hand side of (14.1) would have to be proportional to dr, and in particular, vanish on vectors tangent to the circle through the point $\begin{pmatrix} x \\ y \end{pmatrix}$.

More generally, suppose that $\mathcal{G}$ is group acting on $\mathcal{M}$, and that the function F is invariant under the action of this group, i.e.

$$F(a \cdot \mathbf{x}) = F(\mathbf{x}) \quad \forall a \in \mathcal{G}.$$

Let $\mathcal{O} = \mathcal{G} \cdot \mathbf{x}$ denote the orbit through $\mathbf{x}$, so $\mathcal{G} \cdot \mathbf{x}$ consists of all points of the form $a\mathbf{x}$, $a \in \mathcal{G}$. Then the function F is constant on $\mathcal{O}$ and so $dF_\mathbf{x}$ must vanish when evaluated on the tangent space to $\mathcal{O}$. We may write this symbolically as

$$\ell \in (T\mathcal{O})^0_\mathbf{x} \tag{14.2}$$

if (14.1) holds. Of course, in the infinite dimensional situations where we want to apply this equation, we must use some imagination to understand what is meant by the tangent space to the orbit.

We want to consider what happens when we modify ℓ by adding to it a "small" element, $\mu \in \mathcal{V}^*$. Presumably the solution $\mathbf{x}$ to our "source equation" (14.1) would then be modified by a small amount and so the tangent space to the orbit would change. We would then have to apply (14.2) to $\ell + \mu$ using the modified tangent space. [One situation where disregarding this change in $\mathbf{x}$ could be justified is when $\ell = 0$. Presumably the modification of $\mathbf{x}$ will be of first order in μ, and hence the change in (14.2) will be a second order effect which can be ignored if μ is small.]

A *passive equation* of physics is where we apply (14.2) but disregard the change in the tangent space and so obtain the equation

$$\mu \in (T\mathcal{O})^0_\mathbf{x}. \tag{14.3}$$

The justification for ignoring the non-linear effect of μ of $\mathbf{x}$ may be problematical from our abstract point of view, but the equation we have just obtained for the passive reaction of μ to the presence of $\mathbf{x}$ is a powerful principle of physics. About half the laws of physics are of this form.

We have enunciated two principles of physics, a source equation (14.1) which amounts to inverting a Legendre transformation, and the passive equation (14.3) which is a consequence of symmetry. We now turn to how Hilbert and Einstein implemented these principles for gravity.

14.4 The Hilbert "function".

David Hilbert

**Born: 23 Jan. 1862 in Konigsberg, Prussia
(now Kaliningrad, Russia)
Died: 14 Feb. 1943 In Gottingen, Germany**

The space $\mathcal{M}$ is the space of Lorentzian metrics on a given manifold, M. Hilbert chooses as his function

$$F(\mathbf{g}) = -\int_M Sg, \quad S = \hat{\mathbf{g}} \bullet \text{Ric}(\mathbf{g}).$$

As discussed above, this "function" need not be defined since the integral in question need not converge. But the differential

$$dF_\mathbf{g}[\mathbf{S}]$$

will be defined when evaluated on a variation $\mathbf{S}$ of compact support. I apologize for the notation: S denotes the scalar curvature of the given metric $\mathbf{g}$ and $\mathbf{S}$ denotes a variation of this metric (i.e. an element of $T\mathcal{M}_\mathbf{g}$ of compact support).

The integral defining F involves **g** at three locations: in the definition of the density g, in the dual metric $\hat{\mathbf{g}}$ and in Ric. Thus, by Leibniz's rule

$$-dF_{\mathbf{g}}[\mathbf{S}] = \int_M \hat{\mathbf{g}} \bullet \mathrm{Ric}(\mathbf{g}) dg[\mathbf{S}] + \int_M d\hat{\mathbf{g}}[\mathbf{S}] \bullet \mathrm{Ric}(\mathbf{g}) g + \int_M \hat{\mathbf{g}} \bullet d\mathrm{Ric}_{\mathbf{g}}[\mathbf{S}] g.$$

We have already done the hard work involved in showing that the third integral vanishes in the preceding chapter, see equation (13.22). So we are left with the first two terms.

The first term.

The first term involves the infinitesimal change $dg[\mathbf{S}]$ of the volume density. We recall a computation we did in local coordinates in the preceding chapter: With apologies for the overuse of Δ, we have

$$g = Gdx, \quad G = |\det(\langle \partial_i, \partial_j \rangle)|^{\frac{1}{2}} = (\epsilon \det(\langle \partial_i, \partial_j \rangle))^{\frac{1}{2}}$$

where

$$\epsilon := \mathrm{sgn}\det(\langle \partial_i, \partial_j \rangle).$$

Write

$$\Delta := \det(\langle \partial_i, \partial_j \rangle) = \det(g_{ij}).$$

To see how Δ depends differentially on g_{ij} expand the determinant along the i-th row:

$$\Delta = \det(g_{ij}) = \sum_j g_{ij} \Delta^{ij}$$

where Δ^{ij} are the corresponding cofactors. If we think of Δ as a function of the n^2 variables, g_{ij} then, since none of the Δ_{ik} (for a fixed i) involve g_{ij}, we conclude from the above cofactor expansion that

$$\frac{\partial \Delta}{\partial g_{ij}} = \Delta^{ij}.$$

By Cramer's rule,

$$\Delta^{ij} = \Delta g^{ij}$$

where (g^{ij}) is the inverse "matrix" of (g_{ij}). So from $G = (\epsilon \Delta)^{\frac{1}{2}}$ we obtain

$$\frac{\partial G}{\partial g_{ij}} = \frac{1}{2}(\epsilon \Delta)^{-\frac{1}{2}} \epsilon \Delta g^{ij}$$

so

$$\frac{\partial G}{\partial g_{ij}} = \frac{1}{2} G g^{ij}. \tag{13.11}$$

The coordinate free way or writing this is

$$dg_{\mathbf{g}}[\mathbf{S}] = \frac{1}{2}\hat{\mathbf{g}} \bullet \mathbf{S} g.$$

14.4. THE HILBERT "FUNCTION".

This takes care of the first term in

$$-dF_{\mathbf{g}}[\mathbf{S}] = \int_M \hat{\mathbf{g}} \bullet \text{Ric}(\mathbf{g}) dg[\mathbf{S}] + \int_M d\hat{\mathbf{g}}[\mathbf{S}] \bullet \text{Ric}(\mathbf{g}) g.$$

The second term.

As to the second term,

$$\int_M d\hat{\mathbf{g}}[\mathbf{S}] \bullet \text{Ric}(\mathbf{g}) g,$$

recall that in local coordinates, $\hat{\mathbf{g}}$ is given by $\sum g^{ij}\partial_i\partial_j$ where (g^{ij}) is the inverse matrix of g_{ij}. So we recall a formula we derived for the differential of the inverse function of a matrix that you derived in the first problem set: If inv denotes the inverse function, so

$$\text{inv}(B) = B^{-1},$$

then it follows from differentiating the identity

$$BB^{-1} \equiv I$$

using Leibniz's rule that

$$d\,\text{inv}_B[C] = -B^{-1}CB^{-1}.$$

It follows that the differential of the function $\mathbf{g} \mapsto \hat{\mathbf{g}}$ when evaluated at $\mathbf{S}$ is $-\mathbf{S}\uparrow\uparrow$, the contravariant symmetric tensor obtained from $-\mathbf{S}$ by applying the raising operator (coming from $\mathbf{g}$) twice. Now

$$(\mathbf{S}\uparrow\uparrow) \bullet \text{Ric} = \mathbf{S} \bullet \text{Ric}\uparrow\uparrow.$$

So if we define

$$\text{RIC} := \text{Ric}\uparrow\uparrow$$

to be the contravariant form of the Ricci tensor we obtain

$$dF_{\mathbf{g}}[\mathbf{S}] = \int_M (\text{RIC} - \frac{1}{2}S\hat{\mathbf{g}}) \cdot \mathbf{S}g. \tag{14.4}$$

This is left hand side of the source equation (14.1). The right hand side is a linear function on the space $T(\mathcal{M})_{\text{compact}}$. We know that if $\mathbf{T}$ is a smooth symmetric second order contravariant tensor field, then it defines a linear function on $T(\mathcal{M})_{\text{compact}}$ given by

$$\ell_{\mathbf{T}}(\mathbf{S}) = \int_M \mathbf{S} \bullet \mathbf{T} g.$$

Thus for $\ell = \ell_{\mathbf{T}}$ equation (14.1) becomes the celebrated Einstein field equations

$$\text{RIC} - \frac{1}{2}S\hat{\mathbf{g}} = \mathbf{T} \tag{14.5}$$

So if we regard the physical objects as Lorentzian metrics, and if we believe that matter determines the metric, by a source type equation, then matter should be considered as a linear function on $T(\mathcal{M})_{\text{compact}}$. In particular a "smooth" matter distribution is a contravariant symmetric tensor field. If we believe that the laws of physics are described by the function given by Hilbert, we get the Einstein field equations. Modifying the function would change the source equations. For example, if we replace S by $S + c$ where c is a constant, this would have the effect of adding a term $\frac{1}{2}c\hat{\mathbf{g}}$ to the left hand side of the field equations. This is the "cosmological constant" term that entered into the De Sitter solution, as we discussed in Section 4.13.1, and which leads to the current theories of the expanding universe as discussed in Section 6.7.1 .

We will take our group of symmetries to be the group of diffeomorphisms of M of compact support - diffeomorphisms which are the identity outside some compact set. Such transformations preserve the function F.

If V is a vector field of compact support which generates a one parameter group, ϕ_s of transformations, then these transformations have compact support, and the fact that the function F is invariant under these transformations translates into the assertion that $dF_{\mathbf{g}}[L_V \mathbf{g}] = 0$.
In other words, the "tangent space to the orbit through $\mathbf{g}$" is the subspace of $T(\mathcal{M})_{\text{compact}}$ consisting of all $L_V \mathbf{g}$ where V is a vector field of compact support. From the results obtained above we now know that the passive equation translates into

$$\text{div } \mathbf{T} = 0$$

for a smooth tensor field and into

$$T = \pm \gamma' \otimes \gamma', \quad \gamma \text{ a geodesic}$$

for a continuous tensor field concentrated along a curve.

I repeat: These results are independent of the choice of F satisfying our invariance condition. In this sense, the "passive equations" (14.3) are more general than the source equations (14.1).

To summarize: We have two sorts of laws of physics:

- Source equations (14.1)
$$dF_{\mathbf{x}} = \ell$$
where we are inverting a Legendre transformation. The Einstein field equations are of this type.

- Passive equations which describe the motion of a small object in the presence of a "force field" where we ignore the effect on the field produced by the small object. They are given by the "group theoretical" equation (14.2)
$$\ell \in (T\mathcal{O})_{\mathbf{x}}^0.$$

14.5 Harmonic maps and minimal immersions as solutions to a passive equation.

Let us return to equation (14.3) in the setting of the group of diffeomorphisms of compact support of a manifold M acting on the semi-Riemannian metrics. In the case that our linear function μ was given by a "delta function tensor field supported along a curve" we saw that condition (14.3) implies that the curve γ is a geodesic and the tensor field is $\pm \gamma' \otimes \gamma'$ (under suitable reparametrization of the curve and assuming that the tensor field does not vanish anywhere on the curve). We now examine what condition (14.3) says for a "delta function tensor field" on a more general submanifold. So we are interested in the condition

$$\mu(\mathcal{S}(\nabla V \downarrow)) = 0$$

for all V of compact support where μ is provided by the following data:

1. A k dimensional manifold Q and a proper map $f : Q \to M$,

2. A smooth section $\mathbf{t}$ of $f^*S^2(TM)$, so $\mathbf{t}$ assigns to each $q \in Q$ an element $\mathbf{t}(q) \in S^2 TM_{f(q)}$, and

3. A density ω on Q.

For any section s of $S^2 T^*M$ and any $q \in Q$ we can form the "double contraction" $s(q) \bullet \mathbf{t}(q)$ since $s(q)$ and $\mathbf{t}(q)$ take values in dual vector spaces, and since f is proper, if s has compact support then so does the function $q \mapsto s(q) \bullet \mathbf{t}(q)$ on Q. We can then form the integral

$$\mu[s] := \int_Q s(\cdot) \bullet \mathbf{t}(\cdot) \omega. \tag{14.6}$$

We observe (and this will be important in what follows) that μ depends on the tensor product $\mathbf{t} \otimes \omega$ as a section of $f^*S^2 TM \otimes \mathbf{D}$ where $\mathbf{D}$ denotes the line bundle of densities of Q rather than on the individual factors.

We apply the equation $\mu(\mathcal{S}(\nabla V \downarrow)) = 0$ to this μ and to $v = \phi W$ where ϕ is a function of compact support and W a vector field of compact support on M. Since

$$\nabla(\phi W) = d\phi \otimes W + \phi \nabla W$$

and $\mathbf{t}$ is symmetric, this becomes

$$\int_Q \mathbf{t} \bullet (d\phi \otimes W \downarrow + \phi \nabla W \downarrow) \omega = 0. \tag{14.7}$$

We first apply this to a ϕ which vanishes on $f(Q)$, so that the term $\phi \nabla W$ vanishes when restricted to Q. We conclude that the "single contraction" $\mathbf{t} \cdot \theta$ must be tangent to $f(Q)$ at all points for all linear differential forms θ and hence that

$$\mathbf{t} = df_* \mathbf{h}$$

for some section **h** of $S^2(TQ)$.

Again, let us apply condition (14.7), but no longer assume that ϕ vanishes on $f(Q)$. For any vector field Z on Q let us, by abuse of language, write

$$Z\phi \quad \text{for} \quad Zf^*\phi,$$

for any function ϕ on M, write

$$\langle Z, W \rangle \quad \text{for} \quad \langle df_* Z, W \rangle_M$$

where W is a vector field on M, and

$$\nabla_Z W \quad \text{for} \quad \nabla_{df_* Z} W.$$

Write

$$\mathbf{h} = \sum h^{ij} e_i e_j$$

in terms of a local frame field $e_1, \ldots, e_k$ on Q. Then

$$\mathbf{t} \bullet (\nabla V \downarrow) = \sum h^{ij} \left[e_i(\phi) \langle e_j, W \rangle + \phi \langle \nabla_{e_i} W, e_j \rangle \right].$$

Now

$$\langle \nabla_{e_i} W, e_j \rangle = e_i \langle W, e_j \rangle - \langle W, \nabla_{e_i} e_j \rangle$$

so

$$\mathbf{t} \bullet \nabla V \downarrow = \sum_{ij} \left[h^{ij} e_i(\phi \langle e_j, W \rangle) - \phi \langle W, h^{ij} \nabla_{e_i} e_j \rangle \right].$$

Also,

$$\int_Q \sum h^{ij} e_i (\phi \langle e_j, W \rangle) \omega = - \int_Q \phi \langle e_j, W \rangle L_{\sum_i h^{ij} e_i} \omega.$$

Let us write

$$z^j = \text{div}_\omega (\sum h^{ij} e_i)$$

so

$$L_{\sum_i h^{ij} e_i} \omega = w^j \omega.$$

If we set

$$Z := \sum z^j e_j$$

then condition (14.7) becomes

$$\sum_{ij} h^{ij} {}^M\nabla_{e_i} e_j = -Z, \tag{14.8}$$

where we have used ${}^M\nabla$ to emphasize that we are using the covariant derivative with respect to the Levi-Civita connection on M, i.e.

$$ {}^M\nabla_{e_i} e_j := \nabla_{df_* e_i}(df_* e_j).$$

14.5. HARMONIC MAPS AS SOLUTIONS TO A PASSIVE EQUATION.

To understand (14.8) suppose that we assume that $\mathbf{h}$ is non-degenerate, and so induces a semi-Riemannian metric $\check{\mathbf{h}}$ on Q, and let us *assume* that ω is the volume form associated with $\check{\mathbf{h}}$. (In all dimensions except $k = 2$ this second assumption is harmless, since we can rescale $\mathbf{h}$ to arrange it to be true.) Let $^{\mathbf{h}}\nabla$ denote covariant differential with respect to $\check{\mathbf{h}}$. Let us choose the frame field $e_1, \ldots, e_k$ to be "orthonormal" with respect to $\check{\mathbf{h}}$, i.e.

$$h^{ij} = \epsilon_j \delta_{ij}, \quad \text{where } \epsilon_j = \pm 1$$

so that

$$\sum_i h^{ij} e_i = \epsilon_j e_j.$$

Then

$$L_{e_j}\omega = C(^{\mathbf{h}}\nabla e_j)\omega$$

and

$$C(^{\mathbf{h}}\nabla e_j) = \sum_i \epsilon_i \langle ^{\mathbf{h}}\nabla_{e_i} e_j, e_i \rangle_{\check{\mathbf{h}}} = -\sum_i \langle e_j, \epsilon_i^{\mathbf{h}}\nabla_{e_i} e_i \rangle_{\check{\mathbf{h}}},$$

so

$$Z = -\sum_j \sum_i \epsilon_j \langle e_j, \epsilon_i^{\mathbf{h}}\nabla_{e_i} e_i \rangle_{\check{\mathbf{h}}} e_j = -\sum_i \epsilon_i^{\mathbf{h}}\nabla_{e_i} e_i = -\sum_{ij} {^{\mathbf{h}}}\nabla_{e_i} e_j.$$

Given a metric $\check{\mathbf{h}}$ on Q, a metric $\mathbf{g}$ on M, the **second fundamental form of a map** $f : Q \to M$, is defined as

$$B_f(X, Y) := {^{\mathbf{g}}}\nabla_{df(X)}(df(Y)) - df(^{\mathbf{h}}\nabla_X Y). \tag{14.9}$$

Here X and Y are vector fields on Q and $df(X)$ denotes the "vector field along f" which assigns to each $q \in Q$ the vector $df_q(X_q) \in TM_{f(q)}$.

The **tension field** $\tau(f)$ of the map f (relative to a given $\mathbf{g}$ and $\check{\mathbf{h}}$) is the trace of the second fundamental form so

$$\tau(f) = \sum_{ij} h^{ij} {^{\mathbf{g}}}\nabla_{df(e_i)}(df(e_j)) - df(^{\mathbf{h}}\nabla_{e_i} e_j)$$

in terms of local frame field.

A map f such that $\tau(f) \equiv 0$ is called **harmonic**. We thus see that under the above assumptions about $\mathbf{h}$ and ω

Theorem 31. *Condition (14.7) says that f is harmonic relative to $\mathbf{g}$ and $\check{\mathbf{h}}$.*

Suppose that we make the further assumption that $\check{\mathbf{h}}$ is the metric induced from $\mathbf{g}$ by the map f. Then

$$df(^{\mathbf{h}}\nabla_X Y) = (^{\mathbf{g}}\nabla_{df(X)} df(Y))^{\tan},$$

the tangential component of ${^{\mathbf{g}}}\nabla_{df(X)} df(Y)$ and hence

$$B_f(X, Y) = (^{\mathbf{g}}\nabla_{df(X)} df(Y))^{\text{nor}},$$

the normal component of $^{\mathbf{g}}\nabla_{df(X)} df(Y)$. This is just the classical second fundamental form vector of Q regarded as an immersed submanifold of M. Taking its trace gives kH where H is the mean curvature vector of the immersion. Thus if in addition to the above assumptions we make the assumption that the metric $\check{h}$ is induced by the map f, then we conclude that (14.3) says that $H = 0$, i.e. that the immersion f must be a minimal immersion.

14.6 Schrodinger's equation as a passive equation.

In quantum mechanics, the background is a complex Hilbert space. In order to avoid technicalities, let us assume that $\mathcal{H}$ is a finite dimensional complex vector space with an Hermitian scalar product. Let $\mathcal{M}$ denote the space of all self adjoint operators on $\mathcal{H}$. Let $\mathcal{G}$ be the group of all unitary operators, and let $\mathcal{G}$ act on $\mathcal{M}$ by conjugation: $U \in \mathcal{G}$ acts on $\mathcal{M}$ by sending

$$A \mapsto UAU^{-1}.$$

Since $\mathcal{M}$ is a vector space, its tangent bundle is automatically trivialized. We may also identify the space of linear functions on $\mathcal{M}$ with $\mathcal{M}$ by assigning to $B \in \mathcal{M}$ the linear function ℓ_B defined by

$$\ell_B(A) = \operatorname{tr} AB.$$

If C is a self adjoint matrix, the tangent to the curve

$$\exp(itC) A \exp(-itC)$$

at $t = 0$ is $i[C, A]$. So the "tangent space to the orbit through A" consists of all $i[C, A]$.
The passive equation (14.3) becomes

$$[A, B] = 0$$

for $\mu = \ell_B$. A linear function is called a *pure state* if it is of the form ℓ_B where B is projection onto a one dimensional subspace. This means that there is a unit vector $\phi \in \mathcal{H}$ (determined up to phase) so that

$$Bu = (u, \phi)\phi \quad \forall u \in \mathcal{H}$$

where (,) denotes the scalar product on $\mathcal{H}$.
A pure state satisfies (14.3) if and only if ϕ is an eigenvector of H:

$$H\phi = \lambda\phi$$

for some real number λ. This is the (time independent) Schrodinger equation.

Chapter 15

The Frobenius theorem.

Ferdinand Georg Frobenius

Born: 26 Oct. 1849 in Berlin-Charlottenburg, Prussia (now Germany)
Died: 3 Aug. 1917 in Berlin, Germany

In this chapter I want to prove and apply a theorem of Frobenius. Strictly speaking, this important theorem belongs in a text on the general theory of differentiable manifolds. But since it has so many applications in differential

15.1 The Frobenius theorem.

Let X be a vector field on manifold M. The fundamental existence theorem of ordinary differential equations tells us, at least locally, that X determines a flow, i.e. a locally defined one parameter group. Through each $p \in M$ there is a curve $t \mapsto \phi_t p$, the "trajectory through the point p". If we replace X by fX where f is a smooth nowhere vanishing function, the new flow has the "same" trajectories, the change is in the "speed" along each trajectory. Thus the curves, up to reparametrization, are determined by a "field of lines" rather than a vector field.

15.1.1 Differential systems.

We can study a more general situation: a "field of k-dimensional subspaces". We assume that we are given at each point a k-dimensional subspace of the tangent space, and that this subspace varies smoothly from point to point. Such a family will be denoted by $\mathfrak{D}$. We say that $\mathfrak{D}$ is a "sub-bundle of the tangent bundle" or we say that $\mathfrak{D}$ is a **differential system** or we say that $\mathfrak{D}$ is a **distribution**. These are all different names for the same thing.

In some subjects, such as control theory, one allows the dimension k to vary from point to point. I explicitly exclude this possibility from the current chapter.

Integral submanifolds of a differential system.

A manifold N together with an immersion $\iota : N \to M$ is called an **integral manifold** of $\mathfrak{D}$ if at each $q \in N$

$$d\iota_q(T_q N) \subset \mathfrak{D}(\iota(q)) \tag{15.1}$$

where $\mathfrak{D}(m)$ is the subspace of $T_m(M)$ given by $\mathfrak{D}$.

The example of a one form.

For example, if α a nowhere vanishing one form on M, then at each $x \in M$ we get a differential system of codimension one where $\mathfrak{D}_x \subset T_x M$ consists of those $v \in T_x M$ such that $\langle \alpha_x, v \rangle = 0$.

If there are functions (locally defined) T and S such that $\alpha = TdS$ then at each x_0 the space $\mathfrak{D}_{x_0}$ is tangent to the hypersurface $S = S(x_0)$. If $\alpha = TdS$ then $d\alpha = dT \wedge dS$ and so

$$\alpha \wedge d\alpha \equiv 0.$$

On other hand, consider the form $\alpha = dz + xdy$ (say on three dimensional space). Then

$$\alpha \wedge d\alpha = dx \wedge dy \wedge dz$$

15.1. THE FROBENIUS THEOREM.

is nowhere zero. So we can not write $\alpha = TdS$ on any neighborhood. As we will see, the planes of the differential system $\mathfrak{D}$ defined by α do not "fit together" to be tangent to surfaces.

15.1.2 Foliations, submersions, and fibrations.

Submersions.

Let Q and B be manifolds. A **submersion** is a smooth map $f : Q \to B$ such that for all $q \in Q$ the differential

$$df_q : T_q Q \to T_{f(q)} B$$

is surjective. For any submersion the "fibers"

$$f^{-1}(a), \quad a \in B$$

are smooth embedded submanifolds of Q of dimension $k = n - \dim B$. This follows from the implicit function theorem.

Foliations.

In fact Q is **foliated** by such submanifolds in the following sense: Every point $q \in Q$ has a coordinate neighborhood U with coordinates $x_1, \ldots, x_n$ with q corresponding to $x = 0$, and $f(q)$ a coordinate neighborhood with coordinates $y_1, \ldots y_{n-k}$ with $f(q)$ corresponding to $y = 0$ such that in terms of these coordinates, the map f is given by by projection to the first $n - k$ coordinates. Of course, the set of tangent spaces (at all points of Q) to the fibers of a fibration form a differential system.

More generally, we say that a differential system $\mathfrak{D}$ is a **foliation** or is **completely integrable** if it is tangent to to a foliation in the preceding sense: Every point $q \in Q$ has a coordinate neighborhood U with coordinates $x_1, \ldots, x_n$ with q corresponding to $x = 0$ and a map $f : U \to \mathbb{R}^{n-k}$ with

$$f(x_1, \ldots, x_n) = (x_1, \ldots, x_{n-k})$$

and such that at every point q of U, the space $\mathfrak{D}_q$ is the tangent space to the fiber of f through q.

For example, as we have seen, every one dimensional differential system is completely integrable. This was an immediate consequence of the existence theorem for ordinary differential equations. But not every one dimensional foliation is a submersion. For example, the irrational line foliation on a two dimensional torus.

Also, as we will see, the two dimensional differential system on $\mathbb{R}^3$ defined by the null planes of $dz + xdy$ is *not* completely integrable. The Frobenius theorem to be stated below (in two versions) gives a useful necessary and sufficient condition for a differential system to be a foliation.

Fibrations.

A more stringent condition on a submersion is for $f : Q \to B$ to be a **fibration**. We say that f is a fibration if it is surjective and has the local triviality condition: There exists a manifold F (called the standard fiber) such that about every point in B there is a neighborhood W and a diffeomorphism $\phi : f^{-1}(W) \to W \times F$ which conjugates f into projection to the first factor.

15.1.3 The vector fields of a differential system, the Frobenius theorem.

Let $\mathfrak{D}$ be a differential system on some manifold M. We let $\mathfrak{X}(\mathfrak{D})$ denote the space of vector fields X on M with the property that $X_p \in \mathfrak{D}_p$ at all p.

Theorem 32. Theorem of Frobenius. *A necessary and sufficient condition for $\mathfrak{D}$ to be completely integrable is that $\mathfrak{X}(\mathfrak{D})$ be "involutive" meaning that it is closed under Lie bracket, i.e.*

$$X, Y \in \mathfrak{X}(\mathfrak{D}) \Rightarrow [X, Y] \in \mathfrak{X}(\mathfrak{D}).$$

Proof of necessity.

Proof. If $\mathfrak{D}$ is completely integrable of dimension k, every point has a coordinate neighborhood with coordinates $x_1, \ldots, x_n$ such that

$$\frac{\partial}{\partial x_i} \in \mathfrak{X}(\mathfrak{D}) \text{ for } i = 1, \ldots k.$$

So the values of these vector fields span $\mathfrak{D}_p$ at each point of the neighborhood and hence any element of $\mathfrak{X}(\mathfrak{D})$ is a sum of elements of $\mathfrak{X}(\mathfrak{D})$ of the form $a\frac{\partial}{\partial x_i}$ for some $1 \leq i \leq k$ and where a is smooth function. But

$$\left[a\frac{\partial}{\partial x_i}, b\frac{\partial}{\partial x_j} \right] = a\frac{\partial b}{\partial x_i}\frac{\partial}{\partial x_j} - b\frac{\partial a}{\partial x_j}\frac{\partial}{\partial x_i}$$

lies in $\mathfrak{X}(\mathfrak{D})$. □

For example, consider the family $\mathfrak{D}$ of two planes in $\mathbb{R}^3$ determined by the equation $\alpha = 0$, $\alpha = dz + xdy$. The corresponding $\mathfrak{X}(\mathfrak{D})$ is spanned at every point by the vector fields

$$\frac{\partial}{\partial x} \text{ and } \frac{\partial}{\partial y} - x\frac{\partial}{\partial z}.$$

But the Lie bracket of these two vector fields is $-\frac{\partial}{\partial z}$ which does not belong to $\mathfrak{X}(\mathfrak{D})$. So this $\mathfrak{D}$ is **not** completely integrable.

We will see that our general definition of "curvature" will be a measure of the failure of the Frobenius condition to hold.

15.1. THE FROBENIUS THEOREM.

The proof of the sufficiency is a bit more involved and will be by induction on k. We know the theorem to be true for $k = 1$ where the condition is vacuous and all differential systems are completely integrable.

Proof of sufficiency.

Proof. Let $x \in M$ and let $X_1, \ldots, X_k$ be vector fields defined in a neighborhood of x and whose values at each point p in the neighborhood span $\mathfrak{D}_p$. Let S be a submanifold of dimension $n-1$ passing through x and such that X_1 is nowhere tangent to S. Solve the ordinary differential equation corresponding to X_1 with initial conditions for the trajectories to lie in S at time 0. If $y_2, \ldots y_n$ are coordinates on S, and if $y_1(p)$ denotes the time at which a trajectory starting at S reaches p, then the variables $y_1, \ldots y_n$ are coordinates about x in terms of which

$$X_1 = \frac{\partial}{\partial y_1}.$$

Let

$$f_i := X_i y_1$$

and set

$$Y_1 = X_1, \quad Y_i := X_i - f_i X_1, \quad i = 2, \ldots, k.$$

Then the vector fields $Y_1, \ldots Y_k$ are still linearly independent at all points in a neighborhood about x and span $\mathfrak{D}$ in this neighborhood, and, in addition,

$$Y_i y_1 = 0, \quad i = 2, \ldots, k.$$

The preceding equation implies that the Y_i are all tangent to the initial hypersurface S given by $y_1 = 0$.

So there are vector fields $Z_2, \ldots, Z_k$ on S such that

$$d\iota_q(Z_{iq}) = Y_{iq}$$

at all $q \in S$, where $\iota : S \to M$ denotes the injection of S into M. The vector fields $Z_2, \ldots, Z_k$ define a differential system of dimension $k-1$ on S which satisfies Frobenius's criterion that $[Z_i, Z_j]$ lie in the subspace spanned by the Z_i at all points. Indeed, if this were not true at some point q of S then the the same would hold for $[Y_i, Y_j]$ since none of the Y_i have any component in $\frac{\partial}{\partial y_1}$ direction.

So by induction we conclude that we can find coordinates $w_2, \ldots, w_n$ on S near x so that the differential system spanned by $Z_2, \ldots, Z_k$ is tangent to the foliation given by projection onto the last $n-k$ coordinates. Then

$$(x_1, \ldots, x_n) = (y_1, w_2, \ldots, w_n)$$

is a system of coordinates about x in M and

$$\frac{\partial x_i}{\partial y_1} = 0, \quad i = 2, \ldots n$$

so
$$Y_1 = \frac{\partial}{\partial y_1} = \frac{\partial}{\partial x_1}.$$

In particular,
$$Y_1 x_s = 0 \quad \text{for } s = k+1, \ldots, n.$$
Therefore there are functions c_{ij} such that
$$\begin{aligned}
\frac{\partial}{\partial x_1}(Y_i x_s) &= Y_1(Y_i x_s) \\
&= Y_1(Y_i x_s) - Y_i(Y_1 x_s) \\
&= [Y_1, Y_i] x_s \\
&= c_{i1} Y_1 x_s + \sum_{j=2}^{k} c_{ij} Y_j x_s \quad \text{since } [Y_1, Y_i] \in \mathfrak{X}(\mathfrak{D}) \\
&= \sum_{j=2}^{k} c_{ij} Y_j x_s.
\end{aligned}$$

So the $Y_i x_s$ satisfy the system of (ordinary) homogeneous linear differential equations
$$\frac{\partial}{\partial x_1}(Y_i x_s) = \sum_{j=2}^{k} c_{ij} Y_j x_s$$
with the initial conditions at $x_1 = 0$ given by
$$Y_i x_s = Z_i x_s = 0.$$
The uniqueness theorem for differential equations then implies that
$$Y_i x_s \equiv 0, \quad i \leq k, \quad s > k.$$
So the vector fields Y_i are function linear combinations of
$$\frac{\partial}{\partial x_1}, \ldots, \frac{\partial}{\partial x_k}$$
and since the $Y_1, \ldots, Y_k$ are linearly independent everywhere, this shows that $\mathfrak{D}$ is completely integrable

$\square$

15.1.4 Connected and maximal leaves of an integrable system.

Let $\mathfrak{D}$ be a completely integrable differential system of dimension k on a manifold M. We know that there is a covering of M by coordinate neighborhoods $(U_\alpha, x_\alpha^1, \ldots, x_\alpha^n)$ such that in U_α the integral manifolds of $\mathfrak{D}$ are given by
$$x_\alpha^{k+1} = \text{const.}, \quad \ldots, \quad x_\alpha^n = \text{const.} \,.$$

15.2. MAPS INTO A LIE GROUP.

Suppose that N is a connected integral manifold of $\mathfrak{D}$. More precisely, suppose that N is a connected manifold, together with a smooth map $f : N \to M$ such that at each $q \in N$ equation (15.1) holds:

$$df_q(T_q N) \subset \mathfrak{D}(f(q)).$$

Any $x \in N$ has a neighborhood V such that $f(V) \subset U_\alpha$ for some α. Thus $f(V)$ is contained in a subset of U_α given by $x_\alpha^{k+1} = $ const., ..., $x_\alpha^n = $ const. for suitable constants. Let y be another point of N. If C is a curve in N joining x to y, we can cover C by a finite number of neighborhoods of type V. Thus x and y can be joined by a finite number of curves C_i with $C_i(1) = C_{i+1}(0)$ such that

$$f(C_i) \text{ is an integral curve of } \mathfrak{D}.$$

This gives us a clue as how to construct a *maximal* integral manifold of $\mathfrak{D}$: Let p be a point of M. Let $K \subset M$ be the set of all $q \in M$ such that there exist a finite number of curves C_i with $C_i(1) = C_{i+1}(0)$ such that each C_i is a one dimensional integral submanifold of $\mathfrak{D}$. We sketch how to make K into a differentiable manifold:

Let $q \in K$. Then $q \in U_\alpha$ for some α and the set of points $z \in U_\alpha$ given by

$$x_\alpha^{k+1}(z) = x_\alpha^{k+1}(q), \ldots, x_\alpha^n(z) = x_\alpha^n(q)$$

also belong to K. We take this collection of points z as a coordinate neighborhood of q with coordinates $x_\alpha^1, \ldots, x_\alpha^k$. It is routine to check that this makes K into a manifold. This manifold is *not*, in general, an embedded submanifold of M even in the case $k = 1$ as evidenced by the irrational flow on the torus. But from the previous discussion it is clear that any integral manifold of $\mathfrak{D}$ passing through p maps into a subset of K. So we have proved:

Theorem 33. *Let $\mathfrak{D}$ be a completely integrable differential system on a manifold M. Through each $p \in M$ there passes a maximal connected integral manifold K_p (called the **maximal leaf** of the foliation determined by $\mathfrak{D}$). Any other connected integral manifold of $\mathfrak{D}$ passing though p is a submanifold of K_p*

15.2 Maps into a Lie group.

Let G be a Lie group with Lie algebra $\mathfrak{g}$ and let $\mathfrak{h}$ be a subalgebra of $\mathfrak{g}$. We may think of $\mathfrak{h}$ as the left invariant differential system whose value at the identity e is the subspace given by $\mathfrak{h}$. Then $\mathfrak{h}$ determines a (left invariant) completely integrable system by Frobenius's theorem. Let H be the maximal leaf passing through e. If $y \in H$, then $L_{y^{-1}} y = e$ so $L_{y^{-1}}$ takes H into H. In other words, if $x, y \in H$ then $y^{-1} x \in H$. this implies that H is a subgroup of G and the manifold structure on H (as given by the theorem on maximal leaves) shows that H is a Lie group in its own right.

Notice that any maximal leaf K of the differential system determined by $\mathfrak{h}$ is a coset of H: Indeed, any left translate of K is again a maximal leaf since our differential system is left invariant. If we left translate by the inverse of an element of K we get H.

15.2.1 Applying the above to the diagonal.

We are going to apply the preceding discussion with G replaced by $G \times G$, so that the full Lie algebra is now $\mathfrak{g} \oplus \mathfrak{g}$ and the subalgebra $\mathfrak{h}$ is the "diagonal" subalgebra consisting of all
$$\xi \oplus \xi, \quad \xi \in \mathfrak{g}.$$
Let π_1 and π_2 denote the projections of $G \times G$ onto the first and second components. Let ω range over all (or a basis of) the left invariant linear differential forms on G. Then the differential system given by our diagonal subalgebra $\mathfrak{h}$ is clearly determined by the equations
$$\pi_1^* \omega - \pi_2^* \omega = 0.$$

Theorem 34. *Let f_1 and f_2 be two differentiable maps of a connected manifold M into a Lie group G such that*
$$f_1^* \omega = f_2^* \omega \tag{15.2}$$
as ω ranges over (a basis of) the left invariant forms of G. Then f_1 and f_2 differ by a left translation. That is, there is an $a \in G$ such that
$$f_2 = L_a \circ f_1.$$

Proof. Let $\phi : M \to G \times G$,
$$\phi(m) := (f_1(m), f_2(m)).$$
Then (15.2) asserts that ϕ makes M into a connected integral submanifold of $G \times G$ for the differential system determined by the diagonal subalgebra $\mathfrak{h}$. So $\phi(M)$ lies in a coset of the diagonal. In other words, there exist elements $x, y \in G$ such that $L_x \circ f_1 = L_y \circ f_2$. $\square$

15.2.2 The induced metric and the Weingarten map determine a hypersurface up to a Euclidean motion.

Theorem 35. *Let ϕ_1 and ϕ_2 be two immersions of an oriented n-dimensional manifold M int $\mathbb{R}^{n+1}$. If they induce the same Riemann metric and the same Weingarten map they differ by a Euclidean motion.*

Proof. For the proof we recall some of the discussion in Problem set **1**:

Given an oriented submanifold of dimension n in $\mathbb{R}^{n+1}$, we obtain the manifold of adapted frames whose last vector is the unit normal, and whose first n-vectors are tangent to the submanifold and form an orthonormal frame in the induced metric. Since the induced metrics by the immersion are the same, we

can identify the two bundles of frames. So we can think of a single $\mathcal{O}(M)$ and two maps f_1 and f_2 of $\mathcal{O}(M)$ into the set of all Euclidean frames (which we identify with the Euclidean group (by a choice of an initial frame)) and such that
$$f_1^*\theta_i = \vartheta_i = f_2^*\theta_i, \quad 1 = 1,\ldots,n$$
and
$$f_2^*\Theta_{ij} = \Theta_{ij} = f_2^*\Theta_{ij} \quad \text{for } i,j \leq n$$
The θ_i and the Θ_{ij} (for $i,j = 1,\ldots,n+1$) span the left invariant forms on the group of Euclidean motions. We know that $f_1^*\theta_{n+1} = f_2^*\theta_{n+1} = 0$. So we could apply Theorem 34 if we knew that
$$f_1^*\Theta_{i,n+1} = f_2^*\Theta_{i,n+1}, \quad i = 1,\ldots,n.$$
Now on the space of all Euclidean frames we have
$$de_{n+1} = \sum_{i=1}^{n} \Theta_{n+1,i} e_i.$$
The pull-back $f_i^* e_{n+1}$ is just $\nu_i \circ \pi$ where ν_i is the Gauss map for f_i and $\pi : \mathcal{O}(M) \to M$ is the projection of the bundle of frames down to M. Thus the $f_i^*\Theta_{n+1,i}$ give the expression of the Weingarten map in terms of a frame. So if the Weingarten maps are the same, so are the $\Theta_{i,n+1}$, proving the theorem. $\square$

15.3 Another application of Frobenius: to reduction.

Let Ω be a **closed** form of any degree. Consider the set of vector fields which satisfy
$$i(X)\Omega = 0.$$
This may or may not define a differential system, since the dimension of the space spanned by such X_p may vary with p. Suppose that this dimension is constant. Then

Proposition 32. *The differential system we get is a foliation.*

Proof. If $i(X)\Omega = 0$ then Weil's formula implies that
$$L_X\Omega = di(X)\Omega + i(X)d\Omega = 0.$$
Hence if $i(Y)\Omega = 0$ then
$$i([X,Y])\Omega = L_X(i(Y)\Omega) = 0.$$

$\square$

We call this foliation, that is the foliation spanned by the vector fields satisfying $i(X)\Omega = 0$ the **null foliation** of Ω.

There is a very pretty generalization of this fact due Cartan:

Let $\mathfrak{I}$ be an ideal in the ring of differential forms on a manifold M which is homogeneous: if $\sigma \in \mathfrak{I}$ then all the homogeneous components of σ belong to $\mathfrak{I}$. Consider the set of vector fields X on M which satisfy

$$i(X)\mathfrak{I} \subset \mathfrak{I}.$$

In other words, the set of all vector fields X with the property that

$$\sigma \in \mathfrak{I} \Rightarrow i(X)\sigma \in \mathfrak{I}.$$

Once again, this may or may not define a differential system, since the dimension of the space spanned by the X_p may vary from one p to another. Suppose that this dimension is constant. So we get a differential system which is called the **characteristic system** of the ideal $\mathfrak{I}$.

Theorem 36. [Cartan.] *If $d\mathfrak{I} \subset \mathfrak{I}$, its characteristic system is completely integrable.*

The condition $d\mathfrak{I} \subset \mathfrak{I}$ means that if $\sigma \in \mathfrak{I}$ then $d\sigma \in \mathfrak{I}$.

Proof. If $i(X)\mathfrak{I} \subset \mathfrak{I}$ and $\sigma \in \mathfrak{I}$ then

$$L_X \sigma = i(X)d\sigma + di(X)\sigma \in \mathfrak{I}.$$

So if $i(Y)\sigma \in \mathfrak{I}$ then $L_X(i(Y)\sigma) \in \mathfrak{I}$. But

$$L_X(i(Y)\sigma) = i([X,Y])\sigma + i(Y)L_X\sigma$$

so if $i(X)\mathfrak{I} \subset \mathfrak{I}$ and $i(Y)\mathfrak{I} \subset \mathfrak{I}$ then $i([X,Y])\mathfrak{I} \subset \mathfrak{I}$. □

15.4 A dual formulation of Frobenius' theorem.

As a special case of Cartan's theorem, we can formulate a version of Frobenius's theorem involving differential forms instead of vector fields:

A differential system $\mathfrak{D}$ can be described locally as the set of all hyperplanes $\omega_1 = 0, \ldots, \omega_{n-k} = 0$ where the ω_i are linear differential forms. If $\mathfrak{D}$ is completely integrable, then we can find, locally, coordinates such that the $\omega_i = \sum_j a_{ij} dx_j$ where $A = (a_{ij})$ is an invertible matrix with function entries. In matrix and vector notation $\quad d\omega = Adx \quad$ where

$$\omega = \begin{pmatrix} \omega_1 \\ \vdots \\ \omega_{n-k} \end{pmatrix} \quad \text{and} \quad dx = \begin{pmatrix} dx_1 \\ \vdots \\ dx_{n-k} \end{pmatrix}.$$

So $d\omega = dA \wedge dx = (dA)A^{-1} \wedge \omega$. In other words, the ideal $\mathfrak{I}$ generated by the ω_i is closed under d. From Cartan's theorem just proved, we know that the converse is true. So

Theorem 37. Frobenius in dual form. *Suppose that $\mathfrak{D}$ is given as $\omega_1 = 0, \ldots, \omega_{n-k} = 0$. Then $\mathfrak{D}$ is completely integrable if and only if the ideal $\mathfrak{I}$ generated by the ω_i is closed under d. In more prosaic language this says that there are linear differential forms c_{ij} such that*

$$d\omega_i = \sum_j c_{ij} \wedge \omega_j.$$

15.5 Horizontal and basic forms of a fibration.

Let $\pi : Q \to B$ be a fibration. If τ is a differential form on B then $\sigma = \pi^*\tau$ is a differential form on Q with the following two properties:

- It is **horizontal** in the sense that if X is a vector field which is everywhere tangent to the fiber (we say that X is **vertical**) then

$$i(X)\sigma = 0$$

and

- It is vertically invariant in the sense that $L_X \sigma = 0$ for every vertical vector field.

Conversely, suppose that σ satisfies the first condition. Let us introduce local product coordinates $(x_1, \ldots, x_f, y_1, \ldots, y_b)$ where $(x_1, \ldots, x_f)$ are local coordinates on the fiber F and $(y_1, \ldots, y_b)$ are local coordinates on the base B. If σ is horizontal then σ must be a linear combination with function coefficients of products of the dy's. These functions a might depend on all the variables. But if σ also vertically invariant, they must satisfy

$$\frac{\partial a}{\partial x_i} \equiv 0. \quad i = 1, \ldots, f.$$

In other words they are locally constant in the fiber direction. If F is connected, this implies that $\sigma = \pi^*\tau$ for some form τ on B.

We say that σ is **basic** in the sense that it come from the base. We have proved:

Proposition 33. *Let $Q \to B$ be a fibration with connected fibers. Then a differential form on Q is basic if and only if it is horizontal and is vertically invariant.*

15.6 Reduction of a closed form.

Let Ω be a closed form on a manifold Q and consider the set of vector fields which satisfy

$$i(X)\Omega = 0.$$

Recall that this may or may not define a differential system, since the dimension of the space spanned by such X_p may vary with p. Suppose that this dimension is constant. We say that Ω has **constant rank**. Then we know that the differential system that we get is a foliation $\mathcal{D}$ called the **null foliation**.

Furthermore, by definition, Ω is horizontal with respect to this foliation in the sense that if $X \in \mathfrak{X}(\mathcal{D})$ then $i(X)\Omega = 0$ and is vertically invariant with respect to this foliation in the sense that if $X \in \mathfrak{X}(\mathcal{D})$ then $L_X\Omega = 0$. This follows from Weil's formula as we have seen.

Suppose that this foliation is a fibration $\pi : Q \to B$ with connected fibers. Then by the preceding proposition, we know that $\Omega = \pi^*\omega$ for a uniquely determined form ω on B. Since π^* is an injection and $0 = d\Omega = d\pi^*\omega = \pi^*d\omega$ we conclude that $d\omega = 0$. Finally, I claim that ω is non-degenerate in the following sense: For any $b \in B$ and $v \in T_bB$,

$$i(v)\omega_b = 0 \;\Rightarrow\; v = 0.$$

Indeed, let us introduce local product coordinates

$$(x_1, \ldots, x_f, y_1, \ldots, y_b)$$

around a point $q = (f, b)$ as above. If $i(v)\omega_b = 0$, then the vector $w \in T_qQ \cong T_fF \oplus T_bB$ given by

$$w = (0, v)$$

satisfies

$$i(w)\Omega_q = 0.$$

So $w \in \mathcal{D}_q$. But all vectors of $\mathcal{D}_q$ are vertical, i.e of the form $(z, 0)$. So $v = 0$.

We have proved:

Proposition 34. *Let Ω be a closed form of constant rank on a manifold Q. Suppose that its null foliation is a fibration $\pi : Q \to B$ with connected fibers. Then*

$$\Omega = \pi^*\omega$$

where ω is a non-degenerate closed form on B.

Chapter 16

Connections on principal and associated bundles.

According to the current "standard model" of elementary particle physics, every fundamental force is associated with a kind of curvature. But the curvatures involved are not only the geometric curvatures of space-time, but curvatures associated with the notion of a connection on a geometrical object (a "principal bundle") which is a generalization of the bundle of frames studied in Problem set **1**. In this chapter I will give a gentle introduction to this subject, starting with the Cartan formalism for semi-Riemannian geometry in a frame field.

16.1 Connection and curvature forms in a frame field.

I want to begin by reviewing frame fields and the Cartan calculus in semi-Riemannian geometry. So I start with a repeat of the material presented at the beginning of Chapter 6. Let $E = (E_1, \ldots, E_n)$ be a(n "orthonormal") frame field and

$$\theta = \begin{pmatrix} \theta^1 \\ \vdots \\ \theta^n \end{pmatrix}$$

the dual frame field so

$$\text{id} = E_1 \otimes \theta^1 + \cdots + E_n \otimes \theta^n$$

or

$$\text{id} = (E_1, \ldots, E_n) \otimes \begin{pmatrix} \theta^1 \\ \vdots \\ \theta^n \end{pmatrix}$$

315

where id is the tautological tensor field which assigns the identity map to each tangent space.

We write this more succinctly as

$$\text{id} = E\theta.$$

The (matrix of) connection form(s) in terms of the frame field is then determined by

$$d\theta + \varpi \wedge \theta = 0$$

and the (anti-)symmetry properties of ϖ via Cartan's lemma.

I am now using ϖ instead of ω for the matrix of connection forms because I want to reserve ω for a related but different meaning. I am also now using $\overline{\Omega}$ for the associated curvature in a frame field. The curvature is determined by by

$$d\varpi + \varpi \wedge \varpi = \overline{\Omega}.$$

We now repeat an argument that we gave when discussing the general Maurer Cartan form:

Recall that for any two form τ and a pair of vector fields X and Y we write $\tau(X,Y) = i(Y)i(X)\tau$. Thus

$$(\varpi \wedge \varpi)(X,Y) = \varpi(X)\varpi(Y) - \varpi(Y)\varpi(X),$$

the commutator of the two matrix valued functions, $\varpi(X)$ and $\varpi(Y)$. Consider the commutator of two matrix valued one forms, ϖ and σ,

$$\varpi \wedge \sigma + \sigma \wedge \varpi$$

(according to our usual rules of superalgebra). We denote this by

$$[\varpi \wedge, \sigma].$$

In particular we may take $\varpi = \sigma$ to obtain

$$[\varpi \wedge, \varpi] = 2\varpi \wedge \varpi.$$

We can thus also write the curvature as

$$\overline{\Omega} = d\varpi + \frac{1}{2}[\varpi \wedge, \varpi].$$

This way of writing the curvature has useful generalizations when we want to study connections on principal bundles later on in this chapter.

16.2 Change of frame field.

Suppose that E' is a second frame field whose domain of definition overlaps with the domain of definition of E. On the intersection of their domains of definition

16.2. CHANGE OF FRAME FIELD.

we must have

$$\begin{aligned} E'_1 &= c_{11}E_1 + \cdots + c_{n1}E_n \\ E'_2 &= c_{12}E_1 + \cdots + c_{n2}E_n \\ &\vdots \qquad \vdots \\ E'_n &= c_{1n}E_1 + \cdots + c_{nn}E_n \end{aligned}$$

where the c_{ij} are functions which fit together to form an "orthogonal" matrix. So we can write this more succinctly as

$$E' = EC$$

where C is a(n "orthogonal") matrix valued function.

Let θ' be the dual frame field of E'. On the common domain of definition we have

$$EC\theta' = E'\theta' = \mathrm{id} = E\theta$$

so

$$\theta = C\theta'.$$

Let ϖ' be the connection form associated to θ', so ϖ' is determined (using Cartan's lemma) by the anti-symmetry condition and

$$d\theta' + \varpi' \wedge \theta' = 0.$$

Then

$$d\theta = d(C\theta') = dC \wedge \theta' + C d\theta' = dCC^{-1} \wedge \theta - C\varpi'C^{-1} \wedge \theta$$

implying that

$$\varpi = -dCC^{-1} + C\varpi'C^{-1}$$

or

$$\varpi' = C^{-1}\varpi C + C^{-1}dC. \tag{16.1}$$

We have

$$\varpi' \wedge \varpi' = C^{-1}\varpi \wedge \varpi C + C^{-1}\varpi \wedge dC + C^{-1}dCC^{-1} \wedge \varpi C + C^{-1}dC \wedge C^{-1}dC$$

while

$$d\varpi' = d(C^{-1}) \wedge \varpi C + C^{-1}d\varpi C - C^{-1}\varpi \wedge dC + d(C^{-1}) \wedge dC.$$

Now it follows from

$$C^{-1}C \equiv I$$

that

$$d(C^{-1}) = -C^{-1}dCC^{-1}$$

and hence from the expression
$$\overline{\Omega}' = \varpi' \wedge \varpi' + d\varpi'$$
we get
$$\overline{\Omega}' = C^{-1}\overline{\Omega}C. \tag{16.2}$$

Notice that this equation contains the assertion that the curvature is a tensor. Indeed, recall that for any pair of tangent vectors $\xi, \eta \in TM_p$ the matrix $\Omega(\xi, \eta)$ gives the matrix of the operator $R_{\xi\eta} : TM_p \to TM_p$ relative to the orthonormal basis $E_1(p), \ldots, E_n(p)$. Let $\zeta \in TM_p$ be a tangent vector at p and let z^i be the coordinates of ζ relative to this basis so $\zeta = z^1 E_1 + \cdots z^n E_n$ which we can write as
$$\zeta = E(p)z \quad \text{where} \quad z = \begin{pmatrix} z^1 \\ \vdots \\ z^n \end{pmatrix}.$$

Then
$$R_{\xi\eta}\zeta = E(p)\overline{\Omega}(\xi, \eta)z.$$

If we use a different frame field $E' = EC$ then $\zeta = E'(p)z'$ where $z' = C^{-1}(p)z$. Equation (16.2) implies that
$$\overline{\Omega}'(\xi, \eta)z' = C^{-1}(p)\Omega(\xi, \eta)z$$
which shows that
$$E'(p)\overline{\Omega}'(\xi, \eta)z' = E(p)\overline{\Omega}(\xi, \eta)z.$$
Thus the transformation $\zeta \mapsto E(p)\overline{\Omega}(\xi, \eta)z$ is a well defined linear transformation. So if we did not yet know that $R_{\xi\eta}$ is a well defined linear transformation, we could conclude this fact from (16.2).

16.3 The bundle of frames.

We will now make a reinterpretation of the arguments of the preceding section which will have far reaching consequences. Let $\mathcal{O}(M)$ denote the set of all "orthonormal" bases of all TM_p. So a point, $\mathcal{E}$, of $\mathcal{O}(M)$ is an "orthonormal" basis of TM_p for some point $p \in M$, and we will denote this point by $\pi(\mathcal{E})$. So
$$\pi : \mathcal{O}(M) \to M, \quad \mathcal{E} \text{ is an o.n. basis of } TM_{\pi(\mathcal{E})}$$
assigns to each $\mathcal{E}$ the point at which it is the orthonormal basis.

Suppose that E is a frame field defined on an open set $U \subset M$. If $p \in U$, and $\pi(\mathcal{E}) = p$, then there is a unique "orthogonal" matrix A such that
$$\mathcal{E} = E(p)A.$$

16.3. THE BUNDLE OF FRAMES.

We will denote this matrix A by $\phi(\mathcal{E})$. (If we want to make the dependence on the frame field explicit, we will write ϕ_E instead of ϕ.) Thus

$$\mathcal{E} = E(\pi(\mathcal{E}))\phi(\mathcal{E}).$$

This gives an identification

$$\psi : \pi^{-1}(U) \to U \times G, \quad \psi(\mathcal{E}) = (\pi(\mathcal{E}), \phi(\mathcal{E})) \tag{16.3}$$

where G denotes the group of all "orthogonal" matrices.

It follows from the definition that

$$\phi(\mathcal{E}B) = \phi(\mathcal{E})B, \quad \forall B \in G. \tag{16.4}$$

Let E' be a second frame field defined on an open set U'. We have a map

$$C : U \cap U' \to G$$

such that

$$E' = EC$$

as before.

Thus

$$\mathcal{E} = E\phi_E(\mathcal{E}) = EC\left(\pi(\mathcal{E})\right)\phi_{E'}(\mathcal{E})$$

so

$$\phi_E \phi_{E'}^{-1} = C \circ \pi. \tag{16.5}$$

This shows that the identifications given by (16.3) define, in a consistent way, a manifold structure on $\mathcal{O}(M)$. The manifold $\mathcal{O}(M)$ together with the action of the "orthogonal group" G by "multiplication on the right"

$$R_A : \mathcal{E} \mapsto \mathcal{E} \circ A^{-1}$$

and the differentiable map $\pi : \mathcal{O}(M) \to M$ is called the **bundle of (orthonormal) frames**.

We will now define forms

$$\vartheta = \begin{pmatrix} \vartheta^1 \\ \vdots \\ \vartheta^n \end{pmatrix} \quad \text{and} \quad \omega = (\omega^i_j)$$

on $\mathcal{O}(M)$ such that

$$d\vartheta + \omega \wedge \vartheta = 0, \quad \epsilon_i \omega^i_j = -\epsilon_j \omega^j_i,$$

where $\epsilon_i = \langle E_i, E_i \rangle \, (= \pm 1)$.

16.3.1 The form ϑ.

Let $\xi \in T(\mathcal{O}(M))_{\mathcal{E}}$ be a tangent vector at the point $\mathcal{E} \in \mathcal{O}(M)$. Then $d\pi_{\mathcal{E}}(\xi)$ is a tangent vector to M at the point $\pi(\mathcal{E})$:

$$d\pi_{\mathcal{E}}(\xi) \in TM_{\pi(\mathcal{E})}.$$

As such, the vector $d\pi_{\mathcal{E}}(\xi)$ has coordinates relative to the basis, $\mathcal{E}$ of $TM_{\pi(\mathcal{E})}$ and these coordinates depend linearly on ξ. So we may write

$$d\pi_{\mathcal{E}}(\xi) = \vartheta^1(\xi)\mathcal{E}_1 + \cdots \vartheta^n(\xi)\mathcal{E}_n$$

defining the forms ϑ^i. As usual, we write this more succinctly as

$$d\pi = \mathcal{E}\vartheta.$$

16.3.2 The form ϑ in terms of a frame field.

Let $v \in T(\mathcal{O}(M))_{\mathcal{E}}$ be a tangent vector at the point $\mathcal{E} \in \mathcal{O}(M)$. Assume that $\pi(\mathcal{E})$ lies in the domain of definition of a frame field E and that $\mathcal{E} = E(p)A$ where $p = \pi(\mathcal{E})$. Let us write $d\pi(v)$ instead of $d\pi_{\mathcal{E}}(v)$ so as not to overburden the notation. We have

$$d\pi(v) = E(p)\theta(d\pi(v)) = \mathcal{E}\vartheta(v) = E(p)A\vartheta(v)$$

so

$$A\vartheta(v) = \theta(d\pi(v)). \tag{16.6}$$

Notice that θ is a vector valued linear differential form on U, the domain of definition of the frame field, E, while A is a matrix valued function on G. So $A^{-1}\theta$ is a vector valued linear differential form on $U \times G$.

So we can write (16.6) as

$$\vartheta = \psi^* \left[A^{-1}\theta \right] \tag{16.7}$$

where

$$A^{-1}\theta$$

is the one form defined on $U \times G$ by

$$A^{-1}\theta(\eta + \zeta) = A^{-1}(\theta(\eta)), \quad \eta \in TM_x, \quad \zeta \in TG_A.$$

Here we have made the standard identification of $T(U \times G)_{(x,A)}$ as a direct sum,

$$T(U \times G)_{(x,A)} \sim TM_x \oplus TG_A,$$

valid on any product space.

16.3.3 The definition of ω.

Next we will define ω in terms of the identification

$$\psi : \pi^{-1}(U) \to U \times G$$

given by a local frame field, and check that it satisfies

$$d\vartheta + \omega \wedge \vartheta = 0, \quad \epsilon_i \omega^i_j = -\epsilon_j \omega^j_i.$$

By Cartan's lemma, this uniquely determines ω, so the definition must be independent of the choice of frame field, and so ω is globally defined on $\mathcal{O}(M)$.

Let ϖ be the connection form (of the Levi-Civita connection) of the frame field E.

Define
$$\omega := \psi^* \left[A^{-1} \varpi A + A^{-1} dA \right] \tag{16.8}$$

where the expression in brackets on the right is a matrix valued one form defined on $U \times G$. Then on $U \times G$ we have

$$\begin{aligned} d[A^{-1}\theta] &= -A^{-1}dA \wedge A^{-1}\theta + A^{-1}d\theta \\ &= -A^{-1}dA \wedge A^{-1}\theta - A^{-1}\varpi A \wedge A^{-1}\theta \quad \text{so} \\ 0 &= d[A^{-1}\theta] + \left[A^{-1}\varpi A + A^{-1}dA \right] \wedge A^{-1}\theta. \end{aligned}$$

Applying ψ^* yields

$$d\vartheta + \omega \wedge \vartheta = 0.$$

as desired. The "anti-symmetry" condition says that ϖ takes values in the Lie algebra of G. Hence so does $A\varpi A^{-1}$ for any $A \in G$. We also know that $A^{-1}dA$ takes values in the Lie algebra of G. Hence so does ω.

16.4 Connection forms in a frame field as a pull-backs.

We now have a reinterpretation of the connection form, ϖ, associated to a frame field. Indeed, the form ω is a matrix valued linear differential form defined on all of $\mathcal{O}(M)$. A frame field, E, defined on an open set U, can be thought of as a map, $x \mapsto E(x)$ from U to $\mathcal{O}(M)$:

$$E : U \to \mathcal{O}(M), \quad x \mapsto E(x).$$

Then the pull-back of ω under this map is exactly ϖ, the connection form associated to the frame field! In symbols

$$E^* \omega = \varpi.$$

Claim: $E^*\omega = \varpi$.

To see this, observe that under the map $\psi : \pi^{-1}(U) \to U \times G$, we have $\psi(E(x)) = (x, I)$ where I is the identity matrix. Thus

$$\psi \circ E = (id, I)$$

where $id : U \to U$ is the identity map and I means the constant map sending every point x into the identity matrix. By the chain rule

$$\begin{aligned} E^*\omega &= E^*\psi^* \left[A^{-1}\varpi A + A^{-1} dA \right] \\ &= (\psi \circ E)^* \left[A^{-1}\varpi A + A^{-1} dA \right] \\ &= \omega. \qquad \square \end{aligned}$$

Thus, for example, the frame field E is parallel relative to a vector field, X on M if and only if $\nabla_X(E) \equiv 0$ which is the same as

$$i(X)\varpi \equiv 0$$

where ϖ is the connection form of the frame field. In view of the preceding result this is the same as

$$i\left[dE(X)\right]\omega \equiv 0.$$

Here $dE(X)$ denotes the vector field along the map $E : U \to \mathcal{O}(M)$ which assigns to each $x \in U$ the vector $dE_x(X(x))$.

Let me repeat this important point in a slightly different version. Suppose that $C : [0, 1] \to M$ is a curve on M, and we start with an initial frame $\mathcal{E}(0)$ at $C(0)$. We know that there is a unique curve $t \mapsto \mathcal{E}(t)$ in $\mathcal{O}(M)$ which gives the parallel transport of $\mathcal{E}(0)$ along the curve C. We have "lifted" the curve C on M to the curve $\gamma : t \mapsto \mathcal{E}(t)$ on $\mathcal{E}(M)$. The curve γ is completely determined by

- its initial value $\gamma(0)$,

- the fact that it is a lift of C, i.e. that $\pi(\gamma(t)) = C(t)$ for all t, and

- $$i(\gamma'(t))\omega = 0. \qquad (16.9)$$

I now want to describe two important properties of the form ω. For $B \in G$, recall that R_B denotes the transformation

$$R_B : \mathcal{O}(M) \to \mathcal{O}(M), \quad \mathcal{E} \mapsto \mathcal{E}B^{-1}.$$

We will use the same letter, R_B to denote the transformation

$$R_B : U \times G \to U \times G, \quad (x, A) \mapsto (x, AB^{-1}).$$

Because of (16.4), we may use this ambiguous notation since

$$\psi \circ R_B = R_B \circ \psi.$$

16.4. CONNECTION FORMS IN A FRAME FIELD AS A PULL-BACKS.

It then follows from the local definition (16.8) that

$$R_B^* \omega = B\omega B^{-1}. \tag{16.10}$$

Indeed

$$R_B^* \omega = R_B^* \psi^* \left[A^{-1} \varpi A + A^{-1} dA \right] = \psi^* R_B^* \left[A^{-1} \varpi A + A^{-1} dA \right]$$

and

$$R_B^* (A^{-1} \varpi A) = B(A^{-1} \varpi A) B^{-1}$$

since ϖ does not depend on G and

$$R_B^* (A^{-1} dA) = B(A^{-1} dA) B^{-1}.$$

We can write (16.10) as

$$R_B^* \omega = \mathrm{Ad}_B(\omega). \tag{16.11}$$

For the second property, I introduce some notation: Let ξ be a matrix which is "anti-symmetric" in the sense that

$$\epsilon_i \xi_j^i = -\epsilon_j \xi_i^j.$$

This implies that the one parameter group

$$t \mapsto \exp -t\xi = I - t\xi + \frac{1}{2} t^2 \xi^2 - \frac{1}{3!} t^3 \xi^3 + \cdots$$

lies in our group G for all t. Then the one parameter group of transformations

$$R_{\exp -t\xi} : \mathcal{O}(M) \to \mathcal{O}(M)$$

has as its infinitesimal generator a vector field, which we shall denote by X_ξ.

The one parameter group of transformations

$$R_{\exp -t\xi} : U \times G \to U \times G$$

also has an infinitesimal generator: Identifying the tangent space to the space of matrices with the space of matrices, we see that the vector field generating this one parameter group of transformations of $U \times G$ is

$$Y_\xi : (x, A) \mapsto A\xi.$$

So the vector field Y_ξ takes values at each point in the TG component of the tangent space to $U \times G$ and assigns to each point (x, A) the matrix $A\xi$. In particular $\varpi(Y_\xi) = 0$ since ϖ is only sensitive to the TU component. Also dA is by definition the tautological matrix valued differential form which assigns to any tangent vector Z the matrix Z. Hence

$$A^{-1} dA(Y_\xi) = \xi.$$

From
$$R_B \circ \psi = \psi \circ R_B$$
it follows that
$$\psi^*(Y_\xi) = X_\xi$$
and hence that
$$\omega(X_\xi) \equiv \xi. \tag{16.12}$$

Finally, the curvature form from the point of view of the bundle of frames is given as usual as
$$\Omega := d\omega + \frac{1}{2}[\omega \wedge, \omega]. \tag{16.13}$$

16.5 Submersions, fibrations, and connections.

16.5.1 Submersions.

Recall that a smooth map $\pi : Y \to X$ is called a **submersion** if $d\pi_y : TY_y \to TX_{\pi(y)}$ is surjective for every $y \in Y$. Suppose that X is n-dimensional and that Y is $n + k$ dimensional. The implicit function theorem implies the following for a submersion:

If $\pi : Y \to X$ is a submersion, then about any $y \in Y$ there exist coordinates $z^1, \ldots, z^n; y^1, \ldots, y^k$ (such that y has coordinates $(0, \ldots, 0; 0 \ldots, 0)$) and coordinates $x^1, \ldots, x^n$ about $\pi(y)$ such that in terms of these coordinates π is given by
$$\pi(z^1, \ldots, z^n; y^1, \ldots, y^k) = (z^1, \ldots, z^n).$$

In other words, locally in Y, a submersion looks like the standard projection from $\mathbb{R}^{n+k}$ to $\mathbb{R}^n$ near the origin. For the next few paragraphs I will let $\pi : Y \to X$ denote a submersion.

Vertical tangent vectors to a submersion.

For each $y \in Y$ we define the **vertical** subspace Vert_y of the tangent space TY_y to consist of those $\eta \in TY_y$ such that
$$d\pi_y(\eta) = 0.$$

In terms of the local description, the vertical subspace at any point in the coordinate neighborhood of y given above is spanned by the values of the vector fields
$$\frac{\partial}{\partial y^1}, \ldots, \frac{\partial}{\partial y^k}$$
at the point in question. This shows that the Vert_y fit together to form a smooth sub-bundle, call it Vert, of the tangent bundle TY.

16.5. SUBMERSIONS, FIBRATIONS, AND CONNECTIONS.

General connections on submersions.

Charles Ehresmann

Born: 19 April 1905 in Strasbourg, France
Died: 22 Sept. 1979 in Amiens, France

A **general connection** (or "Ehresmann connection") on the given submersion is a choice of complementary subbundle Hor to Vert. This means that at each $y \in Y$ we are given a subspace $\text{Hor}_y \subset TY_y$ such that

$$\text{Vert}_y \oplus \text{Hor}_y = TY_y$$

and that the Hor_y fit together smoothly to form a sub-bundle of TY. It follows from the definition that Hor_y has the same dimension as $TX_{\pi(y)}$ and, in fact, that the restriction of $d\pi_y$ to Hor_y is an isomorphism of Hor_y with $TX_{\pi(y)}$. We should emphasize that the vertical bundle Vert comes along with the notion of the submersion π. A connection Hor, on the other hand, is an additional piece of geometrical data above and beyond the submersion itself.

CHAPTER 16. CONNECTIONS ON PRINCIPAL BUNDLES.

This notion of a general connection and much of the material in the next few sections is due to Ehresmann.

Local descriptions of general connections.

Let us describe a (general) connection in terms of the local coordinates given above. The local coordinates $x^1, \ldots, x^n$ on X give rise to the vector fields

$$\frac{\partial}{\partial x^1}, \ldots, \frac{\partial}{\partial x^n}$$

which form a basis of the tangent spaces to X at every point in the coordinate neighborhood on X. Since $d\pi$ restricted to Hor is a bijection at every point of Y, we conclude that there are functions a_{ri}, $r = 1, \ldots k$, $i = 1, \ldots n$ on the coordinate neighborhood on Y such that

$$\frac{\partial}{\partial z^1} + \sum_{r=1}^{k} a_{r1} \frac{\partial}{\partial y^r}, \ldots, \frac{\partial}{\partial z^n} + \sum_{r=1}^{k} a_{rn} \frac{\partial}{\partial y^r}$$

span Hor at every point of the neighborhood.

Horizontal lifts of curves in a general connection.

Let $C : [0, 1] \to X$ be a smooth curve on X. We say that a smooth curve γ on Y is a **horizontal lift** of C if

- $\pi \circ \gamma = C$ and
- $\gamma'(t) \in \text{Hor}_{\gamma(t)}$ for all t.

For the first condition to hold, each point $C(t)$ must lie in the image of π. (The condition of being a submersion does not imply, without some additional hypotheses, that π is surjective.) Let us examine the second condition in terms of our local coordinate description. Suppose that $x = C(0)$, that $x = \pi(y)$, and we look for a horizontal lift with $\gamma(0) = y$. We can write

$$C(t) = (x^1(t), \ldots, x^n(t))$$

in terms of the local coordinate system on X. So if γ is any lift (horizontal or not) of C, we have

$$\gamma(t) = (x^1(t), \ldots, x^n(t); y^1(t), \ldots, y^k(t))$$

in terms of the local coordinate system.

For γ to be horizontal, we must have

$$\gamma'(t) = \sum_{i=1}^{n} x^{i\prime}(t) \frac{\partial}{\partial z^i} + \sum_{r} \sum_{i} a_{ri}(\gamma(t)) x^{i\prime}(t) \frac{\partial}{\partial y^r}.$$

Thus the condition that γ be a horizontal lift amount to the system of ordinary differential equations

$$\frac{dy^r}{dt} = \sum_r a_{ri}(x^1(t), \ldots, x^n(t); y^1(t), \ldots, y^k(t))x^{i'}(t)$$

where the x^i and $x^{i'}$ are given functions of t. This is a system of (possibly) non-linear ordinary differential equations.

The existence and uniqueness theorem for ordinary differential equations says that for a given initial condition $\gamma(0)$ there is some $\epsilon > 0$ for which there exists a unique solution of this system of differential equations for $0 \leq t < \epsilon$. Standard examples in the theory of differential equations show that the solutions can "blow up" in a finite amount of time; that in general one can not conclude the existence of the horizontal lift γ over the entire interval of definition of the curve C.

In the case of linear differential equations, we do have existence for all time, and therefore in the case of linear connections, or the connection that we studied on the bundle of orthogonal frames, there was global lifting.

16.5.2 Fibrations.

We will now impose some restrictive conditions. We will say that the map $\pi : Y \to X$ is a locally trivial **fibration** if there exists a manifold F such that every $x \in X$ has a neighborhood U such that there exists a diffeomorphism

$$\psi_U : \pi^{-1}(U) \to U \times F$$

such that
$$\pi_1 \circ \psi = \pi$$

where
$$\pi_1 : U \times F \to U$$

is projection onto the first factor.

The implicit function theorem asserts that a submersion $\pi : Y \to X$ looks like a projection onto a first factor locally in Y. The more restrictive condition of being a fibration requires that π look like projection onto the first factor locally on X, with a second factor F which is fixed up to a diffeomorphism. If the map $\pi : Y \to X$ is a surjective submersion and is proper (meaning that the inverse image of a compact set is compact) then we shall prove below that π is a fibration if X is connected.

Globality of lifts.

A second condition that we will impose is on the connection Hor . We will assume that every smooth curve C has a global horizontal lift γ. We saw that this is the case when local coordinates can be chosen so that the equations for the lifting are linear, we shall see that it is also true when π is proper. But let us take this global lifting condition as a *hypothesis* for the moment.

Transport.

Let $C : [a, b] :\to X$ by a smooth curve. For any $y \in \pi^{-1}(C(a))$ we have a unique lifting $\gamma : [a, b] \to Y$ with $\gamma(a) = y$, and this lifting depends smoothly on y by the smooth dependence of solutions of differential equations on initial conditions. We thus have a smooth diffeomorphism associated with any smooth curve $C : [a, b] \to X$ sending

$$\pi^{-1}(C(a)) \to \pi^{-1}(C(b)).$$

If $c \in [a, b]$ if follows from the definition (and the existence and uniqueness theorem for differential equations) that the composite of the map

$$\pi^{-1}(C(a)) \to \pi^{-1}(C(c))$$

associated with the restriction of C to $[a, c]$ with the map

$$\pi^{-1}(C(c)) \to \pi^{-1}((C(b))$$

associated with the restriction of the curve C to $[c, b]$ is exactly the map

$$\pi^{-1}(C(a)) \to \pi^{-1}(C(b))$$

above. This then allows us to define a map $\pi^{-1}(C(a)) \to \pi^{-1}(C(b))$ associated to any piecewise differentiable curve, and the diffeomorphism associated to the concatenation of two curves which form a piecewise differentiable curve is the composite diffeomorphism.

Suppose that X has a smooth retraction to a point. This means that there is a smooth map $\phi : [0, 1] \times X \to X$ satisfying the following conditions where

$$\phi_t : X \to X$$

denotes the map

$$\phi_t(x) = \phi(t, x)$$

as usual. Here are the conditions:

- $\phi_0 = \text{id}$.
- $\phi_1(x) = x_0$, a fixed point of X.
- $\phi_t(x_0) = x_0$ for all $t \in [0, 1]$.

Suppose also that the submersion $\pi : Y \to X$ is surjective and has a connection with global lifting. We claim that this implies that that the submersion is a trivial fibration; that there is a manifold F and a diffeomorphism

$$\Phi : Y \to X \times F \quad \text{with } \pi_1 \circ \Phi = \pi$$

where π_1 is projection onto the first factor. Indeed, take

$$F = \pi^{-1}(x_0).$$

16.5. SUBMERSIONS, FIBRATIONS, AND CONNECTIONS.

For each $x \in X$ define
$$\Phi_x : \pi^{-1}(x) \to F$$
to be given by the lifting of the curve
$$t \mapsto \phi_t(x).$$

Then define
$$\Phi(y) = (\pi(y), \Phi_{\pi(y)}(y)).$$

The fact that Φ is a diffeomorphism follows from the fact that we can construct the inverse of Φ by doing the lifting in the opposite direction on each of the above curves. Every point on a manifold has a neighborhood which is diffeomorphic to a ball around the origin in Euclidean space. Such a ball is retractible to the origin by shrinking along radial lines. This proves that any surjective submersion which has a connection with a global lifting is locally trivial, i.e. is a fibration.

Existence of connections.

For any submersion we can always construct a connection. Simply put a Riemann metric on Y and let Hor be the orthogonal complement to Vert relative to this metric.

So to prove that if $\pi : Y \to X$ is a surjective submersion which is proper then it is a fibration, it is more than enough to prove that every connection has the global lifting property in this case.

So let $C : [0,1] \to X$ be a smooth curve. Extend C so it is defined on some slightly larger interval, say $[-a, 1+a], a > 0$. For any $y \in \pi^{-1}(C(t)), t \in [0,1]$ we can find a neighborhood U_y and an $\epsilon > 0$ such that the liftng of $C(s)$ exists for all $z \in U_y$ and $t - \epsilon < s < t + \epsilon$. This is what the local existence theorem for differential equations gives. But $C([0,1])$ is a compact subset of X, and hence $\pi^{-1}(C([0,1])$ is compact since π is proper. This means that we can cover $\pi^{-1}(C([0,1])$ by finitely many such neighborhoods, and hence choose a fixed $\epsilon > 0$ that will work for all $y \in \pi^{-1}(C([0,1])$. But this clearly implies that we have global lifting, since we can do the lifting piecemeal over intervals of length less than ϵ and patch the local liftings together.

16.5.3 Projection onto the vertical.

Projections.

Suppose that T is a vector space and we are given two complimentary subspaces V and H of T. This means that every vector $w \in T$ can be written uniquely as $w = v + h$ with $v \in V$ and $h \in H$. So we obtain a **projection** $\mathbf{V} : T \to V$ where
$$\mathbf{V}w = v \quad \text{if} \quad w = v + h.$$

The word "projection" in this context means that $\mathbf{V}^2 = \mathbf{V}$.

Conversely, if we are given a linear map $\mathbf{V} : T \to V$ of a vector space onto a subspace which is a projection in the above sense, and define

$$H := \ker \mathbf{V},$$

then $H \cap V = \{0\}$ since if $h \in V$ then $h = \mathbf{V}h$ and if $h \in H$ then $\mathbf{V}h = 0$. Also, we can write every $w \in T$ as

$$w = \mathbf{V}w + (w - \mathbf{V}w)$$

and

$$\mathbf{V}(w - \mathbf{V}w) = \mathbf{V}w - \mathbf{V}^2 w = 0$$

so $(w - \mathbf{V}w) \in H$. In short, if we are given the subspace $V \subset T$ then choosing a complement H to V is the same as choosing a projection $\mathbf{V} : T \to V$.

Applying this to a connection on a submersion.

We can apply all this to the case to the tangent space at each point $y \in Y$ where we have a submersion $\pi : Y \to X$. We see that given a general connection on Y is the same as giving a tensor $\mathbf{V}$ of type (1,1) on Y where at each $y \in Y$

$$\mathbf{V}_y : TY_y \to \mathrm{Vert}_y$$

is a projection onto the vertical subspace.

16.5.4 Frobenius, generalized curvature, and local triviality.

A connection Hor on a submersion $\pi : Y \to X$ is just a special kind of distribution. So we may apply Frobenius: We can consider the set of vector fields $\mathfrak{X}(\mathrm{Hor})$ which are everywhere horizontal. If the Frobenius condition:

$$X, Y \in \mathfrak{X}(\mathrm{Hor}) \Rightarrow [X, Y] \in \mathfrak{X}(\mathrm{Hor})$$

is satisfied, then we know that in a neighborhood of any point $y \in Y$ there exist functions $w_1, \ldots w_k$ which are functionally independent and such that the subspaces Hor are tangent to the submanifolds obtained by setting the w's constant.

On the other hand, if $x^1, \ldots, x^n$ are coordinates on a neighborhood of $\pi(y)$, we know from the fact that π is a submersion, that the functions $z^i = \pi^* x^i$ are defined and functionally independent near y. Furthermore, the restriction of $dz^1, \ldots, dz^n$ to Hor_p are linearly independent for p near y. So the differentials $dz^1, \ldots, dz^n, dw^1, \ldots, dw^k$ are linearly independent at all points near y and so $z^1, \ldots, z^n, w^1, \ldots, w^k$ form a system of coordinates near y in terms of which $\pi(z^1, \ldots, z^n, w^1, \ldots, w^k) = (z^1, \ldots, z^n)$ and parallel transport simply means keeping the w's constant.

16.6. PRINCIPAL BUNDLES AND INVARIANT CONNECTIONS.

In other words, if we restrict to a small neighborhood of y, then parallel transport along a curve joining

$$a = \begin{pmatrix} a^1 \\ \vdots \\ a^n \end{pmatrix} \text{ to } b = \begin{pmatrix} b^1 \\ \vdots \\ b^n \end{pmatrix}$$

is

$$\begin{pmatrix} a^1 \\ \vdots \\ a^n \\ w^1 \\ \vdots \\ w^k \end{pmatrix} \mapsto \begin{pmatrix} b^1 \\ \vdots \\ b^n \\ w^1 \\ \vdots \\ w^k \end{pmatrix}.$$

In particular it is independent of the path.

Measuring the failure of the Frobenius condition.

So it is useful to have an expression which measures by how much the Frobenius condition fails to hold. For this we use $\mathbf{V}$: Let $X, Y \in \mathfrak{X}(\text{Hor})$. Consider the expression

$$\mathbf{V}([X, Y]).$$

For complete integrability to hold, this must vanish identically for all $X, Y \in \mathfrak{X}(\text{Hor})$. Notice that

$$\mathbf{V}([fX, gY]) = fg\mathbf{V}([X, Y])$$

for any smooth functions f, g and for $X, Y \in \mathfrak{X}(\text{Hor})$.

As usual, this implies that if $\xi, \eta \in \text{Hor}_p$ and $X, Y \in \mathfrak{X}(\text{Hor})$ are such that $X(p) = \xi$ and $Y(p) = \eta$ then $\mathbf{V}([X, Y])$ is independent of the choice of X and Y. So we get an antisymmetric bilinear map from Hor_p to Vert_p at each $p \in Y$ given by

$$\xi, \eta \mapsto \mathbf{V}([X, Y])$$

where $X, Y \in \mathfrak{X}(\text{Hor})$ with $X(p) = \xi$ and $Y(p) = \eta$.

This is our most general formulation of curvature.

16.6 Principal bundles and invariant connections.

16.6.1 Principal bundles.

We now embark on a far reaching generalization of the bundle of frames:

Let G be a Lie group with Lie algebra $\mathfrak{g}$. Let P be a space on which G acts. To tie in with our earlier notation, and also for later convenience, we will denote this action by

$$(p, a) \mapsto pa^{-1}, \quad p \in P, \quad a \in G$$

so $a \in G$ acts on P by a diffeomorphism that we will denote by r_a:

$$r_a : P \to P, \qquad r_a(p) = pa^{-1}.$$

If $\xi \in \mathfrak{g}$, then $\exp(-t\xi)$ is a one parameter subgroup of G, and hence

$$r_{\exp(-t\xi)}$$

is a one parameter group of diffeomorphisms of P, and for each $p \in P$, the curve

$$r_{\exp(-t\xi)} p = p(\exp t\xi)$$

is a smooth curve starting at p at $t = 0$. The tangent vector to this curve at $t = 0$ is a tangent vector to P at p. In this way we get a linear map

$$u_p : \mathfrak{g} \to TP_p, \qquad u_p(\xi) = \frac{d}{dt} p(\exp t\xi)_{|t=0}. \qquad (16.14)$$

Freedom of an action.

For example, if we take $P = G$ with G acting on itself by right multiplication, and if we assumed that G is a subgroup of $Gl(n)$, so that we may identify TP_p as a subspace of the the space of all $n \times n$ matrices, then we have seen that

$$u_p(\xi) = p\xi$$

where the meaning of $p\xi$ on the right hand side is the product of the matrix p with the matrix ξ. For this case, if $r_a(p) = p$ for some $p \in P$, this implies that $a = e$, the identity element.

In general, we say that the group action of G on P is **free** if no point of P is fixed by any element of G other than the identity. So "free" means that if $r_a(p) = p$ for some $p \in P$ then $a = e$. Clearly, if the action is free, then the map u_p is injective for all $p \in P$.

If we have an action of G on P and on Q, then we automatically get an action of G (diagonally) on $P \times Q$, and if the action of P is free then so is the action on $P \times Q$.

For example (to change the notation slightly), if X is a space on which G acts trivially, and if we let G act on itself by right multiplication, then we get a free action of G on $X \times G$. This is what we encountered when we began to construct the manifold structure on the bundle of orthogonal frames out of a local frame field. We now generalize this construction:

Definition of a principal G-bundle.

If we are given an action of G on P we have a projection $\pi : P \to P/G$ which sends each $p \in P$ to its G-orbit. We make the following assumptions:

- The action of G on P is free.

16.6. PRINCIPAL BUNDLES AND INVARIANT CONNECTIONS.

- The space P/G is a differentiable manifold M and the projection $\pi : P \to M$ is a smooth fibration.

- The fibration π is locally trivial consistent with the G action in the sense that every $m \in M$ has a neighborhood U such that

there exists a diffeomorphism

$$\psi_U : \pi^{-1}(U) \to U \times G$$

such that

$$\pi_1 \circ \psi = \pi$$

where

$$\pi_1 : U \times F \to U$$

is projection onto the first factor and if $\psi(p) = (m, b)$ then

$$\psi(r_a p) = (m, ba^{-1}).$$

When all this happens, we say that $\pi : P \to M$ is a **principal fiber bundle** over M with **structure group** G or a **principal G-bundle**.

Suppose that $\pi : P \to M$ is a principal fiber bundle with structure group G. Since π is a submersion, we have the sub-bundle Vert of the tangent bundle TP, and from its construction, the subspace $\text{Vert}_p \subset TP_p$ is spanned by the tangents to the curves $p(\exp t\xi)$, $\xi \in \mathfrak{g}$. In other words, u_p given by

$$u_p : \mathfrak{g} \to TP_p, \qquad u_p(\xi) = \frac{d}{dt} p(\exp t\xi)_{|t=0} \qquad (18.2)$$

is a surjective map from $\mathfrak{g}$ to Vert_p. Since the action of G on P is free, we know that u_p is injective. Putting these two facts together we conclude that

Proposition 35. *If $\pi : P \to M$ is a principal fiber bundle with structure group G then u_p is an isomorphism of $\mathfrak{g}$ with Vert_p for every $p \in P$.*

Let us compare the isomorphism u_p with the isomorphism $u_{r_b(p)} = u_{pb^{-1}}$. The action of $b \in G$ on P preserves the fibration and hence

$$d(r_b)_p : \text{Vert}_p \to \text{Vert}_{pb^{-1}}.$$

Let $v = u_p(\xi) \in \text{Vert}_p$. This means that $v = \frac{d}{dt}(p \exp t\xi)_{t=0}$. By definition

$$d(r_b)_p v = \frac{d}{dt} (r_b(p \exp t\xi))_{|t=0} = \frac{d}{dt}((p \exp t\xi)b^{-1})_{|t=0}.$$

We have

$$\begin{aligned} p(\exp t\xi)b^{-1} &= pb^{-1}(b(\exp t\xi)b^{-1}) \\ &= pb^{-1} \exp t \, \text{Ad}_b \, \xi \end{aligned}$$

where Ad is the adjoint action of G on its Lie algebra. We have thus shown that

$$d(r_b)_p u_p(\xi) = u_{r_b(p)}(\text{Ad}_b \, \xi). \qquad (16.15)$$

16.6.2 Connections on principal bundles.

Connections on principal bundles.

Let $\pi : P \to M$ be a principal bundle with structure group G. Recall that in the general setting, we defined a (general) connection to be a sub-bundle Hor of the tangent bundle TP which is complementary to the vertical sub-bundle Vert. Given the group action of G, we can demand that Hor be invariant under G. So by a **connection on a principal bundle** we will mean a sub-bundle Hor of the tangent bundle such that

$$TP_p = \text{Vert}_p \oplus \text{Hor}_p \quad \text{at all } p \in P$$

and

$$d(r_b)_p (\text{Hor}_p) = \text{Hor}_{r_b(p)} \quad \forall\, b \in G,\ p \in P. \tag{16.16}$$

The connection form.

At any p we can, as above, define the projection

$$\mathbf{V}_p : TP_p \to \text{Vert}_p$$

along Hor_p, i.e. $\mathbf{V}_p$ is the identity on Vert_p and sends all elements of Hor_p to 0. Giving Hor_p is the same as giving $\mathbf{V}_p$ and condition (16.16) is the same as the condition

$$d(r_b)_p \circ \mathbf{V}_p = \mathbf{V}_{r_b(p)} \circ d(r_b)_p \quad \forall\, b \in G,\ p \in P. \tag{16.17}$$

Let us compose $u_p^{-1} : \text{Vert}_p \to \mathfrak{g}$ with $\mathbf{V}_p$. So we define the $\mathfrak{g}$ valued form ω by

$$\omega_p := u_p^{-1} \circ \mathbf{V}_p. \tag{16.18}$$

Then it follows from (16.15) and (16.17) that

$$r_b^* \omega = \text{Ad}_b\, \omega. \tag{16.19}$$

The curvature form.

Let ξ_P be the vector field on P which is the infinitesimal generator of $r_{\exp t\xi}$. In view of definition of u_p as identifying ξ with the tangent vector to the curve $t \mapsto p(\exp t\xi) = r_{\exp -t\xi} p$ at $t = 0$, we see that

$$i(\xi_P)\omega = -\xi. \tag{16.20}$$

The infinitesimal version of (16.19) is

$$L_{\xi_P}\omega = [\xi, \omega]. \tag{16.21}$$

Define the curvature by our formula

$$\Omega := d\omega + \frac{1}{2}[\omega \wedge, \omega]. \tag{16.22}$$

16.6. PRINCIPAL BUNDLES AND INVARIANT CONNECTIONS.

It follows from (16.19) that

$$r_b^*\Omega = \mathrm{Ad}_b\,\Omega \quad \forall b \in G. \tag{16.23}$$

Now
$$i(\xi_P)d\omega = L_{\xi_P}\omega - di(\xi_P)\omega$$

by Weil's formula for the Lie derivative. By (16.20) the second term on the right vanishes because it is the differential of the constant $-\xi$. So by (16.21) we get

$$i(\xi_P)d\omega = [\xi,\omega].$$

On the other hand

$$i(\xi_P)[\omega,\omega] = [i(\xi_P)\omega,\omega] - [\omega, i(\xi_P)\omega] = -2[\xi,\omega]$$

where we used (16.20) again. So

$$i(v)\Omega = 0 \quad \text{if } v \in \mathrm{Vert}_p. \tag{16.24}$$

The meaning of the curvature form.

To understand the meaning of Ω when evaluated on a pair of horizontal vectors, let X and Y be pair of horizontal vector fields, that is vector fields whose values at every point are elements of Hor. Then $i(X)\omega = 0$ and $i(Y)\omega = 0$. So

$$\Omega(X,Y) = i(Y)i(X)\Omega = i(Y)i(X)d\omega = d\omega(X,Y).$$

But by our general formula for the exterior derivative we have

$$d\omega(X,Y) = X(i(Y)\omega) - Y(i(X)\omega) - \omega([X,Y]).$$

The first two terms vanish and so

$$\Omega = -\omega([X,Y]). \tag{16.25}$$

This shows how the curvature measures the failure of the bracket of two horizontal vector fields to be horizontal.

16.6.3 Associated bundles.

Let $\pi: P \to M$ be a principal bundle with structure group G, and let F be some manifold on which G acts. We will write this action as multiplication on the left; i.e. we will denote the action of an element $a \in G$ on an element $f \in F$ as af. We then have the diagonal action of G on $P \times F$: For $a \in G$ we define

$$\mathrm{diag}(a): P \times F \to P \times F, \quad \mathrm{diag}(a)(p,f) = (pa^{-1}, af).$$

Since the action of G on P is free, so is its diagonal action on $P \times F$. We can form the quotient space of this action, i.e. identify all elements of $P \times F$ which lie on the same orbit; so we identify the points (p,f) and (pa^{-1}, af).

The quotient space under this identification will be denoted by

$$P \times_G F$$

or by

$$F(P).$$

It is a manifold and the projection map $\pi : P \to M$ descends to a projection of $F(P) \to M$ which we will denote by π_F or simply by π when there is no danger of confusion. The map $\pi_F : F(P) \to M$ is a fibration. The bundle $F(P)$ is called the bundle associated to P by the G-action on F.

Let

$$\rho : P \times F \to F(P)$$

be the map which send (p, f) into its equivalence class.

Suppose that we are given a connection on the principal bundle P. Recall that this means that at each $p \in P$ we are given a subspace $\mathrm{Hor}_p \subset TP_p$ which is complementary to the vertical, and that this assignment is invariant under the action of G in the sense that

$$\mathrm{Hor}_{pa^{-1}} = dr_a(\mathrm{Hor}_p).$$

Given an $f \in F$, we can consider Hor_p as the subspace

$$\mathrm{Hor}_p \times \{0\} \subset T(P \times F)_{(p,f)} = TP_p \oplus TF_f$$

and then form

$$d\rho_{(p,f)} \mathrm{Hor}_p \subset T(F(P))_{\rho(p,f)}$$

which is complementary to the vertical subspace

$$V(F(P))_{\rho(p,f)} \subset T(F(P))_{\rho(p,f)}.$$

The invariance condition of Hor implies that $d\rho_{(p,f)}(\mathrm{Hor}_p)$ is independent of the choice of (p, f) in its equivalence class.

So a connection on a principal bundle induces a connection on each of its associated bundles.

16.6.4 Sections of associated bundles.

If $\pi : Y \to X$ is a submersion, then a **section** of this submersion is a map

$$s : X \to Y$$

such that

$$\pi \circ s = \mathrm{id}.$$

In other words, s is a map which associates to each $x \in X$ an element

$$s(x) \in Y_x = \pi^{-1}(x).$$

16.6. PRINCIPAL BUNDLES AND INVARIANT CONNECTIONS.

Naturally, we will be primarily interested in sections which are smooth.

For example, we might consider the tangent bundle TM. A section of the tangent bundle then associates to each $x \in M$ a tangent vector $s(x) \in TM_x$. In other words, s is a vector field. Similarly, a linear differential form on M is a section of the cotangent bundle T^*M.

Suppose that $\pi = \pi_F : F(P) \to M$ is an associated bundle of a principal bundle P, and that $s : M \to F(P)$ is a section of this bundle. Let x be a point of M, and let $p \in P_x = \pi^{-1}(x)$ be a point in the fiber of the principal bundle $P \to M$ lying in the fiber over x. Then there is a unique $f \in F$ such that

$$\rho((p, f)) = s(x).$$

We thus get a function $\phi_s : P \to F$ by assigning to p this element $f \in F$. In other words, ϕ_s is uniquely determined by

$$\rho((p, \phi_s(p))) = s(\pi(p)). \tag{16.26}$$

Suppose we replace p by $r_a(p) = pa^{-1}$. Since $\rho((pa^{-1}, af)) = \rho((p, f))$ we see that ϕ_s satisfies the condition

$$\phi \circ r_a = a\phi \quad \forall\, a \in G. \tag{16.27}$$

Conversely, suppose that $\phi : P \to F$ satisfies (16.27). Then

$$\rho((p, \phi(p))) = \rho((pa^{-1}, \phi(pa^{-1})))$$

and so defines an element $s(x)$, $x = \pi(p)$. So a $\phi : P \to F$ satisfying (16.27) determines a section $s : M \to F(P)$ with $\phi = \phi_s$. It is routine to check that s is smooth if and only if ϕ is smooth. We have thus proved

Proposition 36. *There is a one to one correspondence between (smooth) sections $s : M \to F(P)$ and (smooth) functions $\phi : P \to F$ satisfying (16.27). The correspondence is given by (16.26).*

An extremely special case of this proposition is where we take F to be the real numbers with the trivial action of G on $\mathbb{R}$. Then $\mathbb{R}(P) = M \times \mathbb{R}$ since the map ρ does not identify two distinct elements of $\mathbb{R}$ but merely identifies all elements of P_x. A section s of $M \times \mathbb{R}$ is of the form $s(x) = (x, f(x))$ where f is a real valued function. The proposition then asserts that we can identify real valued functions on M with real valued functions on P which are constant on the fibers P_x.

16.6.5 Associated vector bundles.

We now specialize to the case that F is a vector space, and the action of G on F is linear. In other words, we are given a linear representation of G on the vector space F. If $x \in M$ we can add two elements v_1 and v_2 of $F(P)_x$ by choosing $p \in P_x$ which then determines f_1 and f_2 in F such that

$$\rho((p, f_1)) = v_1 \quad \text{and} \quad \rho((p, f_2)) = v_2.$$

We then define
$$v_1 + v_2 := \rho((p, f_1 + f_2)).$$
The fact that the action of G on F is linear guarantees that this definition is independent of the choice of p. In a similar way, we define multiplication of an element of $F(P)_x$ by a scalar and verify that all the conditions for $F(P)_x$ to be a vector space are satisfied.

Let $V \to M$ be a vector bundle. So $V \to M$ is a fibration for which each V_x has the structure of a vector space. (As a class of examples of vector bundles we can consider the associated vector bundles $F(P)$ just considered.) We can then consider V valued differential forms on M. For example, a V valued linear differential form τ will be a rule which assigns a linear map
$$\tau_x : TM_x \to V_x$$
for each $x \in M$, and similarly we can talk of V valued k-forms.

For the case that $V = F(P)$ is an associated vector bundle we have a generalization of Proposition 36 to the case of differential forms. That is, we can describe $F(P)$ valued differential forms as certain kinds of F-valued forms on P. To see how this works, suppose that τ is an $F(P)$-valued k-form on M. Let $x \in M$ and let $p \in P_x$. Now
$$\tau_x : \wedge^k(TM_x) \to F(P)_x$$
and p gives an identification map which we will denote by
$$\mathrm{ident}_p$$
of $F(P)_x$ with F - the element $f \in F$ being identified with $\rho((p, f)) \in F(P)_x$. Also,
$$d\pi_p : TP_p \to TM_x$$
and so induces map (which we shall also denote by $d\pi_p$)
$$d\pi_p : \wedge^k(TP_p) \to \wedge^k(TM_x).$$
So
$$\sigma_p := \mathrm{ident}_p \circ \tau_x \circ d\pi_p$$
maps $\wedge^k(TP_p) \to F$. Thus we have defined an F-valued k-form σ on P. If v is a vertical tangent vector at any point p of P we have $d\pi_p(v) = 0$, so
$$i(v)\sigma = 0 \quad \text{if} \quad v \in \mathrm{Vert}\,(P). \tag{16.28}$$

Let us see what happens when we replace p by $r_a(p) = pa^{-1}$ in the expression for σ. Since $\pi \circ r_a = \pi$, we conclude that
$$d\pi_{pa^{-1}} \circ d(r_a)_p = d\pi_p.$$

16.6. PRINCIPAL BUNDLES AND INVARIANT CONNECTIONS. 339

Also,
$$\text{ident}_{pa^{-1}} = a \circ \text{ident}_p$$
where the a on the right denotes the action of a on F. We thus conclude that
$$r_a^* \sigma = a \circ \sigma. \tag{16.29}$$

Conversely, suppose that σ is an F-valued k-form on P which satisfies (16.28) and (16.29). It defines an $F(P)$ valued k-form τ on M as follows: At each $x \in M$ choose a $p \in P_x$. For any k tangent vectors $v_1, \ldots, v_k \in TM_x$ choose tangent vectors $w_1, \ldots, w_k \in TP_p$ such that
$$d\pi_p(w_j) = v_j, \quad j = 1, \ldots, k.$$

Then consider
$$\sigma_p(w_1 \wedge \cdots \wedge w_k) \in F.$$

Condition (16.28) guarantees that this value is independent of the choice of the w_i with $d\pi_p(w_j) = v_j$. In this way we define a map
$$\wedge^k(TM_x) \to F.$$

If we now apply $\rho(p, \cdot)$ to the image, we get a map
$$\wedge^k(TM_x) \to F(P)_x$$
and condition (16.29) guarantees that this map is independent of the choice of $p \in P_x$.

From the construction it is clear that the assignments $\tau \to \sigma$ and $\sigma \to \tau$ are inverses of one another. We have thus proved:

Proposition 37. *There is one to one correspondence between $F(P)$ valued forms on M and F valued forms on P which satisfy (16.28) and (16.29).*

Forms on P which satisfy (16.28) and (16.29) are called **basic** forms because (according to the proposition) F-valued forms on P which satsfy (16.28) and (16.29) correspond to forms on the base manifold M with values in the associated bundle $F(P)$.

For example, equations (16.23) and (16.24) say that the curvature of a connection on a principal bundle is a basic $\mathfrak{g}$ valued form relative to the adjoint action of G on $\mathfrak{g}$. According to the proposition, we can consider this curvature as a two form on the base M with values in $\mathfrak{g}(P)$, the vector bundle associated to P by the adjoint action of G on its Lie algebra.

Here is another important illustration of the concept. Equation (16.19) says that a connection form ω satisfies (16.29), but it certainly does *not* satisfy (16.28). Indeed, the interior product of a vertical vector with the linear differential form ω is given by (16.20). However, suppose that we are given two connection forms ω_1 and ω_2. Then their difference $\omega_1 - \omega_2$ *does* satisfy (16.28) and, of course, (16.29). We can phrase this by saying that the difference of two connections is a basic $\mathfrak{g}$ valued one-form.

16.6.6 Exterior products of vector valued forms.

Suppose that F_1 and F_1 are two vector spaces on which G acts, and suppose that we are given a bilinear map

$$\mathbf{b} : F_1 \times F_2 \to F_3$$

into a third vectors space F_3 on which G acts, and suppose that $\mathbf{b}$ is consistent with the actions of G in the sense that

$$\mathbf{b}(af_1, af_2) = a\mathbf{b}(f_1, f_2).$$

Examples of such a situation that we have come across before are:

1. G is a subgroup of $Gl(n)$ and F_1, F_2 and F_3 are all the vector space of $n \times n$ matrices on which G acts by conjugation, and $\mathbf{b}$ is matrix multiplication.

2. G is a subgroup of $Gl(n)$, F_1 is the space of all $n \times n$ matrices on which Gstill acts by conjugation, F_2 and F_3 are $\mathbf{R}^n$ with its standard action of $Gl(n)$, and $\mathbf{b}$ is multiplication of a matrix times a vector.

3. G is a general Lie group, $F_1 = F_2 = F_3 = \mathfrak{g}$, the Lie algebra of G and $\mathbf{b}$ is Lie bracket.

In each of these cases we have had occasion to form the exterior product of an F_1 valued differential form with an F_2 valued differential form to obtain an F_3 valued form.

We can do this construction in general: form the exterior product of an F_1 valued k-form with an F_2-valued ℓ form to get an F_3 valued $k + \ell$ form. For example, if $f_1^1, \ldots, f_m^1$ is a basis of F_1 and $f_1^2, \ldots, f_n^2$ is a a basis of F_2 then the most general F_1-valued k-form α can be written as

$$\alpha = \sum \alpha^i f_i^1$$

where the α^i are real valued k-forms, and the most general F_2-valued ℓ-form β can be written as

$$\beta = \sum \beta^j f_j^2$$

where the β^j are real valued ℓ forms.

Let $f_1^3, \ldots, f_q^3$ be a basis of F_3 and define the numbers B_{ij}^k by

$$\mathbf{b}(f_i^1, f_j^2) = \sum_k B_{ij}^k f_k^3.$$

Then you can check that $\alpha \wedge \beta$ defined by

$$\alpha \wedge \beta := \sum B_{ij}^k (\alpha^i \wedge \beta^j) f_k^3$$

is independent of the choice of bases.

16.7. COVARIANT DIFFERENTIALS AND DERIVATIVES.

In a similar way we can define the exterior derivative of a vector valued form, the interior product of a vector valued form with a vector field, the pull back of a vector valued form under a map etc. There should be little problem in understanding the concept involved. There is a bit of a notational problem - how explicit do we want to make the map **b** in writing down a symbol for this exterior product. In example 1) we simply wrote $\wedge$ for the exterior product of two matrix valued forms. This forced us to use the rather ugly $[\alpha \wedge, \beta]$ for the exterior product of two Lie algebra valued forms, where the **b** was commutator or Lie bracket. We shall retain this ugly notation for the sake of the clarity it gives.

A situation that we will want to discuss in the next section is: we are given an action of G on a vector space F, and unless forced to be more explicit, we have chosen to denote the action of an element $a \in G$ on an element $f \in F$ simply by af. This determines a bilinear map

$$\mathbf{b}: \mathfrak{g} \times F \to F$$

by

$$\mathbf{b}(\xi, f) := \frac{d}{dt}(\exp t\xi)f_{|t=0}.$$

We therefore get an exterior multiplication of a $\mathfrak{g}$-valued form with an F-valued form. We shall denote this particular type of exterior mutlplication by $\diamond$. So if α is a $\mathfrak{g}$-valued k-form and β is an F-valued ℓ form then $\alpha \diamond \beta$ is an F-valued $(k+\ell)$-form.

We point out that conditions (16.28) and (16.29) make perfectly good sense for vector valued forms, and so we can talk of basic vector valued forms on P, and the exterior product of two basic vector valued forms is again basic.

16.7 Covariant differentials and derivatives.

In this section we consider a fixed connection on a principal bundle P. This means that we are given a projection **V** of TP onto the vertical bundle and therefore a connection form ω. Of course we also have a projection

$$\mathbf{id} - \mathbf{V}$$

onto the horizontal bundle Hor of the connection, where **id** is the identity operator. This projection kills all vertical vectors.

16.7.1 The horizontal projection of forms.

If α is a (possibly vector valued) k-form on P, we will define the horizontal projection $\mathbf{H}\alpha$ of α by

$$\mathbf{H}\alpha(v_1, \ldots, v_k) = \alpha((\mathbf{id} - \mathbf{V})v_1, \ldots, (\mathbf{id} - \mathbf{V})v_k). \tag{16.30}$$

The following properties of **H** follow immediately from its definition and the invariance of the horizontal bundle under the action of G:

1. $\mathbf{H}(\alpha \wedge \beta) = \mathbf{H}\alpha \wedge \mathbf{H}\beta$.

2. $r_a^* \circ \mathbf{H} = \mathbf{H} \circ r_a^* \quad \forall\, a \in G$.

3. If α has the property that $i(w)\alpha = 0$ for any horizontal vector w then $\mathbf{H}\alpha = 0$. In particular,

4. $\mathbf{H}\bar{\omega} = 0$.

5. If α has the property that $i(v)\alpha = 0$ for any vertical vector v then $\mathbf{H}\alpha = \alpha$. In particular,

6. $\mathbf{H}$ is the identity on basic forms.

In 1) α and β could be vector valued forms if we have the bilinear map $\mathbf{b}$ which allows us to multiply them.

The map $\mathbf{H}$ is clearly a projection in the sense that

$$\mathbf{H}^2 = \mathbf{H}.$$

16.7.2 The covariant differential of forms on P.

Define $\mathbf{d}$ mapping k-forms into $(k+1)$ forms by

$$\mathbf{d} := \mathbf{H} \circ d. \tag{16.31}$$

The following facts are immediate:

- $\mathbf{d}(\alpha \wedge \beta) = \mathbf{d}\alpha \wedge \mathbf{H}\beta + (-1)^k \mathbf{H}\alpha \wedge \mathbf{d}\beta$ if α is a k-form.
- $i(v)\mathbf{d} = 0$ for any vertical vector v.
- $r_a^* \circ \mathbf{d} = \mathbf{d} \circ r_a^* \quad \forall\, a \in G$.

It follows from the second and third items that $\mathbf{d}$ carries basic forms into basic forms. The map $\mathbf{d}$ is sometimes called the "exterior covariant derivative".

If F is a vector space on which G acts linearly, we can form the associated vector bundle $F(P)$, and we know from Proposition 37 that k-forms on M with values in $F(P)$ are the same as basic k-forms on P with values in F. So giving a connection on P induces an operator $\mathbf{d}$ mapping k-forms on M with values in $F(P)$ to $(k+1)$-forms on M with values in $F(P)$. For example, a section s of $F(P)$ is just a zero form on M with values in the vector bundle $F(M)$. Giving the connection on P allows us to construct the one form $\mathbf{d}s$ with values in $F(P)$. If X is a vector field on M, then we can define

$$\nabla_X s := i(X)\mathbf{d}s,$$

the covariant derivative of s in the direction X.

16.7.3 A formula for the covariant differential of basic forms.

Let α be a basic form on P with values in the vector space F on which G acts linearly. Let $\mathbf{d}$ be the covariant differential associated with the connection form ω. We claim that
$$\mathbf{d}\alpha = d\alpha + \omega \diamond \alpha. \tag{16.32}$$
In order to prove this formula, it is enough to prove that when we apply $i(v)$ to the right hand side we get zero, if v is vertical. For then applying $\mathbf{H}$ does not change the right hand side. But applying $\mathbf{H}$ to the right hand side yields $\mathbf{d}\alpha$ since $\mathbf{d}\alpha := \mathbf{H}(d\alpha)$ and
$$\mathbf{H}\omega = 0$$
so
$$\mathbf{H}(\omega \diamond \alpha) = 0.$$
So it is enough to show that for any $\xi \in \mathfrak{g}$ we have
$$i(\xi_P)d\alpha = -i(\xi_P)(\omega \diamond \alpha).$$
Since α is basic, we have $i(\xi_P)\alpha = 0$, so by Weil's identity we have
$$i(\xi_P)d\alpha = L_{\xi_P}\alpha = \xi \diamond \alpha$$
by the infinitesimal version of the invariance condition (16.29). On the other hand, since $i(\xi_P)\alpha = 0$ and $i(\xi_P)\omega = -\xi$, we have proved our formula.

There are a couple of special cases of (16.32) worth mentioning. If F is $\mathbf{R}$ with the trivial representation then (16.32) says that $\mathbf{d} = d$. This implies, that if s is a section of an associated vector bundle $F(P)$, and if ϕ is a function on M, so that ϕs is again a section of $F(P)$ then
$$\mathbf{d}(\phi s) = (d\phi) \wedge s + s\mathbf{d}s$$
implying that for any vector field X on M we have
$$\nabla_X(\phi s) = (X\phi)s + \phi(\nabla_X s).$$

Another important special case is where we take $F = \mathfrak{g}$ with the adjoint action. Then (16.32) says that
$$\mathbf{d}\alpha = d\alpha + [\omega \wedge, \alpha].$$

The curvature is $\mathbf{d}\omega$.

We wish to prove that
$$\mathbf{d}\omega = d\omega + \frac{1}{2}[\omega \wedge, \omega]. \tag{16.33}$$
Both sides vanish when we apply $i(v)$ where v is a vertical vector - this is true for the left hand side by definition, and we have already verified this for the right hand side, see equation (16.24). But if we apply $\mathbf{H}$ to both sides, we get $\mathbf{d}\omega$ on the left, and also on the right since $\mathbf{H}\omega = 0$. $\square$

Bianchi's identity.

In our setting this says that
$$\mathbf{d}\Omega = 0. \tag{16.34}$$

Proof. We have
$$d\Omega = d(d\omega) + d\frac{1}{2}[\omega\wedge,\omega] = [d\omega\wedge,\omega].$$

Applying **H** yields zero because $\mathbf{H}\omega = 0$. □

The curvature and $\mathbf{d}^2$.

We wish to show that
$$\mathbf{d}^2\alpha = \Omega \diamond \alpha. \tag{16.35}$$

In this equation α is a basic form on P with values in the vector space F where G acts, and we know that Ω is a basic form with values in $\mathfrak{g}$, so the right hand side makes sense and is a basic F valued form. To prove this we use our formula
$$\mathbf{d}\alpha = d\alpha + \omega \diamond \alpha$$

and apply it again to get
$$\mathbf{d}^2\alpha = d(d\alpha + \omega \diamond \alpha) + \omega \diamond (d\alpha + \omega \diamond \alpha).$$

We have $d^2 = 0$ so the first expression (under the d) becomes
$$d(\omega \diamond \alpha) = d\omega \diamond \alpha - \omega \diamond d\alpha.$$

The second term on the right here cancels the term $\omega \diamond d\alpha$ so we get
$$\mathbf{d}^2\alpha = d\omega \diamond \alpha + \omega \diamond (\omega \diamond \alpha).$$

So to complete the proof we must check that
$$\frac{1}{2}[\omega\wedge,\omega] \diamond \alpha = \omega \diamond (\omega \diamond \alpha).$$

This is a variant of a computation we have done several times before:

Since interior product with vertical vectors sends α to zero, while interior product with horizontal vectors sends ω to zero, it suffices to verify that the above equation is true after we take the interior product of both sides with two vertical vectors, say η_P and ξ_P. Now
$$i(\xi_P)[\omega\wedge,\omega] = -[\xi,\omega] + [\omega,\xi] = -2[\xi,\omega]$$

and so
$$i(\eta_p)i(\xi_P)(\frac{1}{2}[\omega\wedge,\omega] \diamond \alpha) = [\xi,\eta] \diamond \alpha.$$

A similar computation shows that
$$i(\eta_P)i(\xi_P)(\omega \diamond (\omega \diamond \alpha)) = \xi \diamond (\eta \diamond \alpha) - \eta \diamond (\xi \diamond \alpha).$$

16.7. COVARIANT DIFFERENTIALS AND DERIVATIVES.

But the equality of these two expressions follows from the fact that we have an action of G on F which implies that for any $\xi, \eta \in \mathfrak{g}$ and any $f \in F$ we have

$$[\xi, \eta]f = \xi(\eta f) - \eta(\xi f). \quad \square$$

It is a useful exercise to work out how all this relates to the material on linear and semi-Riemannian connections in the earlier chapters of this book: The bundle of frames is a principal bundle with structure group $G = Gl(n)$ (or $O(n)$ in the Riemannian case) and the tangent bundle is the bundle associated to the standard representation of G on $\mathbb{R}^n$, etc.

Chapter 17

Reduction of principal bundles, gauge theories and the Higgs mechanism.

In this chapter I give a rough sketch of how the theory of connections on principal bundles is related to the current gauge theories of forces. I also discuss Cartan's theory of soldering forms and "Cartan connections".

I begin with some history.

17.1 A brief history of gauge theories.

Soon after Einstein's theory of general relativity, Hermann Weyl suggested that the true objects of general relativity should not be (semi-)Riemann metrics, but rather the associated Levi-Civita connection. And if we generalize this connection to be a conformal connection (i.e. if we enlarge the group from $O(1,3)$ to $\mathbb{R}^+ \times O(1,3)$) then we can incorporate electromagnetism.(See his classic *Raum Zeit Materie*, Springer, Berlin (1918)). The word "gauge" derives from Weyl's theory in which the length is changed by a conformal transformation.

Einstein rejected Weyl's proposal of considering a conformal connection as the underlying physical field, although Einstein himself considered the possibility that Riemannian geometry be replaced by conformal geometry as a basis for unified theories - see his article in *Preuss Akad.* 261 (1921) as well as the following notes on the "unified field theory": *loc. cit.* (1925) p. 414, (1928) p. 3, (1929) p. 3.

After the advent of quantum mechanics, Fritz London, in a short note in early 1927(F. London, "Die Theorie von Weyl und die Quantenmechanik", *Naturwiss.* **15** 187) and soon after in a longer paper, ("Quantenmechanische Deutung der Theorie von Weyl," *Zeit. für Physik* **42**, 375-389 (1927)), proposed a quantum mechanical interpretation of Weyl's attempt to unify electromagnetism and gravitation. The essential idea is to replace Weyl's $\mathbb{R}^+$ by $U(1)$ acting as phase transformations of the quantum mechanical state vector. The group $U(1)$ does not act on the tangent space of space time. It is "internal". The London theory for $U(1)$ was generalized to $SU(2)$ by Yang and Mills in 1954.

The "field" in a Yang-Mills theory on space time is a connection on a principal bundle P. One then obtains equations analogous to Maxwell's equations for the curvature of this connection.

Giving a connection on a principal bundle is the same as giving (consistently) the notion of covariant derivative on any associated bundle. The covariant derivative language is more popular in the standard physics texts. I will give a self contained review of the notions of connection and curvature in the more general setting of superconnections and supercurvature later on.

17.2 Cartan's approach to connections.

I now interrupt the discussion of Gauge theories to go back to some work of Cartan in the 1920's.

In our discussion of semi-Riemannian geometry so far in this book, we followed the approach of Levi-Civita - that a connection on a differentiable manifold is essentially a means of parallel transport along curves. This notion of parallel transport then gives rise to the concept of covariant derivative, etc. The principal bundle connected with this approach is the bundle of orthonormal frames.

In the classical (19th century) theory of surfaces, there was another notion intimately related to parallel transport, and that is the development of a surface on a plane along a curve. We briefly described this idea in the discussion after Problem **14** of Chapter 2. Intuitively speaking, one rolled the surface on the plane along the curve, maintaining first-order contact. This gives an identification of the tangent space to the surface at each point of the curve with the plane. Then parallel translation in the Euclidean geometry of the plane gives the parallel transport in the sense of Levi-Civita along the curve. From this point of view the notion of development is central, and the crucial property of the plane is that it is a homogeneous space in the sense that it admits a transitive group of isometries that includes the full rotation group as the isotropy group of a point.

The plane is $E(2)/O(2)$, where $E(2)$ denotes the group of Euclidean motions of the plane. From this point of view, we might also want to study the development of a surface onto a sphere along a curve. The sphere is $O(3)/O(2)$. Or we might want to study the development of a surface onto hyperbolic space $Sl/(2,\mathbb{R})/SO(2)$. This is the starting point for Cartan's approach. We must not only be given the structure group H of a principal bundle, but also a larger group G containing H as a closed subgroup so that G/H is a homogeneous "model" space with which we would like to compare our manifold M. Thus, for example, in general relativity, where H is the Lorentz group, we would take G

17.2. CARTAN'S APPROACH TO CONNECTIONS.

to be the Poincaré group if we were comparing space-time with Minkowski space and G to be the de Sitter group $0(1,4)$ if we were comparing space-time with de Sitter space. In the theory of spontaneous symmetry breaking we are also given a pair of principal bundles. For example, in the Weinberg-Salam model we would have the group $G = SU(2) \times U(1)$ and $H = U(1)$. In this case, the quotient space $G/H = SU(2) \sim S^3$ is not be related in any way to the base.

If the space G/H *does* have the same dimension as M, then we can demand more from the connection form on P. We can demand that its restriction to P, have no kernel. This has the effect of identifying the tangent space to M, at any point, with the tangent space to G/H –giving a precise mathematical formulation to the intuitive idea of "rolling" M along G/H. This was Cartan's idea of "soldering". Although Cartan wrote his fundamental papers in the 1920s the precise modern definition of a Cartan connection was first given by Ehresmann in 1950.

Recalling some notation and results from Chapter 16.

Let G be a Lie group and M a differentiable manifold. A principal G bundle over M is a manifold P on which G acts freely and such that the quotient of this G action is M. Thus we have a smooth map $\pi : P \to M$ and $\pi^{-1}(x)$ is a G orbit for each $x \in M$. One also assumes that P is locally trivial in the sense that about each $x \in M$ there is a neighbourhood, U, such that $\pi^{-1}(U)$ is isomorphic to $U \times G$ (with the obvious definition of isomorphism, see Chapter 16). We denoted the action of an $a \in G$ on a $p \in P$ by

$$r_a(p) = pa^{-1}.$$

Let F be some differentiable manifold on which G acts. We can then form the quotient of the product space $P \times F$ by the G action; call it $F(P)$. Let $\rho : P \times F \to F(P)$ denote the passage to the quotient. Then $F(P)$ is also fibered over M where $\pi_F(\rho(p, f)) = \pi(p)$. So we have the commutative diagram

$$\begin{array}{ccc} P \times F & \xrightarrow{\rho} & F(P) \\ & \searrow \quad \swarrow & \\ & M & \end{array}$$

where the left diagonal map $P \times F \to M$ is $(p, f) \mapsto \pi(p)$ and the right diagonal is the map π_F.

As we saw in Chapter 16,

We may identify the space of sections of F(P) with the space of maps
$$f : P \to F \text{ satisfying}$$

$$f(pa^-1) = af(p). \tag{17.1}$$

Reduction of a principal bundle.

Here is an important case of the above assertion: Suppose that $F = G/H$ where H is a closed subgroup of G. So $F(P)$ is a bundle of homogeneous spaces. Let $f : P \to F$ satisfy (17.1) and so be equivalent to a section, s, of $F(P)$. Consider
$$f^{-1}(H) = \{p \in P | f(p) = H \in G/H\}.$$
If $p \in f^{-1}(H)$, then $f(pa^{-1}) = af(p) \in H$ if and only if $a \in H$. In other words,
$$P_H := f^{-1}(H)$$
is an H sub-bundle of P. It is, a **reduction** of the principal G-bundle P to an H bundle.

Conversely, suppose that P_H is an H sub-bundle of P. Define $f : P \to G/H$ by
$$f(P_H) = H$$
and if $p = qa^{-1}$, $q \in P_H$ then
$$f(p) = aH \in G/H.$$
If $p = q'b^{-1}$ with $q' \in P_H$ then $q' = pb = qa^{-1}b$ and $q' = qh^{-1}$ for some $h \in H$ so
$$a^{-1}b = h^{-1}$$
i.e.
$$b = ah^{-1}$$
so $bH = aH$. This shows that f is well defined. Also, $f(qa^{-1}c^{-1}) = f(q(ca)^{-1}) = (ca)H = cf(qa^{-1})$ showing that f satisfies (17.1). Thus

Proposition 38. *A section s of the bundle $(G/H)(P)$ is the same as a reduction of P to an H-sub-bundle P_H. The restriction of s to the sub-bundle P_H corresponds to the constant function $g \equiv H$ as a map from P_H to G/H. We can consider $(G/H)(P)$ as the bundle associated to P_H by the H action on G/H.*

We can carry this identification further: Let $\mathfrak{g}$ be the Lie algebra of G and $\mathfrak{h}$ the Lie algebra of H. The group G has its adjoint action on $\mathfrak{g}$, and the restriction of this adjoint action to H preserves $\mathfrak{h} \subset \mathfrak{g}$. So we get a representation of H on the quotient space $\mathfrak{g}/\mathfrak{h}$ which I will continue to denote by Ad. At $H \in G/H$ we have an identification of $T(G/H)$ with $\mathfrak{g}/\mathfrak{h}$ consisent with the H action. So we may identify the vector bundle $[\mathfrak{g}/\mathfrak{h}](P_H)$ with the vertical tangent vectors to $G/H)(P)$ along s.

Suppose that $\sigma : TP_H \to \mathfrak{g}/\mathfrak{h}$ is a one form which satisfies (16.28) and (16.29) (relative to the group H). Then σ can be thought of as a one form on M with values in $[\mathfrak{g}/\mathfrak{h}](P_H)$. So we have

17.2. CARTAN'S APPROACH TO CONNECTIONS.

Proposition 39. *Let s be a section of $(G/H)(P)$ and P_H the corresponding reduced bundle. Let σ be a one form on P_H with values in $\mathfrak{g}/\mathfrak{h}$ which satisfies (16.28) and (16.29) (relative to the group H). Then σ can be regarded as a one form on M with values in the bundle of vertical tangent vectors to $(G/H)(P)$ along the section s.*

In the particular case that $\dim G/H = \dim M$, we can further demand that the one form σ on M give an isomorphism (at all points) between TM and the bundle of vertical tangent vectors. This is the situation considered by Cartan. The form σ is called a **soldering form** in this case. For example, suppose that $H = O(V)$ is the "orthogonal" group of a vector space with non-degenerate scalar product and G is the corresponding group of affine transformations, so that $G/H = V$. A soldering form then gives an identification of TM with $V(P_H)$. In particular, this puts a (pseudo) Riemann metric on M and also allows us to identify P_H with the bundle of "orthonormal" frames.

17.2.1 Cartan connections.

Suppose we are in the situation of the preceding section, so we have a restriction of a principal G bundle P to an H-bundle P_H. Let ω_G be a connection form on P and consider its restriction to P_H. This restriction is a $\mathfrak{g}$ valued one form on P_H which satisfies (16.20) for $\xi \in \mathfrak{h}$ and (16.19) for $a \in H$. As $\mathfrak{h}$ is an invariant subspace of $\mathfrak{g}$ under the adjoint action of H on $\mathfrak{g}$, we can define the form

$$\sigma := (\text{restriction of } \omega_G \text{ to } P_H)/\mathfrak{h}$$

as a $\mathfrak{g}/\mathfrak{h}$ valued one form on P_H satisfying the conditions of the preceding section.

If the group H is reductive - or, more generally, if $\mathfrak{h}$ has an H invariant complement, $\mathfrak{n}$ in $\mathfrak{g}$ -we can decompose

$$\mathfrak{g} = \mathfrak{h} \oplus \mathfrak{n}$$

as an H invariant decomposition and $\mathfrak{n}$ becomes identified with $\mathfrak{g}/\mathfrak{h}$. Then the restriction of ω_G to P_H has the decomposition

$$\omega_G|_H = \omega_H + \sigma \qquad (17.2)$$

where ω_H satisfies the conditions for a connection on P_H and σ fulfills the conditions of the preceding section.

In particular, if $\dim G/H = \dim M$ we can add the condition that σ be a soldering form. If this holds, then ω_G is called a **Cartan connection**. If G is the group of affine motions and H the "orthogonal" subgroup then a Cartan connection is called an affine connection.

Recall that the curvature form Ω_G of the connection ω_G is given by

$$\Omega_G = \text{Curv}(\Omega_G) = d\omega_G + \frac{1}{2}[\omega_G \wedge, \omega_G].$$

In the case of a reduced bundle such that the decomposition (17.2) holds, the restriction of Ω_G to P_H is given by

$$\text{Curv}(\omega_H) + d\sigma + \omega_H \cdot \sigma + \frac{1}{2}[\sigma \wedge, \sigma] = \text{Curv}(\omega_H) + d_{\omega_H}\sigma + \frac{1}{2}[\sigma \wedge, \sigma].$$

In particular, for the case of an affine connection $[\mathfrak{n}, \mathfrak{n}] = 0$ so the last term disappears. The term $d_{\omega_H}\sigma$ is the torsion of the linear connection ω_H. In the Levi-Civita theory of connections the torsion entered as a more or less formal object. It was given, in local coordinates, as the antisymmetric components of the Christoffel symbols. But in the Cartan theory the torsion enters as a component of the curvature. Its geometric meaning is clear: it gives the translational component of parallel transport around an infinitesimal closed curve.

17.3 Spontaneous symmetry breaking and "mass acquisition".

In this section we show how the notion of a reduction of a principal bundle arises in the general theory of "spontaneous symmetry breaking . Let $P_G \to M$ be a principal bundle with structure group G. There are various geometrical objects that we have associated with P_G: For a vector space F which is a representation space of G, we have the associated vector bundle $F(P)$ and then the spaces of differential forms on M with values in $F(P)$. We denote the space of k-forms on M with values in $F(P)$ by $A^k(F)$ and shall denote the sum of all these spaces- and for various possible F s simply $A^*(P_G)$. We can also consider the space of all connections on P, which we shall denote by $\text{Conn}(P_G)$. We shall denote the covariant differential with respect to $\omega \in \text{Conn}(P_G)$ by d_ω.

Suppose that M has a (semi) Riemann metric and that $\mathfrak{g}$ is given a G invariant scalar product as is F. Also suppose that we are given some G invariant function V on F. We can then consider various scalar valued functions built out of the geometric objects such as

$$\begin{array}{ll} \|\sigma\|^2 & \sigma \in A^k(F) \\ \|d_\omega \sigma\|^2 & \sigma \in A^k(F), \ \omega \in \text{Conn}(P_G) \\ \|\text{Curv}(\omega)\|^2 & \omega \in \text{Conn}(P_G) \\ V(f) & f \in A^0(F). \end{array}$$

If we have a projection of P_G onto the bundle of orthonormal frames of M, so that $T(M)$ and $T^*(M)$ are associated bundles to P_G then we might also have expressions such as

$$W(f, \sigma)$$

where W is some G invariant polynomial. All of these are G invariant functions on P_G and hence define functions on M. Each of the above expressions thus defines a map L_1, L_2 etc., from geometrical objects to functions on M

$$L_i : \text{Conn}(P_G) \times A^*(P_G) \to C^\infty(M).$$

17.3. SYMMETRY BREAKING AND "MASS ACQUISITION".

If we then integrate these expressions, or some (linear) combination, L, of them over M relative to the volume form we obtain a functional on the space of geometric objects (at least on the set where this integral converges). The study of this "function"
$$\mathcal{F} = \int_M L d\text{vol}$$
is the central object of study in Yang-Mills theories. For example, describing the critical sets of this function is very important, as we have seen.

17.3.1 The Automorphism group and the Gauge group of a principal bundle.

Now $\mathcal{F}$ admits a rather large group of symmetries. Let $\text{Aut}(P_G)$ denote the group of all diffeomorphisms of P_G which commute with the action of G and $\text{Gau}(P_G) \subset \text{Aut}(P_G)$ the subgroup consisting of those which project onto the identity on M. Each of the Lagrangians L_i and hence the function $\mathcal{F}$ is invariant under the action of $\text{Gau}(P_G)$. The idea of spontaneous symmetry breaking is to provide a local partial cross-section for the action of $\text{Gau}(P_G)$ and hence remove some of the redundancy in $\mathcal{F}$.

Warning.

In the physics literature, the phrase "gauge group" is usually taken to mean the group G associated to the principal bundle P_G. So what we are calling the "structure group" of the bundle the physicists call the gauge group. I will stick the the mathematician's terminology using the phrase "gauge group" to mean $\text{Gau}(P_G)$ as above.

Here is a very useful alternative description of $\text{Gau}(P_G)$ in terms of sections of an associated bundle: Let G act on itself by conjugation. Then $G(P_G)$ is an associated bundle, each fiber of which is a group. Let $\Gamma(G(P_G))$ be the space of smooth sections of $G(P_G)$. Pointwise multiplication makes $\Gamma(G(P_G))$ into a group. This group can be identified with $\text{Gau}(P_G)$ as follows: A section of $G(P_G)$ can be viewed as a map
$$\tau : P_G \to G$$
satisfying
$$\tau(pb^{-1}) = b\tau(p)b^{-1} \quad \forall \, p \in P_G, \, b \in G.$$
Consider the transformation $\phi_\tau : P_G \to P_G$ given by
$$\phi_\tau(p) = p\tau(p). \tag{17.3}$$
Then, with $\phi = \phi_\tau$ we have
$$\phi(pb^{-1}) = pb^{-1}\tau(pb^{-1}) = pb^{-1}b\tau(p)b^{-1} = \phi(p)b^{-1}$$

so

$$\phi \circ R_b = R_b \circ \phi \quad \forall b \in G.$$

Conversely, given ϕ satisfying $\phi \circ R_b = R_b \circ \phi \ \forall b \in G$ we can define τ by (17.3) and check that it satisfies $\tau(pb^{-1}) = b\tau(p)b^{-1}$. It is immediate from (17.3) that

$$\phi_{\tau_1 \cdot \tau_2} = \phi_{\tau_1} \circ \phi_{\tau_2}$$

so we may identify $\mathrm{Gau}(P_G)$ with $\Gamma(G(P_G))$ as a group.

In what follows it will be convenient to consider several vector spaces $U, R, L, \ldots$ on which G acts, instead of a single F, and single out one of them, U, for special attention which we shall call the Higgs space. An element $f \in \Gamma(U(P_G))$ (that is, a section of the associated bundle $U(P_g)$) is called a Higgs field.

We assume the following about the action of G on U: That outside of a small G invariant singular set S in U there exists a cross-section, C, to the G action; that is, we assume that $U \setminus S = C \times (G/K)$ as G spaces, where K is the (common) isotropy group of all points of C. For example, let $G = SU(2)$ and $U = \mathbb{C}^2$. Then we can take $S = \{0\}$ and C to be the set of all vectors of the form $\begin{pmatrix} 0 \\ r \end{pmatrix}$ with r real and positive. Here $K = \{e\}$. Let $\Gamma^+(U(P_G)) \subset \Gamma(U(P_G))$ denote the set of all $f : P_G \to U$ which do not intersect S, i.e. such that $f(P_G) \subset U \setminus S$. Any such f determines (by projection onto the second factor) a G equivariant map, $\overline{f} : P_G \to G/K$. But then

$$\overline{f}^{-1}(K) = f^{-1}(C)$$

is a principal K subbundle of P_G, i.e., a reduction of P_G, to a subbundle with structure group K. The group $\mathrm{Gau}(P_G)$ acts naturally on all geometric objects. In particular, it acts on the space of all sections, $\Gamma(G/K)$. We will assume that this space is nonempty and that $\mathrm{Gau}(P_G)$ acts transitively on it.

Note: If $K = \{e\}$ the action is always transitive. Indeed, let f_1 and $f_2 : P_G \to G$ be two functions satisfying

$$f(pb^{-1}) = bf(p).$$

Then $\tau := f_1 f_2^{-1}$ satisfies

$$\tau(pb^{-1}) = bf_1(p)f_2^{-1}(p)b^{-1} = b\tau(p)b^{-1}$$

and hence defines an element of $\mathrm{Gau}(P_G)$. Then

$$f_1(p\tau(p)) = f_2(p)f_1^{-1}(p)f_1(p) = f_2(p)$$

so ϕ_τ carries f_1 into f_2.

Under our transitivity assumption, all the subbundles $\overline{f}^{-1}(K)$ are all conjugate under $\mathrm{Gau}(P_G)$ and thus, in particular, define the same abstract principal

17.3. SYMMETRY BREAKING AND "MASS ACQUISITION".

K bundle which we shall denote by P_K. The choice of a particular f then determines an embedding

$$P_K \xrightarrow{\tilde{f}} P_G$$

as a K subbundle. By construction, the G/K component of $f \circ \tilde{f}$ is identically K, so

$$f \circ \tilde{f} = (\phi, K)$$

where $\phi : P_K \to C$ and K acts trivially on C. So ϕ can be thought of as a function from M to C. It is clear that we can recover f from $\tilde{f}$ and ϕ. Thus giving $f \in \Gamma^+(U(P_G)$ is the same as giving $\tilde{f}$ and ϕ.

Let V be any representation space for G. By restriction, K also acts on V and hence f defines a map, $\tilde{f}^*$, pullback, from V-valued forms on P, to V-valued forms on P_K. It is easy to check that

$$\tilde{f}^* : A^k(V(P_G)) \to A^k(V(P_K)).$$

Notice that

$$\|\tilde{f}^*\sigma\|^2(q) = \|\sigma\|^2(\tilde{f}(q))$$

since σ is basic.

Let us assume that $\mathfrak{l}$ is a K invariant complement in $\mathfrak{g}$ to $\mathfrak{k}$ so

$$\mathfrak{g} = \mathfrak{k} + \mathfrak{l}$$

is a K invariant decomposition of $\mathfrak{g}$ as a vector space. (It is always possible to choose such an $\mathfrak{l}$ if K is compact, for example.) Let $\omega \in \text{Conn}(PG)$. Then we can write

$$\tilde{f}^*\omega = (\tilde{f}^*\omega)_{\mathfrak{k}} + (\tilde{f}^*\omega)_{\mathfrak{l}}.$$

As we have seen in the preceding section, the first term on the right is an element of $\text{Conn}(P_K)$ and the second is an element of $A^1(\mathfrak{l}(P_K))$. We thus have defined a map

$$\text{Conn}(P_G) \xrightarrow{\tilde{f}^*} \text{Conn}(P_K) \times A^1(\mathfrak{l}(P_K)).$$

At $p = \tilde{f}(q)$ we have

$$(\text{Curv}(\omega))(p) = \text{Curv}(\tilde{f}^*\omega)_{\mathfrak{k}}(q) + [d_{(\tilde{f}^*\omega)_{\mathfrak{k}}}(\tilde{f}^*\omega)_{\mathfrak{l}} + \frac{1}{2}[(\tilde{f}^*\omega)_{\mathfrak{l}}, (\tilde{f}^*\omega)_{\mathfrak{l}}]. \quad (17.4)$$

Thus we can write down a geometric function

$$\tilde{L} : \text{Conn}(P_K) \times A^1(\mathfrak{l}(P_K)) \to C^\infty(M)$$

such that

$$\tilde{L} \circ \tilde{f}^*\omega = \|\text{Curv}\,\omega\|^2.$$

In fact, we have proved

Proposition 40. *The assignment $f \mapsto \tilde{f}$ induces a map*

$$\Gamma^+(U) \times \mathrm{Conn}(P_G) \times A^*(P) \times \cdots$$

$$\xrightarrow{\tilde{f}^*} C^\infty(M, C) \times \mathrm{Conn}(P_K) \times A^1(\mathfrak{l}(P_K)) \times A^*(R) \times \cdots.$$

For any geometrical function L on the left hand side there exists a geometrical function L' on the right hand side such that

$$L' \circ \tilde{f}^* = L.$$

In this way, we have reduced the $\mathrm{Gau}(P_G)$ redundancy inherent in L to a $\mathrm{Gau}(P_K)$ redundancy in L. This is known as spontaneous symmetry breaking.

17.3.2 Mass acquisition - the Higgs mechanism.

Let us now explain the idea of "acquisition of mass" under spontaneous symmetry breaking. For this purpose, let us examine the function

$$\mathrm{Conn}(P_G) \times A^0(U) \to C^\infty(M)$$

given by

$$(\omega, f) \mapsto \|d_\omega f\|^2,$$

where, recall, $d_\omega f$ denotes the covariant differential of the section f of U (the Higgs field) with respect to the connection ω.

Along the subbundle $f^{-1}(C)$ we have $f = (\phi, K)$ so

$$df = d\phi : TM \to TC$$

and

$$\omega \cdot f = \omega_\mathfrak{l} \cdot f \in T(O_\phi(m))$$

where O_c denotes the G orbit through $c \in C$. We may identify $\mathfrak{l}$ with the tangent space $T(O_c)$. Suppose the cross section has been chosen so that TC_c, and $TO_{\phi(m)}$ are always perpendicular. Then the above two components of $d_\omega f$ are orthogonal so we get

$$\|d_\omega f\|^2(p) = \|d\phi\|^2(q) + \|\omega_\mathfrak{l}\|^2_{\phi(q)} \tag{17.5}$$

where $\| \ \|_c$ denotes the metric induced on $\mathfrak{l}$ by the identification

$$\mathfrak{l} \sim TO_c \qquad \mathfrak{l} \ni \xi \mapsto \xi \cdot c.$$

Let us assume that $\mathfrak{k}$ and $\mathfrak{l}$ are perpendicular for the scalar product on $\mathfrak{g}$. Then substituting into (17.4) we see that

$$\|\mathrm{Curv}(\omega)\|^2 = \|d(\tilde{f}(\omega)_\mathfrak{k})\|^2 + \|d(\tilde{f}^*\omega)_\mathfrak{l}\|^2$$

$$+\text{terms cubic or higher in } (\tilde{f}^*(\omega)_\mathfrak{k} \text{ and } (\tilde{f}^*\omega)_\mathfrak{l}. \tag{17.6}$$

17.4. THE HIGGS MECHANISM IN THE STANDARD MODEL.

Let us combine (17.6) with (17.5). Consider a Lagrangian L of the form

$$L = \|\operatorname{Curv}\omega\|^2 + a\|d_\omega f\|^2.$$

Then we can write $L = L' \circ \tilde{f}^*$ where

$$L' = \|d(\tilde{f}^*\omega)_{\mathfrak{l}}\|^2 + a\|\omega\mathfrak{l}\|^2_{\phi_0} + \|d\phi\|^2 + \|d(\tilde{f}^*\omega)_{\mathfrak{k}}\|^2$$

+higher order terms.

Here ϕ_0 is some constant value of ϕ usually determined as the minimum of some function.) In L', the quadratic terms in $\omega_{\mathfrak{l}}$, look like the Lagrangian for the Proca equation, with a $a\|\omega\mathfrak{l}\|^2_{\phi_0}$ giving the mass terms. This is the procedure of mass acquisition in spontaneous symmetry breaking.

17.4 The Higgs mechanism in the standard model.

I want to specialize the above discussion to the case of the standard model of electroweak interactions. Before doing so, I should mention some language difficulties I have had over the years in communication between mathematicians and physicists.

17.4.1 Problems of translation between mathematicians and and physicists.

Is there an i in the structure constants of a Lie algebra?

A first barrier between the mathematics literature and most of the physics literature is the ubiquitous factor of i: The mathematical definition of a Lie algebra is that it is a vector space $\mathfrak{g}$ with a bilinear map

$$\mathfrak{g} \times \mathfrak{g} \to \mathfrak{g}$$

which is anti-symmetric and satisfies Jacobi's identity.

So the set of self-adjoint matrices under commutator bracket is **not** a Lie algebra. Indeed the commutator of two self adjoint matrices is skew adjoint. So the Lie algebra of $u(n)$ is not the space of self-adjoint matrices but rather the space of skew adjoint matrices. Indeed, if A is a skew adjoint matrix then $\exp tA$ is a one parameter group of unitary matrices. The physicists prefer to write $\exp itH$ where H is self adjoint. This is of course due to the fact that self adjoint operators are the observables of quantum mechanics, and Noether's theorem suggests that elements of the Lie algebra should correspond to observables. But the price to pay for this is to put an i in front of all brackets.

For example, the three dimensional real vector space consisting of self adjoint two by two matrices of trace zero has, as a basis, τ_i, $i = 1, 2, 3$ the "Pauli matrices" where, to be absolutely sure of the factors $1/2$ etc.,

$$\tau_1 := \begin{pmatrix} 0 & 1 \\ 1 & 0 \end{pmatrix}, \quad \tau_2 := \begin{pmatrix} 0 & -i \\ i & 0 \end{pmatrix}, \quad \tau_3 := \begin{pmatrix} 1 & 0 \\ 0 & -1 \end{pmatrix}. \tag{17.7}$$

The physicists like to think of these as "generators" of $SU(2)$, i.e. as elements of the Lie algebra $su(2)$. Of course, we mathematicians would say that multiplying each of these three matrices by i gives a basis of $su(2)$. This distinction is relatively harmless, but is a nuisance for a mathematician reading a physics book or paper.

Ad invariant metrics on $u(2)$.

If we use the scalar product

$$(A, B) = 2 \operatorname{tr} AB$$

then the elements $\frac{1}{2}\tau_i$, $i = 1, 2, 3$ form an orthonormal basis of our three dimensional space of self-adjoint matrices of trace zero. Since the algebra $su(2)$ is simple, the most general Ad invariant scalar product on our three dimensional space of self-adjoint matrices of trace zero must be a positive multiple of the above scalar product. We will want to consider the four dimensional space of *all* two by two self adjoint matrices. (After multiplication by i this would yield the Lie algebra of $U(2)$.) So we must add the two by two identity matrix to get a basis of this four dimensional real space. The algebra $u(2)$ is not simple, but decomposes into the sum of two ideals consisting of $su(2)$ and all (real) multiples of iI. These ideals must be orthogonal under any Ad invariant metric. So there is a two parameter family of Ad invariant metrics on $u(2)$. We discussed this at the end of Chapter 5.

Ad invariant metrics, coupling constants, and the Weinberg angle.

Indeed, as we saw in Chapter 5, the most general Ad invariant metric on our four dimensional space of self adjoint two by two matrices can be written as

$$\frac{2}{g_2^2} \operatorname{tr}(A - \frac{1}{2}(\operatorname{tr} A)I)(B - \frac{1}{2}(\operatorname{tr} B)I) + \frac{1}{g_1^2} \operatorname{tr} A \operatorname{tr} B. \tag{17.8}$$

Relative to this scalar product the elements

$$\frac{g_2}{2}\tau_1, \quad \frac{g_2}{2}\tau_2, \quad \frac{g_2}{2}\tau_3, \quad \frac{g_1}{2}I \tag{17.9}$$

form an orthonormal basis. Notice that for traceless matrices the second term in (17.8) vanishes, and the first term reduces to a multiple of $2 \operatorname{tr} AB$; similarly, for multiples of I, the first term vanishes.

For mathematicians, the question is "why this strange notation, with the g_2 and g_1 occurring in the denominator?". The answer is that, in the physics literature, these constants are not regarded as parametrizing metrics on $u(2)$, but rather as "universal coupling constants", see the next section.

What are classical fields?

A third difference between the mathematical literature and the physics literature is that in the physics literature all (classical) fields are regarded as scalar

17.4. THE HIGGS MECHANISM IN THE STANDARD MODEL.

valued functions (or vector fields) or n-tuplets of scalar valued functions (or vector fields). One must then discuss the "field transformations" under which, for example, the Lagrangian is invariant. The mathematical literature prefers a "basis free" formulation where many of the invariance properties of the Lagrangian are obvious - they are built into the formulation. The price to pay is that the fields are no longer scalar functions or n-tuplets of scalar functions but vector valued functions, or, more generally, sections of a vector bundle, or differential forms with values in a vector bundle. This means that in the physics literature a basis of the vector space (or a basis of sections of the vector bundle) is chosen. Thus, for example, if we choose a basis $v_1, \ldots, v_n$ of a Lie algebra $\mathfrak{k}$ then the Lie bracket can be given in terms of the "Cartan structure constants" c_{jk}^ℓ where

$$[v_j, v_k] = \sum_\ell c_{jk}^\ell v_\ell.$$

As explained above, in the physics literature there will be an additional factor of i in front of the structure constants as understood by the mathematicians. For example, if we take the orthonormal basis of the space of traceless two by two self adjoint matrices consisting of the first three elements of (17.7), we find by direct computation that

$$\left[\frac{g}{2}\tau_1, \frac{g}{2}\tau_2\right] = i\frac{g^2}{2}\tau_3 = ig\frac{g}{2}\tau_3 \qquad \text{where } g = g_2,$$

with a similar formula for the brackets of the remaining two elements. So relative to this basis, the structure constants are

$$C_{jk\ell} = ig\epsilon jk\ell.$$

Up to an overall sign arising from slightly different conventions this is the statement about the structure constants of $SU(2)_L$ found in S. Weinberg, *The Quantum Theory of Fields*, Cambridge U. Press (1996), vol. 2. page 307 just after equation (21.3.11) giving the expression of the Lagrangian of the Yang-Mills field. So whereas for mathematicians the parameter g describes the scalar product on $su(2)$, for physicists, who write out the fields in terms of an orthonormal basis, the g appears in the structure constants and is interpreted as a "coupling constant", measuring the "strength of the interaction between the fields".

For example, we have seen in Chapter 12 that the electromagnetic field F is a two form whose integral over any surface has units of inverse charge. So $F \wedge \star F$ is a 4-form on space time with units $1/(\text{charge})^2$. Since, in quantum field theory, one takes the exponential of (a multiple of) the integral of this object, we want $F \wedge \star F$ to be a scalar. Thus, in order to get the correct Lagrangian density, we must multiply by ϵ_0, the permittivity of empty space, which (in natural units) has units of $(\text{charge})^2 2$, so that

$$\frac{1}{2}\epsilon_0 F \wedge \star F$$

is the Lagrangian density for the electromagnetic field in empty space. If we want to consider F (strictly speaking iF in the physicist notation) as the curvature of a connection on a $U(1)$ bundle as suggested by London, we see that we must consider ϵ_0 as determining a metric on $u(1)$ (different from the "natural" one regarding $u(1)$ as $i\mathbb{R}$), and this metric has deep physical significance.

More generally, if G is the structure group of the bundle P_G and $\mathfrak{g}_0$ is the Lie algebra of G, the curvature of such a connection is a 2-form on space-time with values in the vector bundle $\mathfrak{g}_0(P)$ associated to the adjoint representation of G. If F is such a curvature form, and if $\star$ denotes the Hodge star operator of space time, then $\star F$ is another 2-form with values in $\mathfrak{g}_0(P)$, so

$$F \wedge \star F$$

is a 4-form with values in $\mathfrak{g}_0(P) \otimes \mathfrak{g}_0(P)$. In order to get a numerical valued 4-form which we can consider as a Lagrangian density, we need an Ad invariant scalar product on $\mathfrak{g}_0$.

In the Standard Model of the electroweak theory, the group under consideration is $U(2)$ or $SU(2) \times U(1)$ with Lie algebra $\mathfrak{g}_0 = u(2)$. As we have seen, there is a two parameter family of invariant metrics on $u(2)$ given by (5.12).

We repeat that we are regarding g_1 and g_2 as parameters describing possible Ad invariant scalar products on the Lie algebra $u(2)$. As such they have physical significance similar to that of the permittivity of free space in electromagnetic theory and are necessary to be able to formulate a Yang-Mills functional. In a general relativistic theory one would expect them to have a space time dependence just as the metric of space time does. The interpretation of g_1 and g_2 as "universal coupling constants" then derives from the interpretation as defining a metric.

17.4.2 The Higgs mechanism in the standard model.

The Higgs mechanism in a nutshell.

The Higgs mechanism in the Standard Model of electroweak interactions is a device for breaking the $u(2) = su(2) \oplus u(1)$ symmetry of a $U(2)$ gauge theory in such a way that the three of the four components of a connection form (originally massless in a pure Yang-Mills theory) become differential forms with values in a vector bundle associated to $U(1)$ and which enter into a Lagrangian whose quadratic terms correspond to particles with positive mass. In mathematical terms this corresponds to a reduction of a principal $U(2)$ bundle to a $U(1)$ bundle.

The ingredients that go into this mechanism and into the computation of the acquired masses are the following:

- An Ad invariant positive definite metric on $u(2)$. This is needed for the original (unbroken) Yang-Mills theory. We have argued that the "universal coupling constants" that enter into the general formulation of this theory are in fact parameters which describe the possible Ad invariant metrics on $u(2)$. In general there is a two parameter family of such metrics. They are related by a certain angle θ_W known as the Weinberg angle. A current measured value gives $\sin^2 \theta_W = 0.2312 \pm 0.003$.

17.4. THE HIGGS MECHANISM IN THE STANDARD MODEL. 361

- A two dimensional Hermitian vector bundle associated to the principal $U(2)$ bundle. In the general presentation of the Standard Model this vector bundle is an extraneous ingredient put in "by hand".

- A degree-four polynomial on this vector bundle. In the general presentation this must also be provided by hand.

- The vector bundle $\mathfrak{g}_1$ is associated to the original $U(2)$ bundle, so $U(2)$ invariance determines the Hermitian metric up to a scalar factor. Once the metric has been fixed, we can write the most general (invariant) degree four polynomial as
$$a \|\cdot\|^4 - b \|\cdot\|^2.$$

For the next three steps of the standard Higgs mechanism, cf. for example A. Derdzinski, *Geometry of the standard model* Section 11. We summarize them here for the reader's convenience. Additional details will be given below.

- If a and b are both positive, then the quadratic polynomial
$$az^2 - bz$$
achieves its minimum at
$$z_0 = \frac{b}{2a}$$
and hence any section ψ of our vector bundle lying on the three-sphere bundle
$$\|\psi\|^2 = z_0$$
is a global minimum. Any such section is called a *vacuum state*. The reduction of the principal $U(2)$ bundle is achieved by fixing one such vacuum. For example, if the bundle is trivial and is given a trivialization which identifies it with the trivial $\mathbb{C}^2$ bundle then we may choose ψ of the form
$$\psi = \psi_0 := \begin{pmatrix} 0 \\ v \end{pmatrix}, v > 0$$
so
$$\|\psi_0\| = \sqrt{\frac{b}{2a}}.$$

- The mass of the W particle is then given as
$$m(W) = \frac{\|\psi_0\|}{\|i\tau_1\|_{u(2)}} \tag{17.10}$$
where
$$\tau_1 = \begin{pmatrix} 0 & 1 \\ 1 & 0 \end{pmatrix}.$$

See the discussion in Section 17.4.2 below. In terms of the parameter g_2 entering into the definition of the metric on $su(2)$ this becomes
$$m(W) = \frac{1}{2}g_2\|\psi_0\| = \frac{1}{2}g_2\sqrt{\frac{b}{2a}}. \tag{17.11}$$

- The mass of the Higgs field (see below) is given by

$$m(\text{Higgs}) = 2\sqrt{b}. \tag{17.12}$$

This gives the value of the Higgs mass in terms of parameters entering into the Higgs model. Notice that only the coefficient of the quadratic term (b) enters into this formula, but if we know the coefficient a of the quartic term, then we can get b from $\|\psi_0\| = \sqrt{b/2a}$.

Other quadratic forms.

Given a positive definite real scalar product $(\cdot, \cdot)$ on a real vector space, any other quadratic form is given by $(S\cdot, \cdot)$ where S is a self-adjoint operator. We can then diagonalize S. If the second quadratic form is positive semi-definite, then these eigenvalues are non-negative, and S has a unique square root $S^{\frac{1}{2}}$. In the Standard Model, i.e. for the case of the reduction of a $U(2)$ bundle to a $U(1)$ bundle via the choice of section of an associated bundle - these eigenvalues are identified with the masses of certain spin 1 particles:

Consider the standard action of $u(2)$ on $\mathbb{C}^2$ and define the "second" quadratic form on $u(2)$ to be

$$q(A) := \|A\psi_0\|_{\mathbb{C}^2}^2 = (A\psi_0, A\psi_0)_{\mathbb{C}^2}$$

where ψ_0 is a fixed element of $\mathbb{C}^2$, and where $(\cdot, \cdot)_{\mathbb{C}^2}$ is some $U(2)$ invariant scalar product on $\mathbb{C}^2$ (and so is some positive multiple of the standard scalar product). The corresponding bilinear form on $u(2)$ is

$$\langle A, B \rangle = \text{Re}(A\psi_0, B\psi_0)_{\mathbb{C}^2}.$$

The masses of the vector bosons.

In fact, let us take

$$\psi_0 = \begin{pmatrix} 0 \\ v \end{pmatrix}, \quad v > 0.$$

Then

$$\tau_1 \psi_0 = \begin{pmatrix} v \\ 0 \end{pmatrix}, \ \tau_2 \psi_0 = \begin{pmatrix} -iv \\ 0 \end{pmatrix}, \ \tau_3 \psi_0 = \begin{pmatrix} 0 \\ -v \end{pmatrix}, \text{ and } I\psi_0 = \begin{pmatrix} 0 \\ v \end{pmatrix}.$$

Then relative to any scalar product $(\cdot, \cdot)$ on $u(2)$ we have

$$(S\tau_1, X) = \langle \tau_1, X \rangle = 0 \quad \text{for } X = \tau_2, \tau_3, I.$$

If $(\cdot, \cdot)$ is any of the invariant metrics (5.12), we have $(\tau_1, X) = 0$ for $X = \tau_2, \tau_3$. This shows that τ_1 is an eigenvector of S with eigenvalue $\|\psi_0\|^2/\|\tau_1\|^2$. Similarly for τ_2. Sections of the line bundles corresponding to these eigenvectors are identified with the W particles. This accounts for the mass of the W as given above.

17.4. THE HIGGS MECHANISM IN THE STANDARD MODEL.

We have $(\tau_3 + I)\psi_0 = 0$, so $\tau_3 + I$ is an eigenvector of S with eigenvalue 0. Expressed in terms of the orthonormal basis (2) and normalized so to have length one gives

$$\frac{1}{(g_1^2 + g_2^2)^{\frac{1}{2}}} \left(g_2 \frac{g_1}{2} I + g_1 \frac{g_2}{2} \tau_3 \right).$$

The corresponding mass zero field is then identified with the electromagnetic field. Taking the orthogonal complement of the three eigenvectors found so far (corresponding to the W's and the electromagnetic field) gives the field of the Z particle.

The mass of the Higgs.

It is assumed that the Higgs field is a section of a Hermitian vector bundle with potential $\mathcal{V}$ which has the form

$$\mathcal{V}(\psi) = f(\langle \psi, \psi \rangle)$$

where

$$f : [0, \infty) \to \mathbb{R}$$

is a smooth function with a minimum at z_0. A particular section ψ_0 is chosen with $\langle \psi_0, \psi_0 \rangle = z_0$. (If, as we shall assume, the Hermitian vector bundle is a two dimensional bundle associated to a principal $U(2)$ or $SU(2) \times U(1)$ bundle this has the effect of reducing the principal bundle to a $U(1)$ bundle.)

The most general section of our vector bundle is then written as $\psi_0 + \eta$ and we consider the quadratic term in the expansion of $f(\psi_0 + \eta)$ as a function of η. It will be given as

$$\frac{1}{2} \operatorname{Hess}(f)(\psi_0)(\eta) = 2f''(\langle \psi_0, \psi_0 \rangle)(\operatorname{Re}\langle \psi_0, \eta \rangle)^2.$$

For η tangent to the orbit of the action of $U(2)$ this vanishes. But for $\eta \in \mathbb{R}\psi_0$ we have $\langle \psi_0, \eta \rangle = \pm \|\psi_0\| \|\eta\|$ so for such η known as the Higgs field) the quadratic term is

$$2z_0 f''(z_0) \|\eta\|^2.$$

We want to consider this as a mass term, which means that we want to write this quadratic expression as $\frac{1}{2} m \|\eta\|^2$.

If

$$f(z) = az^2 - bz$$

with a and b positive constants, then the minimum of f is achieved at

$$z_0 = \frac{b}{2a}$$

and

$$f''(z_0) = 2a.$$

So
$$2z_0 f''(z_0) = 2b.$$
We wish to write $2b\|\eta\|^2$ as $\frac{1}{2}m\|\eta\|^2$ where m is the mass of the Higgs. This gives
$$m(\text{Higgs}) = 2\sqrt{b}.$$

The notion of a connenction can be generalized to that of a "superconnection". See the next chapter. In a series of papers, Yuval Neeman and I argued that the Higgs field be regarded as a piece of the odd component of a superconnection. In our theory, the quartic term of the polynomial giving rise to the Higgs mechanism is the super-Yang-Mills functional. We made some predictions as to the relation between the Higgs mass and the mass of the W boson. We also gave a superalgebra interpretation of the Weinberg angle and an explanation of the charges of the leptons in generations. For details about this, see my Toronto lectures available on the web at

http://www.fields.utoronto.ca/programs/scientific/07-08/geomanalysis/
sternberglectures.pdf.

As of this writing, the Higgs boson has not yet been discovered experimentally.

Chapter 18

Superconnections.

Daniel Gray Quillen

Born: 27 June 1940 in Orange, New Jersey, USA
Died: 30 April 2011 in Gainesville, Florida, USA

In this chapter we give a self contained introduction to the theory of superconnections, a subject due to the late Daniel Quillen. In the main, we follow the exposition given in N. Berline, E. Getzler, M. Vergne: *Heat Kernels and Dirac Operators*, Springer, Berlin 1991 with some changes in convention and notation.

18.1 Superbundles.

18.1.1 Superspaces and superalgebras.

Recall that **superspace** E is just a vector space with a $\mathbb{Z}_2$ grading:

$$E = E^+ \oplus E^-,$$

and that **superalgebra** A is an algebra whose underlying vector space is a superspace and such that

$$A^+ \cdot A^+ \subset A^+, \quad A^- \cdot A^- \subset A^+, \quad A^+ \cdot A^- \subset A^-, \quad A^+ \cdot A^- \subset A^-.$$

The commutator of two homogeneous elements of A is defined as

$$[a, b] := ab - (-1)^{|a| \cdot |b|} ba.$$

We use the notation $|a| = 0$ if $a \in A^+$ and $|a| = 1$ if $a \in A^-$ and we do addition and multiplication mod 2.

A superalgebra is **commutative** if the commutator of any two elements vanishes. For example, the exterior algebra $\wedge(V)$ of a vector space is a commutative superalgebra where

$$\wedge(V)^+ := \wedge^0(V) \oplus \wedge^2(V) \oplus \wedge^4(V) \oplus \cdots,$$

and

$$\wedge(V)^- := \wedge^1(E) \oplus \wedge^3(V) \oplus \cdots.$$

We will be interested in this chapter in superalgebras which are not necessarily commutative.

18.1.2 The tensor product of two superalgebras.

If A and B are superspaces we make $A \otimes B$ into a superspace by

$$|a \otimes b| = |a| + |b|.$$

If A and B are superalgebras we make $A \otimes B$ into a superalgebra by

$$(a \otimes b) \cdot (a' \otimes b') := (-1)^{|b| \cdot |a'|} aa' \otimes bb'.$$

For example, the Clifford algebra of any vector space with a scalar product is a superalgebra, where $C(V)^+$ consists of those elements which can be written as a sum of products of an even number of elements of V and $C(V)^-$ consists of those elements which can be written as a sum of products of an odd number of elements of V. If V and W are two spaces with scalar products then the

18.1. SUPERBUNDLES. 367

Clifford algebra of their orthogonal direct sum is the tensor product of their Clifford algebras:
$$C(V \oplus W) = C(V) \otimes C(W).$$
We will use the convention of the algebraists rather than that of the geometers in the definition of the Clifford algebra. So if V is a vector space with a (not necessarily positive definite) scalar product then $C(V)$ is the universal algebra relative to the relations
$$uv + vu = 2(u,v)\mathbf{1}.$$
(In Berline, Getzler, and Vergne the opposite convention (with a minus sign on the right hand side) is used.)

18.1.3 Lie superalgebras.

If A is an associative superalgebra the commutator of two homogeneous elements of A was defined as
$$[a,b] := ab - (-1)^{|a|\cdot|b|}ba.$$
This commutator satisfies the axioms for a **Lie superalgebra** which are

- $[a,b] + (-1)^{|a|\cdot|b|}[b,a] = 0$, and
- $[a,[b,c]] = [[a,b],c] + (-1)^{|a|\cdot|b|}[b,[a,c]]$.

If A is a commutative superalgebra and L is a Lie superalgebra then $A \otimes L$ is again a Lie superalgebra under the usual definition:
$$[a \otimes X, b \otimes Y] := (-1)^{|X|\cdot|b|} ab \otimes [X,Y].$$

18.1.4 The endomorphism algebra of a superspace.

Let $E = E^+ \oplus E^-$ be a superspace. We make the algebra of all endomorphisms (= linear transformations) of E into a superalgebra by letting $\text{End}(E)^+$ consist of those linear transformations which carry E^+ into E^+ and E^- into E^- while $\text{End}(E)^-$ interchanges the two components. Thus a typical element of $\text{End}(E)^+$ looks like
$$\begin{pmatrix} A & 0 \\ 0 & D \end{pmatrix}, \quad A \in \text{End}(E^+), \quad D \in \text{End}(E^-)$$
while a typical element of $\text{End}(E)^-$ looks like
$$\begin{pmatrix} 0 & B \\ C & 0 \end{pmatrix}, \quad B : E^- \to E^+, \quad C : E^+ \to E^-.$$

An action (or a representation) of an associative algebra A on a superspace E is a (gradation preserving) homomorphism of A into $\text{End}(E)$. We then also say that E is an A module.

Similarly, a representation of a Lie superalgebra L on a superspace E is a homomorphism of L into the commutator Lie superalgebra of $\text{End}(E)$. This is the same as an action of the universal enveloping algebra $U(L)$ on E. We say that E is an L module.

18.1.5 Superbundles.

Let $\mathcal{E} \to M$ be a bundle of superspaces over a manifold M. We call such an object a **superbundle**. So $\mathcal{E} = \mathcal{E}^+ \oplus \mathcal{E}^-$ where $\mathcal{E}^+ \to M$ and $\mathcal{E}^- \to M$ are vector bundles over M. We will call a section of $\mathcal{E}^+$ an even section of $\mathcal{E}$ and a section of $\mathcal{E}^-$ an odd section of $\mathcal{E}$.

If $\mathcal{E}$ and $\mathcal{F}$ are superbundles, then $\mathcal{E} \otimes \mathcal{F}$ is a superbundle. In particular, $\wedge(T^*M)$ is a superbundle where

$$\wedge(T^*M)^+ := \wedge^0(T^*M) \oplus \wedge^2(T^*M) \oplus \wedge^4(T^*M) \oplus \cdots,$$

$$\wedge(T^*M)^- := \wedge^1(T^*M) \oplus \wedge^3(T^*M) \oplus \wedge^5(T^*M) \oplus \cdots.$$

A section of $\wedge(T^*M) \otimes \mathcal{E}$ is called an $\mathcal{E}$-**valued differential form** and the space of all $\mathcal{E}$-valued differential forms will be denoted by $\mathcal{A}(M, \mathcal{E})$. Locally any element of $\mathcal{A}(M, \mathcal{E})$ is a sum of terms of the form $\alpha \otimes s$ where α is a differential form on M and s is a section $\mathcal{E}$.

18.1.6 The endomorphism bundle of a superbundle.

If $\mathcal{E} \to M$ is a superbundle, then we can consider the superbundle $\mathrm{End}(\mathcal{E})$ where, at each $m \in M$ we have $\mathrm{End}(\mathcal{E})_m := \mathrm{End}(\mathcal{E}_m)$. We have an action of any section of $\mathrm{End}(\mathcal{E})$ on any section of $\mathcal{E}$. By tensor product, any element of $\mathcal{A}(M, \mathrm{End}(\mathcal{E}))$ acts on any element of $\mathcal{A}(M, \mathcal{E})$. In particular any element of $\mathcal{A}(M)$, i.e. any differential form acts on $\mathcal{A}(M, \mathcal{E})$ and (super)commutes with all elements of $\mathcal{A}(M, \mathrm{End}(\mathcal{E}))$.

18.1.7 The centralizer of multiplication by differential forms.

Any element of $\mathcal{A}(M)$, i.e. any differential form acts on $\mathcal{A}(M, \mathcal{E})$ and (super)commutes with all elements of $\mathcal{A}(M, \mathrm{End}(\mathcal{E}))$.

There is an important converse to this last assertion. A differential operator on $\mathcal{A}(M, \mathcal{E})$ is by definition an operator which in local coordinates looks like

$$\sum_\gamma a_\gamma \partial^\gamma$$

where a_γ is a section of $\mathcal{A}(M, \mathrm{End}(\mathcal{E}))$ and $\partial^\gamma = \partial_1^{\gamma_1} \cdots \partial_n^{\gamma_n}$ is a partial differentiation operator in terms of the local coordinates. Leibnitz's rule implies that if such an operator commutes with all multiplications by functions, then it can't really involve any differentiations. If furthermore it commutes with the action of all elements of $\mathcal{A}(M)$ it must be given by the action of some element of $\mathcal{A}(M, \mathrm{End}(\mathcal{E}))$. In short: a differential operator on $\mathcal{A}(M, \mathcal{E})$ commutes with the action of $\mathcal{A}(M)$ if and only if it is given by an element of $\mathcal{A}(M, \mathrm{End}(\mathcal{E}))$.

18.1.8 Bundles of Lie superalgebras.

If $\mathfrak{g}$ is a bundle of Lie superalgebras over M then $\mathcal{A}(M,\mathfrak{g})$ is a Lie superalgebra with bracket determined (as we have seen) by

$$[\alpha \otimes X, \beta \otimes Y] = (-1)^{|X|\cdot|\beta|}(\alpha \wedge \beta) \otimes [X,Y].$$

If $\mathcal{E}$ is a superbundle on which $\mathfrak{g}$ acts, meaning that we have a Lie superalgebra homomorphism ρ of $\mathfrak{g}$ into the Lie superalgebra $\operatorname{End}(\mathcal{E})$ (under bracket), then we have an action of $\mathcal{A}(M,\mathfrak{g})$ on $\mathcal{A}(M,\mathcal{E})$ determined by

$$\rho(\alpha \otimes X)(\beta \otimes v) = (-1)^{|X|\cdot|\beta|}(\alpha \wedge \beta) \otimes (\rho(X)v).$$

18.2 Superconnections.

A **superconnection** on a superbundle $\mathcal{E}$ is an odd first order differential operator

$$\mathbb{A} : \mathcal{A}^{\pm}(M,\mathcal{E}) \to \mathcal{A}^{\mp}(M,\mathcal{E})$$

which satisfies

$$\mathbb{A}(\alpha \wedge \theta) = d\alpha \wedge \theta + (-1)^{|\alpha|}\alpha \wedge \mathbb{A}\theta, \quad \forall\, \alpha \in \mathcal{A}(M),\ \theta \in \mathcal{A}(M,\mathcal{E})$$

which we can write as

$$[\mathbb{A}, e(\alpha)] = e(d\alpha) \tag{18.1}$$

where $e(\beta)$ denotes the operation of exterior multiplication by $\beta \in \mathcal{A}(M)$.

Let $\Gamma(\mathcal{E})$ denote the space of smooth sections of $\mathcal{E}$ which we can regard as a subspace of $\mathcal{A}(M,\mathcal{E})$. Then

$$\mathbb{A} : \Gamma(\mathcal{E}^{\pm}) \to \mathcal{A}^{\mp}(M,\mathcal{E})$$

and $\mathbb{A}$ is completely determined by this map since

$$\mathbb{A}(\alpha \otimes s) = d\alpha \otimes s + (-1)^{|\alpha|}\alpha \otimes \mathbb{A}s$$

for all differential forms α and sections s of $\mathcal{E}$.

Conversely, suppose that $\mathbb{A} : \Gamma(\mathcal{E}^{\pm}) \to \mathcal{A}^{\mp}(M,\mathcal{E})$ is a first order differential operator which satisfies

$$\mathbb{A}(fs) = df \otimes s + f \otimes \mathbb{A}s$$

for all functions f and sections s of $\mathcal{E}$. Then we can extend $\mathbb{A}$ to $\mathcal{A}(M,\mathcal{E})$ by setting

$$\mathbb{A}(\alpha \otimes s) = d\alpha \otimes s + (-1)^{|\alpha|}\alpha \otimes \mathbb{A}s$$

without fear of running into a contradiction.

18.2.1 Extending superconnections to the bundle of endomorphisms.

If $\gamma \in \mathcal{A}(M, \text{End}(\mathcal{E}))$ define
$$\mathbb{A}\gamma := [\mathbb{A}, \gamma].$$

We claim that $[\mathbb{A}, \gamma]$ belongs to $\mathcal{A}(M, \text{End}(\mathcal{E}))$. To prove this, we must check that $[\mathbb{A}, \gamma]$ commutes with all $e(\alpha)$, $\alpha \in \mathcal{A}(M)$. For any $\alpha \in \mathcal{A}(M)$ we have

$$\mathbb{A} \circ \gamma \circ e(\alpha) = (-1)^{|\gamma| \cdot |\alpha|} \mathbb{A} \circ e(\alpha) \circ \gamma$$
$$= (-1)^{|\gamma| \cdot |\alpha|} e(d\alpha) \circ \gamma + (-1)^{|\alpha| + |\gamma| \cdot |\alpha|} e(\alpha) \circ \mathbb{A} \circ \gamma$$

while

$$\gamma \circ \mathbb{A} \circ e(\alpha) = \gamma \circ e(d\alpha) + (-1)^{|\alpha|} \gamma \circ e(\alpha) \circ \mathbb{A}$$
$$= (-1)^{|\gamma| + |\gamma| \cdot |\alpha|} e(d\alpha) \circ \gamma + (-1)^{|\alpha| + |\alpha| \cdot |\gamma|} e(\alpha) \circ \gamma \circ \mathbb{A}$$

so

$$[\mathbb{A}, \gamma] \circ e(\alpha) = \mathbb{A} \circ \gamma \circ e(\alpha) - (-1)^{|\gamma|} \gamma \circ \mathbb{A} \circ e(\alpha)$$
$$= (-1)^{|\alpha| + |\alpha| \cdot |\gamma|} e(\alpha) \circ [\mathbb{A}, \gamma]$$

Since $|[\mathbb{A}, \gamma]| = |\gamma| + 1$ this shows that $[[\mathbb{A}, \gamma], e(\alpha)] = 0$ as desired.

18.2.2 Supercurvature.

Consider the even operator $\mathbb{A}^2$. We have,

$$[\mathbb{A}^2, e(\alpha)] = \mathbb{A} \circ [\mathbb{A}, e(\alpha)] + (-1)^{|\alpha|} [\mathbb{A}, e(\alpha)] \circ \mathbb{A} =$$
$$\mathbb{A} \circ e(d\alpha) - (-1)^{|d\alpha|} e(d\alpha) \circ \mathbb{A} = [\mathbb{A}, e(d\alpha)] = e(dd(\alpha)) = 0.$$

So $\mathbb{A}^2 \in \mathcal{A}(M, \text{End}(\mathcal{E}))$. We set

$$\mathbb{F} := \mathbb{A}^2$$

and call it the **curvature** of the superconnection $\mathbb{A}$.

The **Bianchi identity** says that

$$\mathbb{A}\mathbb{F} := 0.$$

Indeed $\mathbb{A}\mathbb{F}$ is defined as $[\mathbb{A}, \mathbb{F}]$ and since $\mathbb{F} := \mathbb{A}^2$ is even we have

$$[\mathbb{A}, \mathbb{A}^2] = \mathbb{A} \circ \mathbb{A}^2 - \mathbb{A}^2 \circ \mathbb{A} = 0$$

by the associative law.

18.2.3 The tensor product of two superconnections.

If $\mathcal{E}$ and $\mathcal{F}$ are superbundles recall that $\mathcal{E} \otimes \mathcal{F}$ is the superbundle with grading

$$(\mathcal{E} \otimes \mathcal{F})^+ = \mathcal{E}^+ \otimes \mathcal{F}^+ \oplus \mathcal{E}^- \otimes \mathcal{F}^-,$$
$$(\mathcal{E} \otimes \mathcal{F})^- = \mathcal{E}^+ \otimes \mathcal{F}^- \oplus \mathcal{E}^- \otimes \mathcal{F}^+.$$

If $\mathbb{A}$ is a superconnection on $\mathcal{E}$ and $\mathbb{B}$ is a superconnection on $\mathcal{F}$ then $\mathbb{A} \otimes \mathbf{1} + \mathbf{1} \otimes \mathbb{B}$ is a superconnection on $\mathcal{E} \otimes \mathcal{F}$. Thus

$$(\mathbb{A} \otimes \mathbf{1} + \mathbf{1} \otimes \mathbb{B})(\alpha \wedge \beta) := \mathbb{A}\alpha \wedge \beta + (-1)^{|\alpha|}\alpha \wedge \mathbb{B}\beta.$$

A bit of computation shows that this definition is consistent and defines a superconnection on $\mathcal{E} \otimes \mathcal{F}$.

18.2.4 The exterior components of a superconnection.

If $\mathbb{A}$ is a superconnection on a superbundle $\mathcal{E}$ we may break $\mathbb{A}$ into its homogeneous components $\mathbb{A}_{[i]}$ which map $\Gamma(M, \mathcal{E})$ into $\mathcal{A}^i(M, \mathcal{E})$, the space of i-forms with values in $\mathcal{E}$:

$$\mathbb{A} = \mathbb{A}_{[0]} + \mathbb{A}_{[1]} + A_{[2]} + \cdots.$$

Let s be a section of $\mathcal{E}$ and f a function. By the above decomposition and the defining property of a superconnection we have

$$\mathbb{A}(fs) = \sum_{i=0}^{n} \mathbb{A}_{[i]}(fs)$$

and

$$\mathbb{A}(fs) = df \otimes s + f \sum_{i=0}^{n} \mathbb{A}_{[i]} s$$

where n is the dimension of M. We see that

$$\mathbb{A}_1(fs) = df \otimes s + f\mathbb{A}_{[1]}s$$

which is the defining property of an ordinary connection. Furthermore, since $\mathbb{A}_{[1]}$ has total odd degree, we see that as an ordinary connection

$$\mathbb{A}_{[1]} : \Gamma(\mathcal{E}^+) \to \Gamma(T^*, \mathcal{E}^+) \quad \text{and} \quad \mathbb{A}_{[1]} : \Gamma(\mathcal{E}^-) \to \Gamma(\mathcal{E}^-).$$

It also follows from the above comparison of the two expressions for $\mathbb{A}(fs)$ that the remaining $\mathbb{A}_{[i]}$, $i \neq 1$ are given by the action of an element of $\mathcal{A}^i(M, \text{End}(\mathcal{E}))$. For example $\mathbb{A}_{[0]}$ is given by an element of $\Gamma(M, \text{End}^-(\mathcal{E}))$.

18.2.5 A local computation.

To see what the supercurvature computation looks like in terms of a local description, let us assume that our bundle $\mathcal{E}$ is trivial, i.e. $\mathcal{E} = M \times E$ where E is a superspace. Let us also assume that $\mathbb{A}$ has only components $\mathbb{A}_{[0]}$ and $\mathbb{A}_{[1]}$. This will be the case in the physical model mentioned at the end of the preceding chapter.

We may thus write $\mathbb{A}_{[0]} = L \in C^\infty(M, \mathrm{End}^-(E))$ so

$$L = \begin{pmatrix} 0 & L^- \\ L^+ & 0 \end{pmatrix}, \quad L^- \in C^\infty(M, \mathrm{Hom}(E^-, E^+)),$$

$$L^+ \in C^\infty(M, \mathrm{Hom}(E^+, E^-)).$$

We may also write

$$\mathbb{A}_{[1]} = d + A, \quad A \in \mathcal{A}^1(M, \mathrm{End}(E)^+).$$

Let ∇ denote the covariant differential corresponding to the ordinary connection $\mathbb{A}_{[1]}$. Then

$$\mathbb{F} := (\mathbb{A})^2 = \mathbb{A}_{[0]}^2 + [\mathbb{A}_{[1]}, \mathbb{A}_{[0]}] + \mathbb{A}_{[1]}^2 = \mathbb{A}_{[0]}^2 + \nabla \mathbb{A}_{[0]} + F$$

where F is the curvature of $\mathbb{A}_{[1]}$. In terms of the matrix decomposition above we have

$$\mathbb{F} = \begin{pmatrix} L^- L^+ + F^+ & \nabla L^- \\ \nabla L^+ & L^+ L^- + F^- \end{pmatrix}$$

where $F^\pm$ is the restriction of F to $E^\pm$. Notice that $\mathbb{F}$ is quadratic in L, and so any quadratic function of $\mathbb{F}$ will involve a quartic function of L.

18.3 Superconnections and principal bundles.

Let $\mathfrak{g} = \mathfrak{g}_0 \oplus \mathfrak{g}_1$ be a Lie superalgebra and G be a Lie group whose Lie algebra is $\mathfrak{g}_0$. Suppose that we have a representation of G as (even) automorphisms of $\mathfrak{g}$ whose restriction to $\mathfrak{g}_0$ is the adjoint representation of G on its Lie algebra.

We will denote the representation of G on all of $\mathfrak{g}$ by Ad.

18.3.1 Recalling some definitions.

Let $P = P_G \to M$ be a principal bundle with structure group G. Recall that this means the following:

- We are given an action of G on P. To tie in with standard notation we will denote this action by

$$(p, a) \mapsto pa^{-1}, \quad p \in P, \quad a \in G$$

18.3. SUPERCONNECTIONS AND PRINCIPAL BUNDLES.

so $a \in G$ acts on P by a diffeomorphism that we will denote by r_a:

$$r_a : P \to P, \qquad r_a(p) = pa^{-1}.$$

If $\xi \in \mathfrak{g}_0$, then $\exp(-t\xi)$ is a one parameter subgroup of G, and hence

$$r_{\exp(-t\xi)}$$

is a one parameter group of diffeomorphisms of P, and for each $p \in P$, the curve

$$r_{\exp(-t\xi)}p = p(\exp t\xi)$$

is a smooth curve starting at t at $t = 0$. The tangent vector to this curve at $t = 0$ is a tangent vector to P at p. In this way we get a linear map

$$u_p : \mathfrak{g}_0 \to TP_p, \qquad u_p(\xi) = \frac{d}{dt}p(\exp t\xi)_{|t=0}. \qquad (18.2)$$

- The action of G on P is free.

- The space P/G is a differentiable manifold M and the projection $\pi : P \to M$ is a smooth fibration.

- The fibration π is locally trivial consistent with the G action in the sense that every $m \in M$ has a neighborhood U such that there exists a diffeomorphism

$$\psi_U : \pi^{-1}(U) \to U \times G$$

such that

$$\pi_1 \circ \psi = \pi$$

where

$$\pi_1 : U \times F \to U$$

is projection onto the first factor and if $\psi(p) = (m, b)$ then

$$\psi(r_a p) = (m, ba^{-1}).$$

Suppose that $\pi : P \to M$ is a principal fiber bundle with structure group G. Since π is a submersion, we have the sub-bundle Vert of the tangent bundle TP where $\text{Vert}_p, p \in P$ consists of those tangent vectors which satisfy $d\pi_p v = 0$. From its construction, the subspace $\text{Vert}_p \subset TP_p$ is spanned by the tangents to the curves $p(\exp t\xi)$, $\xi \in \mathfrak{g}_0$. In other words, u_p is a surjective map from $\mathfrak{g}_0$ to Vert_p. Since the action of G on P is free, we know that u_p is injective. Putting these two facts together we conclude that

If $\pi : P \to M$ is a principal fiber bundle with structure group G then u_p is an isomorphism of $\mathfrak{g}_0$ with Vert_p for every $p \in P$.

An (ordinary) connection on a principal bundle is a choice of a "horizontal" subbundle Hor complementary to the vertical bundle which is invariant under the action of G. At any p we can define the projection

$$\mathbf{V}_p : TP_p \to \text{Vert}_p$$

along Hor_p, i.e. $\mathbf{V}_p$ is the identity on Vert_p and sends all elements of Hor_p to 0. Giving Hor_p is the same as giving $\mathbf{V}_p$ and condition of invariance under G translates into

$$d(r_b)_p \circ \mathbf{V}_p = \mathbf{V}_{r_b(p)} \circ d(r_b)_p \quad \forall b \in G, \; p \in P.$$

This then defines a one form ω on P with values in $\mathfrak{g}_0$:

$$\omega_p := u_p^{-1} \circ \mathbf{V}_p.$$

Invariance of the connection under G translates into

$$r_b^* \omega = \text{Ad}_b \omega.$$

Let ξ_P be the vector field on P which is the infinitesimal generator of $r_{\exp t\xi}$. The the infinitesimal version of the preceding equation is

$$D_{\xi_P} \omega = [\xi, \omega].$$

In view of definition of u_p as identifying ξ with the tangent vector to the curve $t \mapsto p(\exp t\xi) = r_{\exp -t\xi} p$ at $t = 0$, we see that

$$i(\xi_P)\omega = -\xi.$$

18.3.2 Generalizing the above to superconnections.

We now generalize this to superconnections: We define a **superconnection form** A to be an odd element of $\mathcal{A}(P, \mathfrak{g})$ which satisfies

$$r_b^* A = \text{Ad}_b A \quad \forall b \in G \tag{18.3}$$
$$i(\xi_P)A = -\xi \quad \forall \, \xi \in \mathfrak{g}_0. \tag{18.4}$$

The meaning of (18.4) is the following:

$$A = A_{[0]} + A_{[1]} + \cdots + A_{[n]}, \quad n = \dim M$$

where $A_{[i]}$ is an i-form with values in $\mathfrak{g}_0$ if i is odd and with values in $\mathfrak{g}_1$ it i is even. Then $A_{[1]}$ is a connection form and all the other components satisfy

$$i(\xi_P)A_{[i]} = 0.$$

This condition together with (18.3) imply that these other components can be identified with odd i-forms on M with values in $\mathfrak{g}(P)$ the vector bundle over M associated to the representation Ad of G on $\mathfrak{g}$.

More generally, if the superspace E is G module and also a $\mathfrak{g}$ module in a consistent way, then we can form the associated bundle

$$\mathcal{E}(M) = E(P)$$

which is a module for the associated bundle of superalgebras $\mathfrak{g}(P)$. A k-form on M with values in $\mathcal{E}$ is the same thing as a k-form σ on P with values in E which satisfies

1. $i(\xi_P)\sigma = 0 \quad \forall \, \xi \in \mathfrak{g}_0$ and

2. $r_a^*\sigma = \rho(a)\sigma$ where ρ denotes the action of G on E.

The bilinear map

$$\mathfrak{g} \times E \to E$$

given by the action of $\mathfrak{g}$ determines an exterior multiplication

$$\Omega(P, \mathfrak{g}) \times \Omega(P, E) \to \Omega(P, E)$$

which we will denote by $\diamond$. We then obtain a superconnection on $\mathcal{E}$ given by

$$\mathbb{A}\sigma = d\sigma + A \diamond \sigma. \tag{18.5}$$

18.4 Clifford Bundles and Clifford superconnections.

Suppose that M is a semi-Riemannian manifold so that we can form the bundle of Clifford algebras $C(TM)$. Suppose that $\mathcal{F}$ is a bundle of Clifford modules. We denote the action of a section a of $C(TM)$ on a section of $\mathcal{F}$ by $c(a)$. We extend this notation to denote the action of a Clifford bundle valued differential form, i.e. an element of $\mathcal{A}(M, C(TM))$ on $\mathcal{A}(M, \mathcal{F})$ by

$$c(\alpha \otimes a)(\beta \otimes s) = (-1)^{|a| \cdot |\beta|}(\alpha \wedge \beta) \otimes c(a)s$$

on homogeneous elements.

A superconnection $\mathbb{B}$ on $\mathcal{F}$ is called a **Clifford superconnection** if for all sections a of $C(T(M))$ we have

$$[\mathbb{B}, c(a)] = c(\nabla a)$$

where ∇ is the covariant differential on $C(T(M))$ coming from the Levi-Civita connection on M.

Suppose that $\mathbb{B}$ and $\mathbb{B}'$ are Clifford superconnections on $\mathcal{F}$. Then

$$[\mathbb{B} - \mathbb{B}', e(\alpha)] = 0 \quad \forall \, \alpha \in \mathcal{A}(M)$$

so $\mathbb{B} - \mathbb{B}' \in \mathcal{A}^-(M, \operatorname{End}(\mathcal{F}))$. Also

$$[\mathbb{B} - \mathbb{B}', c(a)] = 0$$

implying that
$$\mathbb{B} - \mathbb{B}' \in \mathcal{A}^-(M, \mathrm{End}_{C(M)}(\mathcal{F})).$$

Conversely, if $\tau \in \mathcal{A}^-(M, \mathrm{End}_{C(M)}(\mathcal{F}))$ and $\mathbb{B}'$ is a Clifford superconnection then $\mathbb{B} = \mathbb{B}' + \tau$ is a Clifford superconnection. Thus the collection of all Clifford superconnections is an affine space modeled on the linear space $\mathcal{A}^-(M, \mathrm{End}_{C(M)}(\mathcal{F}))$.

If $\mathcal{E}$ is a superbundle and $\mathcal{F}$ is a bundle of Clifford modules then we can make $\mathcal{E} \otimes \mathcal{F}$ into a Clifford module by letting a section a of $C(TM)$ act as $1 \otimes c(a)$ where $c(a)$ denote the action of a on $\mathcal{F}$. If $\mathbb{A}$ is a superconnection on $\mathcal{E}$ then

$$[\mathbb{A} \otimes \mathbf{1}, \mathbf{1} \otimes c(a)] = 0$$

for all sections a of $C(TM)$ and so

$$[\mathbb{A} \otimes \mathbf{1} + \mathbf{1} \otimes \mathbb{B}, \mathbf{1} \otimes c(a)] = \mathbf{1} \otimes c(\nabla a).$$

In other words, the tensor product of a superconnection with a Clifford superconnection is a Clifford superconnection.

Chapter 19

Semi-Riemannian submersions.

The treatment here is that of a 1966 paper by O'Neill, "The fundamental equations of a submersion", *Michigan Mathematical Journal* **13** 459 - 469, following earlier basic work by Hermann. In a sense, the subject can be regarded as yet another generalization of the notion of a "surface of revolution".

Indeed, a special case of a semi-Riemannian submersion is a "warped product", see below for the definition. A warped product is itself a generalization of the Friedmann-Robertson-Walker metrics which are, as we pointed out, generalizations of the metric of a surface of revolution. In his book, O'Neill restricts his discussion to the case of a warped product.

So in this chapter, we will refer to his paper when discussing the general case except where the page numbers make it clear that we are referring to his book. The references in the discussion of warped products are to his book.

Barrett O'Neill

Born: 26 March 1924 in Washington, D.C., USA
Died: 16 June 2011 in Los Angeles, California, USA

19.1 Submersions.

Let M and B be differentiable manifolds, and $\pi : M \to B$ be a submersion, which means that $d\pi_m : TM_m \to TB_{\pi(m)}$ is surjective for all $m \in M$. The implicit function theorem then guarantees that $\pi^{-1}(b)$ is a submanifold of M for all $b \in B$. These submanifolds are called the *fibers* of the submersion. By the implicit function theorem, the tangent space to the fiber through $m \in M$ just the kernel of the differential of the projection, π. Call this space $V(M)_m$. So

$$V(M)_m := \ker d\pi_m.$$

The set of such tangent vectors at m is called the set of vertical vectors, and a vector field on M whose values at every point are vertical will be called a vertical vector field. We will denote the set of vertical vector fields by $\mathcal{V}(M)$.

If ϕ is a smooth function on B, and V is a vertical vector field, then $V\pi^*\phi = 0$. Conversely, if $V\pi^*\phi = 0$ for all smooth functions, ϕ on B, then V is vertical. In particular, if U and V are vertical vector fields, then so is $[U, V]$.

Now suppose that both M and B are (semi-)Riemann manifolds. Let

$$H(M)_m := V(M)_m^\perp.$$

19.1. SUBMERSIONS.

We assume the following:
$$d\pi_m : H(M)_m \to TB_{\pi(m)}$$

is an isometric isomorphism, i.e. is bijective and preserves the scalar product of tangent vectors. Notice that this implies that $V(M)_m \cap H(M)_m = \{0\}$ so that the restriction of the scalar product to $V(M)_m$ is non-singular. (Of course in the Riemannian case this is automatic.)

We let $\mathcal{H} : T(M)_m \to H(M)_m$ denote the orthogonal projection at each point and also let $\mathcal{H}(M)$ denote the set of "horizontal" vector fields (vector fields which belong to $H(M)_m$ at each point). Similarly, we let $\mathcal{V}$ denote orthogonal projection onto $V(M)_m$ at each point. So if E is a vector field on M, then $\mathcal{V}E$ is a vertical vector field and $\mathcal{H}E$ is a horizontal vector field. We will reserve the letters U, V, W for vertical vector fields, and X, Y, Z for horizontal vector fields.

Among the horizontal vector fields, there is a subclass, the *basic* vector fields. They are defined as follows: Let X_B be a vector field on B. If $m \in M$, there is a unique tangent vector, call it $X(m) \in H(M)_m$ such that
$$d\pi_m X(m) = X_B(\pi(m)).$$

This defines the the basic vector field, X, corresponding to X_B. We can say that X and X_B are π-related.

Notice that if X is the basic vector field corresponding to X_B, and if ϕ is a smooth function on B, then
$$X\pi^*\phi = \pi^*(X_B\phi).$$

Also, by definition,
$$\langle X, Y\rangle_M = \pi^*\langle X_B, Y_B\rangle_B$$

for basic vector fields X and Y. In general, if X and Y are horizontal, or even basic vector fields, their Lie bracket, $[X,Y]$ need not be horizontal. But if X and Y are basic, then we can compute the horizontal component of $[X,Y]$ as follows: If ϕ is any function on B and if X and Y are basic vector fields, then

$$\begin{aligned}(\mathcal{H}[X,Y])\pi^*\phi &= [X,Y]\pi^*\phi \\ &= XY\pi^*\phi - YX\pi^*\phi \\ &= \pi^*(X_BY_B\phi - Y_BX_B\phi) \\ &= \pi^*([X_B,Y_B]\phi)\end{aligned}$$

so $\mathcal{H}[X,Y]$ is the basic vector field corresponding to $[X_B, Y_B]$.

We claim that

$$\mathcal{H}(\nabla_X Y) \text{ is the basic vector field corresponding to } \nabla^B_{X_B}(Y_B) \quad (19.1)$$

where ∇^B denotes the Levi-Civita covariant derivative on B and ∇ denotes the covariant derivative on M. Indeed, let X_B, Y_B, Z_B be vector fields on B and X, Y, Z the corresponding basic vector fields on M. Then

$$X\langle Y, Z\rangle_M = X\left(\pi^*\langle X_B, Y_B\rangle_B\right) = \pi^*(X_B\langle Y_B, Z_B\rangle_B)$$

while
$$\langle X, [Y, Z]\rangle = \langle X, \mathcal{H}[Y, Z]\rangle = \pi^*\left(\langle X_B, [Y_B, Z_B]\rangle_B\right)$$
since $\mathcal{H}[Y, Z]$ is the basic vector field corresponding to $[Y_B, Z_B]$. From the Koszul formula it then follows that
$$\langle \nabla_X Y, Z\rangle_M = \pi^* \langle \nabla^B_{X_B} Y_B, Z_B\rangle_B.$$
Therefore $d\pi_m(\nabla_X Y(m)) = \nabla^B_{X_B} Y_B(\pi(m))$ for all points m which implies (19.1).

Suppose that γ is a horizontal geodesic, so that $\pi\gamma$ is a regular curve, so an integral curve of a vector field X_B on B. Let X be the corresponding basic vector field, so γ is an integral curve of X. The fact that γ is a geodesic implies that $\nabla_X X = 0$ along γ, and hence by (19.1) $\nabla_{X_B} X_B = 0$ along $\pi\gamma$ so $\pi\gamma$ is a geodesic. We have proved

$$\pi(\gamma) \text{ is a geodesic if } \gamma \text{ is a horizontal geodesic.} \qquad (19.2)$$

If V and W are vertical vector fields, then we may consider their restriction to each fiber as a vector field along that fiber, and may also consider the Levi-Civita connection on the fiber considered as a semi-Riemann manifold in its own right. We will denote the covariant derivative of W with respect to V relative to the connection induced by the metric on each fiber by $\nabla^\mathcal{V}_V W$. It follows from the Koszul formula, and the fact that $[V, W]$ is vertical if V and W are that

$$\nabla^\mathcal{V}_V W = \mathcal{V}(\nabla_V W) \qquad (19.3)$$

for vertical vector fields. Here ∇ is the Levi-Civita covariant derivative on M, so that $\nabla_V W$ has both a horizontal and a vertical component.

19.2 The fundamental tensors of a submersion.

19.2.1 The tensor T.

For arbitrary vector fields E and F on M define
$$T_E F := \mathcal{H}[\nabla_{\mathcal{V}E}(\mathcal{V}F)] + \mathcal{V}[\nabla_{\mathcal{V}E}(\mathcal{H}F)],$$
where, in this equation, ∇ denotes the Levi-Civita covariant derivative determined by the metric on M.

If f is any differentiable function on M, then $\mathcal{V}fF = f\mathcal{V}F$ and $\nabla_{\mathcal{V}E}(f\mathcal{V}F) = [(\mathcal{V}E)f]\mathcal{V}F + f\nabla_{\mathcal{V}E}(\mathcal{V}F)$ so
$$\mathcal{H}[\nabla_{\mathcal{V}E}(\mathcal{V}(fF))] = f\mathcal{H}[\nabla_{\mathcal{V}E}(\mathcal{V}F)].$$

Similarly f pulls out of the second term in the definition of T. Also $\mathcal{V}(fE) = f\mathcal{V}E$ and $\nabla_{f\mathcal{V}E} = f\nabla_{\mathcal{V}E}$ by a defining property of ∇.

This proves that T is a tensor of type $(1,2)$: $T_{fE}F = T_E(fF) = fT_E F$.

19.2. THE FUNDAMENTAL TENSORS OF A SUBMERSION. 381

By definition, $T_E = T_{\mathcal{V}E}$ depends only on the vertical component, $\mathcal{V}E$ of E. If U and V are vertical vector fields, then

$$\begin{aligned} T_U V &= \mathcal{H} \nabla_U V \\ &= \mathcal{H} \nabla_V U + \mathcal{H}([U, V]) \\ &= \mathcal{H} \nabla_V U \end{aligned}$$

since $[U, V]$ is vertical. Thus

$$T_U V = T_V U \tag{19.4}$$

for vertical vector fields. Also notice that if U is a vertical vector field then $T_E U$ is horizontal, while if X is a horizontal vector field, then $T_E X$ is vertical.

1. Show that

$$\langle T_E F_1, F_2 \rangle = -\langle F_1, T_E F_2 \rangle \tag{19.5}$$

for any pair of vector fields F_1, F_2.

19.2.2 The tensor A.

This is defined by interchanging the role of horizontal and vertical in T, so

$$A_E F := \mathcal{V} \nabla_{\mathcal{H}E}(\mathcal{H}F) + \mathcal{H} \nabla_{\mathcal{H}E}(\mathcal{V}F).$$

The same proof as above shows that A is a tensor, that A_E sends horizontal vector fields into vertical vector fields and vice versa, and your solution of problem 1 will also show that

$$\langle A_E F_1, F_2 \rangle = -\langle F_1, A_E F_2 \rangle$$

for any pair of vector fields F_1, F_2.

Notice that any horizontal vector field can be written (locally) as a function linear combination of basic vector fields, and if V is vertical and X basic, then

$$[V, X]\pi^*\phi = V\pi^*(X_B \phi) - XV\pi^*\phi = 0,$$

so the Lie bracket of a vertical vector field and a basic vector field is vertical.

2. Show that $A_X X = 0$ for any horizontal vector field, X, and hence that

$$A_X Y = -A_Y X \tag{19.6}$$

for any pair of horizontal vector fields X, Y. Since

$$\mathcal{V}[X, Y] = \mathcal{V}(\nabla_X Y - \nabla_Y X) = A_X Y - A_Y X$$

it then follows that

$$A_X Y = \frac{1}{2}\mathcal{V}[X, Y]. \tag{19.7}$$

(Hint, it suffices to show that $A_X X = 0$ for basic vector fields, and for this, that $\langle V, A_X X \rangle = 0$ for all vertical vector fields since $A_X X$ is vertical. Use Koszul's formula.)

We can express the relations between covariant derivatives of horizontal and vertical vector fields and the tensors T and A:

$$\nabla_V W = T_V W + \nabla_V^\mathcal{V} W \qquad (19.8)$$
$$\nabla_V X = \mathcal{H} \nabla_V X + T_V X \qquad (19.9)$$
$$\nabla_X V = A_X V + \mathcal{V} \nabla_X V \qquad (19.10)$$
$$\nabla_X Y = \mathcal{H} \nabla_X Y + A_X Y \qquad (19.11)$$

If X is a basic vector field, then $\nabla_V X = \nabla_X V + [V, X]$ and $[V, X]$ is vertical. Hence

$$\mathcal{H} \nabla_V X = A_X V \text{ if } X \text{ is basic.} \qquad (19.12)$$

19.2.3 Covariant derivatives of T and A.

The definition of covariant derivative of a tensor field gives

$$(\nabla_{E_1} A)_{E_2} E_3 = \nabla_{E_1} (A_{E_2} E_3) - A_{\nabla_{E_1} E_2} E_3 - A_{E_2} (\nabla_{E_1} E_3)$$

for any three vector fields E_1, E_2, E_3. Suppose, in this equation, we take $E_1 = V$ and $E_2 = W$ to be vertical, and $E_3 = E$ to be a general vector field. Then $A_{E_2} = A_W = 0$ so the first and third terms on the right vanish. In the middle term we have

$$A_{\nabla_V W} = A_{\mathcal{H} \nabla_V W} = A_{T_V W}$$

so that we get

$$(\nabla_V A)_W = -A_{T_V W}. \qquad (19.13)$$

If we take $E_1 = X$ to be horizontal and $E_2 = W$ to be vertical, again only the middle term survives and we get

$$(\nabla_X A)_W = -A_{A_X W}. \qquad (19.14)$$

Similarly,

$$(\nabla_X T)_Y = -T_{A_X Y} \qquad (19.15)$$
$$(\nabla_V T)_Y = -T_{T_V Y}. \qquad (19.16)$$

3. Show that

$$\langle (\nabla_U A)_X V, W \rangle = \langle T_U V, A_X W \rangle - \langle T_U W, A_X V \rangle \qquad (19.17)$$
$$\langle (\nabla_E A)_X Y, V \rangle = -\langle (\nabla_E A)_Y X, V \rangle \qquad (19.18)$$
$$\langle (\nabla_E T)_V W, X \rangle = \langle (\nabla_E T)_W V, X \rangle \qquad (19.19)$$

19.2. THE FUNDAMENTAL TENSORS OF A SUBMERSION.

where U, V, W are vertical, X, Y are horizontal and E is a general vector field.

We also claim that

$$\mathcal{S}\langle (\nabla_Z A)_X Y, V\rangle = \mathcal{S}\langle A_X Y, T_V Z\rangle \tag{19.20}$$

where V is vertical, X, Y, Z horizontal and $\mathcal{S}$ denotes cyclic sum over the horizontal vectors.

Proof. This is a tensor equation, so we may assume that X, Y, Z are basic and that the corresponding vector fields X_B, Y_B, Z_B have all their Lie brackets vanish at $b = \pi(m)$ where m is the point at which we want to check the equation. Thus all Lie brackets of X, Y, Z are vertical at m. We have $\frac{1}{2}[X, Y] = A_X Y$ by (19.7), so

$$\frac{1}{2}[[X, Y], Z] = [A_X Y, Z] = \nabla_{A_X Y}(Z) - \nabla_Z(A_X Y)$$

and the cyclic sum of the leftmost side vanishes by the Jacobi identity. So

$$\mathcal{S}[\nabla_{A_X Y}(Z)] = \mathcal{S}[\nabla_Z(A_X Y)]. \tag{19.21}$$

Taking scalar product with the vertical vector $V(m)$, we have (at the point m) by repeated use of (19.4) and (19.5)

$$\begin{aligned}\langle \nabla_{A_X Y}(Z), V\rangle &= \langle T_{A_X Y}(Z), V\rangle \\ &= -\langle Z, T_{A_X Y}(V)\rangle \\ &= -\langle Z, T_V(A_X Y)\rangle \\ &= \langle T_V Z, A_X Y\rangle\end{aligned}$$

We record this fact for later use as

$$\langle T_{A_X Y}(Z), V\rangle = \langle T_V Z, A_X Y\rangle. \tag{19.22}$$

Using (19.21) we obtain

$$\mathcal{S}\langle \nabla_Z(A_X Y), V\rangle = \mathcal{S}\langle T_V Z, A_X Y\rangle. \tag{19.23}$$

Now

$$\langle \nabla_Z(A_X Y), V\rangle - \langle (\nabla_Z A)_X Y, V\rangle = \langle A_{\nabla_Z X}(Y), V\rangle + \langle A_X(\nabla_Z Y), V\rangle$$

while

$$A_{\nabla_Z X}(Y) = -A_Y(\mathcal{H}\nabla_Z X) = -A_Y(\mathcal{H}\nabla_X Z)$$

using (19.6) for the first equations and the fact that $[X, Z]$ is vertical for the second equation. Taking scalar product with V gives

$$-\langle A_Y(\mathcal{H}\nabla_X Z), V\rangle = -\langle A_Y(\nabla_X Z), V\rangle$$

since $A_Y U$ is horizontal for any vertical vector, and hence

$$\langle A_Y(V\nabla_X Z), V\rangle = 0.$$

We thus obtain

$$\langle \nabla_Z(A_X Y), V\rangle - \langle (\nabla_Z A)_X Y, V\rangle = \langle A_X(\nabla_Z Y), V\rangle - \langle A_Y(\nabla_X Z), V\rangle.$$

The cyclic sum of the right hand side vanishes. So, taking cyclic sum and applying (19.23) establishes (19.20).

19.2.4 The fundamental tensors for a warped product.

A very special case of a semi-Riemann submersion is that of a "warped product" following O'Neill's terminology. Here $M = B \times F$ as a manifold, so π is just projection onto the first factor. We are given a positive function, f on B and metrics $\langle\,,\,\rangle_B$ and $\langle\,,\,\rangle_F$ on each factor. At each point $m = (b,q)$, $b \in B, q \in F$ we have the direct sum decomposition

$$TM_m = TB_b \oplus TF_q$$

as vector spaces, and the warped product metric is defined as the direct sum

$$\langle\,,\,\rangle = \langle\,,\,\rangle_B \oplus f^2 \langle\,,\,\rangle_F.$$

O'Neill writes $M = B \times_f F$ for the warped product, the metrics on B and F being understood. As mentioned above, the notion of warped product can itself be considered as a generalization of a surface of revolution, where B is a plane curve not meeting the axis of revolution, where f is the distance to the axis, and where $F = S^1$, the unit circle with its standard metric.

On a warped product, the basic vector fields are just the vector fields of B considered as vector fields of $B \times F$ in the obvious way, having no F component. In particular, the Lie bracket of two basic vector fields, X and Y on M is just the Lie bracket of the corresponding vector fields X_B and Y_B on B, considered as a vector field on M via the direct product. In particular, $[X, Y]$ has no vertical component, so $A_X Y = 0$. In fact, we can be more precise. For each fixed $q \in F$, the projection π restricted to $B \times \{q\}$ is an isometry of $B \times \{q\}$ with B. Thus

$$\nabla_X Y = \text{the basic vector field corresponding to } \nabla^B_{X_B} Y_B.$$

On a warped product, there is a special class of vertical vector fields, those that are vector fields on F considered as vector fields on $B \times F$ via the direct product decomposition. Let us denote the collection of these vector fields by $\mathcal{L}(F)$, the "lifts" of vector fields on F to use O'Neill's terminology. If $V \in \mathcal{L}(F)$ and X is a basic vector field, then $[X, V] = 0$ since they "depend on different variables" and hence $\nabla_X V = \nabla_V X$. The vector field $\nabla_X V$ is vertical, since $\langle \nabla_X V, Y\rangle = -\langle V, \nabla_X Y\rangle = 0$ for any basic vector field, Y, as $\nabla_X Y$ is horizontal.

19.3. CURVATURE.

This shows that $A_X V = 0$ as well, so $A = 0$. We claim that once again we can be more precise:

$$\nabla_X V = \nabla_V X = \frac{Xf}{f} V \quad \forall \text{ basic } X, \text{and } \forall V \in \mathcal{L}(F). \tag{19.24}$$

Indeed, the only term that survives in the Koszul formula for $2\langle \nabla_X V, W \rangle$, $W \in \mathcal{L}(F)$ is $X\langle V, W \rangle$. We have

$$\langle V, W \rangle = f^2 \langle V_F, W_F \rangle_F$$

where we have written f instead of $\pi^* f$ by the usual abuse of language for a direct product. Now $\langle V_F, W_F \rangle_F$ is a function on F (pulled back to $B \times F$) and so is annihilated by X. Hence

$$X\langle V, W \rangle = 2f(Xf)\langle V_F, W_F \rangle_F = \frac{2Xf}{f}\langle V, W \rangle,$$

proving (19.24). Notice that (19.24) gives us a piece of T, namely

$$T_V X = \frac{Xf}{f} V.$$

We can also derive the "horizontal" piece of T, namely

$$T_V W = -\frac{\langle V, W \rangle}{f} \operatorname{grad} f. \tag{19.25}$$

Indeed

$$\begin{aligned}
\langle \nabla_V W, X \rangle &= -\langle W, \nabla_V X \rangle \\
&= -\frac{Xf}{f}\langle V, W \rangle \quad \text{and} \\
Xf &= \langle \operatorname{grad} f, X \rangle.
\end{aligned}$$

In this formula, it doesn't matter whether we consider f as a function on M and compute its gradient there, or think of f as a function on B and compute its gradient relative to B and then take the horizontal lift. The answer is the same since f has no F dependence. Finally, the vertical component of $\nabla_V W$, $V, W \in \mathcal{L}(F)$ is just the same as the extension to M of $\nabla^F_{V_F} W_F$ since the metric on each fiber differs from that of F by a constant factor, which has no influence on the covariant derivative.

19.3 Curvature.

We want equations relating the curvature of the base and the curvature of the fibers to the curvature of M and the tensors T, A, and their covariant derivatives.

Warning:

I will be following O'Neill's paper and book, where he uses the opposite convention for the Riemann curvature tensor than the one we have been using throughout this text. Undoubtedly, many errors will be introduced if I try to re-arrange his formulas in accordance with our conventions. So I will use the following device (suggested by O'Neill in his book (page 89) but in the opposite direction !). For the rest of this chapter, I will let

$$\mathcal{R} := -R.$$

Thus
$$\mathcal{R}_{XY}Z = \nabla_{[X,Y]}Z - [\nabla_X, \nabla_Y]Z.,$$

We will be considering expressions of the form

$$\langle \mathcal{R}_{E_1 E_2} E_3, E_4 \rangle$$

where $\mathcal{R}$ is the curvature of M (in O'Neill's convention) and the E's are either horizontal or vertical. We let $n = 0, 1, 2, 3,$ or 4 denote the number of horizontal vectors, the remaining being vertical. This gives five cases. So we will get five equations for curvature. For example, $n = 0$ corresponds to all vectors vertical, so we are asking for the relation between the curvature of the fiber and the full curvature. Let $\mathcal{R}^{\mathcal{V}}$ denote the curvature tensor of the fiber (as a semi-Riemann submanifold).

The case **n = 0** is the **Gauss equation** of each fiber:

$$\langle \mathcal{R}_{UV}W, F \rangle =$$

$$\langle \mathcal{R}^{\mathcal{V}}_{UV}W, F \rangle - \langle T_U W, T_V F \rangle + \langle T_V W, T_U F \rangle, \quad U, V, W, F \in \mathcal{V}(M). \quad (19.26)$$

We recall the proof (O'Neill 100). We may assume $[U, V] = 0$ so

$$\mathcal{R}_{UV} = -\nabla_U \nabla_V + \nabla_V \nabla_U$$

and, using (19.3) and the definition of T, if we have

$$\begin{aligned}
\langle \nabla_U \nabla_V W, F \rangle &= \langle \mathcal{V} \nabla_U \nabla^{\mathcal{V}}_V W, F \rangle + \langle \nabla_U (T_V W), F \rangle \\
&= \langle \nabla^{\mathcal{V}}_U \nabla^{\mathcal{V}}_V W, F \rangle + U \langle T_V W, F \rangle - \langle T_V W, \nabla_U F \rangle \\
&= \langle \nabla^{\mathcal{V}}_U \nabla^{\mathcal{V}}_V W, F \rangle - \langle T_V W, T_U F \rangle.
\end{aligned}$$

Substituting the above expression for $\mathcal{R}_{UV}$ into $\langle \mathcal{R}_{UV}W, F \rangle$ then proves (19.26).

The case **n = 1** is the **Codazzi equation** for each fiber: Let U, V, W be vertical vector fields and X a horizontal vector field. Then

$$\langle \mathcal{R}_{UV}W, X \rangle = \langle (\nabla_V T)_U W, X \rangle - \langle (\nabla_U T)_V W, X \rangle \quad (19.27)$$

19.3. CURVATURE.

This is also in O'Neill, page 115. We recall the proof. We assume that $[U,V] = 0$ so $\mathcal{R}_{UV} = -\nabla_U\nabla_V + \nabla_V\nabla_U$ as before. We have

$$\begin{aligned}\langle \nabla_U\nabla_V W, X\rangle &= \langle \nabla_U\nabla_V^{\mathcal{V}} W, X\rangle + \langle \nabla_U(T_V W), X\rangle \\ &= \langle T_U(\nabla_V^{\mathcal{V}} W), X\rangle + \langle \nabla_U(T_V W), X\rangle.\end{aligned}$$

We write
$$\nabla_U(T_V W) = (\nabla_U T)_V W + T_{\nabla_U V} W + T_V \nabla_U W$$

and
$$\langle T_V \nabla_U W, X\rangle = \langle T_V \nabla_U^{\mathcal{V}} W, X\rangle$$

so

$$\langle \nabla_U\nabla_V W, X\rangle = \langle (\nabla_U T)_V W, X\rangle + \langle T_U(\nabla_V^{\mathcal{V}} W), X\rangle + \langle T_V(\nabla_U W), X\rangle + \langle T_{\nabla_U V} W, X\rangle.$$

Interchanging U and V and subtracting, using $\nabla_U V = \nabla_V U$ proves (19.27).

We now turn to the opposite extreme, $n = 4$ and $n = 3$ but first some notation. We let $\mathcal{R}^{\mathcal{H}}$ denote the horizontal lift of the curvature tensor of B: If $h_i \in H(M)_m$ with $v_i := d\pi_m h_i$ define $\mathcal{R}^{\mathcal{H}}_{h_1 h_2} h_3$ to be the unique horizontal vector such that

$$d\pi_m\left(\mathcal{R}^{\mathcal{H}}_{h_1 h_2} h_3\right) = \mathcal{R}^{B}_{v_1 v_2} v_3.$$

The case **n = 4** is given by

$$\langle \mathcal{R}_{XY} Z, H\rangle = \langle \mathcal{R}^{\mathcal{H}}_{XY} Z, H\rangle - 2\langle A_X Y, A_Z H\rangle + \langle A_Y Z, A_X H\rangle + \langle A_Z X, A_Y H\rangle \tag{19.28}$$

for any four horizontal vector fields X, Y, Z, H. As usual, we may assume X, Y, Z are basic and all their brackets are vertical. We will massage each term on the right of
$$\mathcal{R}_{XY} Z = \nabla_{[X,Y]} Z - \nabla_X \nabla_Y Z + \nabla_Y \nabla_X Z.$$

Since $[X,Y]$ is vertical, $[X,Y] = 2A_X Y$. So

$$\nabla_{[X,Y]} Z = 2\mathcal{H}\nabla_{A_X Y} Z + 2T_{A_X Y}(Z).$$

Since Z is basic we can apply (19.12) to the first term giving

$$\nabla_{[X,Y]} Z = 2A_Z(A_X Y) + 2T_{A_X Y}(Z).$$

Let us write
$$\nabla_Y Z = \mathcal{H}\nabla_Y Z + A_Y Z$$

and apply equation (19.1) which we write, by abuse of language as

$$\mathcal{H}\nabla_Y Z = \nabla_Y^B Z.$$

Then

$$\nabla_X \nabla_Y Z = \nabla_X^B \nabla_Y^B Z + A_X(\nabla_Y^B Z) + A_X A_Y Z + \mathcal{V}\nabla_X(A_Y Z).$$

Separating the horizontal and vertical components in the definition of $\mathcal{R}$ gives

$$\mathcal{H}\mathcal{R}_{XY} Z = -[\nabla_X^B, \nabla_Y^B] Z + 2 A_Z A_X Y - A_X A_Y Z + A_Y A_X Z \quad (19.29)$$
$$\mathcal{V}\mathcal{R}_{XY} Z = 2 T_{A_X Y}(Z) - \mathcal{V}\nabla_X(A_Y Z) +$$
$$+ \mathcal{V}\nabla_Y(A_X Z) - A_X(\nabla_Y^B Z) + A_Y(\nabla_X^B Z) \quad (19.30)$$

As we have chosen X, Y such that $[X_B, Y_B] = 0$, the first term on the right of (19.29) is just $\mathcal{R}^{\mathcal{H}}_{XY} Z$. Taking the scalar product of (19.29) with a horizontal vector field (and using the fact that A_E is skew adjoint relative to the metric and $A_X Z = -A_Z X$) proves (19.28).

If we take the scalar product of (19.30) with a vertical vector field, V we get an expression for $\langle \mathcal{R}_{XY} Z, V \rangle$ (and we can drop the projections $\mathcal{V}$). Let us examine what happens when we take the scalar product of the various terms on the right of (19.30) with V. The first term gives

$$\langle T_{A_X Y}(Z), V \rangle = \langle T_V Z, A_X Y \rangle$$

by (19.22). The next two terms give

$$\langle \nabla_Y(A_X Z), V \rangle - \langle \nabla_X(A_Y Z), V \rangle = \langle (\nabla_Y A)_X Z, V \rangle - \langle (\nabla_X A)_Y Z, V \rangle$$
$$+ \langle A_X(\nabla_Y Z), V \rangle - \langle A_Y(\nabla_X Z), V \rangle$$

since $\nabla_X Y - \nabla_Y X = [X, Y]$ is vertical by assumption. The last two terms cancel the terms obtained by taking the scalar product of the last two terms in (19.30) with V and we obtain

$$\langle \mathcal{R}_{XY} Z, V \rangle = 2 \langle A_X Y, T_V Z \rangle + \langle (\nabla_Y A)_X Z, V \rangle - \langle (\nabla_X A)_Y Z, V \rangle. \quad (19.31)$$

We can simplify this a bit using (19.18) and (19.20). Indeed, by (19.18) we can replace the second term on the right by $-\langle (\nabla_T A)_Z X, V \rangle$ and then apply (19.20) to get, for **n = 3**,

$$\langle \mathcal{R}_{XY} Z, V \rangle = \langle (\nabla_Z A)_X Y, V \rangle + \langle A_X Y, T_V Z \rangle - \langle A_Y Z, T_V X \rangle - \langle A_Z X, T_V Y \rangle. \quad (19.32)$$

Finally we give an expression for the case **n = 2**:

19.3. CURVATURE. 389

$$\langle \mathcal{R}_{XV}Y, W\rangle = \langle(\nabla_X T)_V W, Y\rangle + \langle(\nabla_V A)_X Y, W\rangle - \langle T_V X, T_W Y\rangle + \langle A_X V, A_Y W\rangle. \tag{19.33}$$

To prove this, write

$$\mathcal{R}_{XV} = \nabla_{\nabla_X V} - \nabla_{\nabla_V X} - \nabla_X \nabla_V + \nabla_V \nabla_X$$

and

$$\begin{aligned}
\langle \nabla_{\nabla_X V} Y, W\rangle &= -\langle Y, T_{\nabla_X V} W\rangle + \langle A_{\nabla_X V} Y, W\rangle \\
-\langle \nabla_{\nabla_V X} Y, W\rangle &= -\langle T_{\nabla_V X} Y, W\rangle - \langle A_{\nabla_V X} Y, W\rangle \\
-\langle \nabla_X \nabla_V Y, W\rangle &= -\langle \nabla_X(T_V Y), W\rangle + \langle \nabla_V Y, A_X W\rangle \\
\langle \nabla_V \nabla_X Y, W\rangle &= \langle \nabla_V A_X Y, W\rangle - \langle \nabla_X Y, T_V W\rangle
\end{aligned}$$

where, for example, in the last equation we have written $\nabla_X Y = A_X Y + \mathcal{H}\nabla_X Y$ and

$$\langle \nabla_V \mathcal{H} \nabla_X Y, W\rangle = -\langle \mathcal{H}\nabla_X Y, \nabla_V W\rangle = \langle \mathcal{H}\nabla_X Y, T_V W\rangle = \langle \nabla_X Y, T_V W\rangle.$$

We have

$$\begin{aligned}
\langle(\nabla_X T)_V W, Y\rangle &= -\langle W, (\nabla_X T)_V Y\rangle \\
&= -\langle W, \nabla_X(T_V Y)\rangle + \langle W, T_{\nabla_X V} Y\rangle + \langle W, T_V \nabla_X Y\rangle \\
\langle(\nabla_V A)_X Y, W\rangle &= \langle \nabla_V(A_X Y), W\rangle - \langle A_{\nabla_V X} Y, W\rangle - \langle A_X \nabla_V Y, W\rangle.
\end{aligned}$$

The six terms on the right of the last two equations equal six of the eight terms on the right of the preceding four leaving two remaining terms,

$$\langle A_{\nabla_X V} Y, W\rangle - \langle T_{\nabla_V X} Y, W\rangle.$$

But

$$\begin{aligned}
\langle A_{\nabla_X V} Y, W\rangle &= -\langle A_Y \mathcal{H}\nabla_X V, W\rangle \\
&= -\langle A_Y A_X V, W\rangle \\
&= \langle A_X V, A_Y W\rangle
\end{aligned}$$

and a similar argument deals with the second term.

We repeat our equations. In terms of increasing values of n we have

$$\begin{aligned}
\langle \mathcal{R}_{UV} W, F\rangle &= \langle \mathcal{R}^{\mathcal{V}}_{UV} W, F\rangle - \langle T_U W, T_V F\rangle + \langle T_V W, T_U F\rangle, \\
\langle \mathcal{R}_{UV} W, X\rangle &= \langle(\nabla_V T)_U W, X\rangle - \langle(\nabla_U T)_V W, X\rangle \\
\langle \mathcal{R}_{XV} Y, W\rangle &= \langle(\nabla_X T)_V W, Y\rangle + \langle(\nabla_V A)_X Y, W\rangle - \langle T_V X, T_W Y\rangle + \langle A_X V, A_Y W\rangle, \\
\langle \mathcal{R}_{XY} Z, V\rangle &= \langle(\nabla_Z A)_X Y, V\rangle + \langle A_X Y, T_V Z\rangle - \langle A_Y Z, T_V X\rangle - \langle A_Z X, T_V Y\rangle, \\
\langle \mathcal{R}_{XY} Z, H\rangle &= \langle \mathcal{R}^{\mathcal{H}}_{XY} Z, H\rangle - 2\langle A_X Y, A_Z H\rangle + \langle A_Y Z, A_X H\rangle + \langle A_Z X, A_Y H\rangle.
\end{aligned}$$

We have stated the formula for $n = 2$, i.e. two vertical and two horizontal fields for the case $\langle \mathcal{R}_{XV}Y, W\rangle$, i.e. where one horizontal and one vertical vector occur in the subscript $\mathcal{R}_{E_1 E_2}$. But it is easy to check that all other arrangements of two horizontal and two vertical fields can be reduced to this one by curvature identities. Similarly for $n = 1$ and $n = 3$.

19.3.1 Curvature for warped products.

The curvature formulas simplify considerably in the case of a warped product where $A = 0$ and

$$T_V X = \frac{Xf}{f}V, \quad T_V W = -\frac{\langle V, W\rangle}{f}\operatorname{grad} f.$$

We will give the formulas where X, Y, Z, H are basic and $U, V, W, F \in \mathcal{L}(F)$. We have $Vf = 0$ and $\langle \nabla_V \operatorname{grad} f, X\rangle = VXf - \langle \operatorname{grad} f, \nabla_V X\rangle = 0$. We conclude that the right hand side of (19.27) vanishes, so $\mathcal{R}_{UV}W$ is vertical and we conclude from (19.26) that

$$\mathcal{R}_{UV}W = \mathcal{R}^F_{UV}W - \frac{\langle \operatorname{grad} f, \operatorname{grad} f\rangle}{f^2}(\langle U, W\rangle V - \langle V, W\rangle U) \qquad (19.34)$$

The Hessian of a function f on a semi-Riemann manifold is defined to be the bilinear form on the tangent space at each point defined by

$$H^f(X, Y) = \langle \nabla_X \operatorname{grad} f, Y\rangle.$$

In fact, we have

$$\begin{aligned}\langle \nabla_X \operatorname{grad} f, Y\rangle &= XYf - \langle \operatorname{grad} f, \nabla_X Y\rangle \\ &= \nabla_X(df(Y)) - df(\nabla_X Y) \\ &= [\nabla\nabla f](X, Y)\end{aligned}$$

which gives an alternative definition of the Hessian as

$$H^f = \nabla\nabla f$$

and shows that it is indeed a (0,2) type tensor field. Also

$$\begin{aligned}H^f(X, Y) &= XYf - (\nabla_X Y)f \\ &= [X, Y]f + YXf - [\nabla_X Y - \nabla_Y X + \nabla_Y X]f \\ &= YXF - (\nabla_Y X)f \\ &= H^f(Y, X)\end{aligned}$$

showing that H^f is a symmetric tensor field.
We have

$$\nabla_X V = \frac{Xf}{f}V = T_V X, \quad T_V W = -\frac{\langle V, W\rangle}{f}\operatorname{grad} f,$$

19.3. CURVATURE.

if X is basic and $V, W \in \mathcal{L}(F)$. So

$$\begin{aligned}(\nabla_X T)_V W &= \nabla_X(T_V W) - T_{\nabla_X V} - T_V(\nabla_X W)\\ &= \langle V, W\rangle \left(\frac{Xf}{f^2}\operatorname{grad} f - \frac{1}{f}\nabla_X \operatorname{grad} f\right) + \frac{Xf}{f^2}\end{aligned}$$

and $\langle \operatorname{grad} f, Y\rangle = Yf$. Therefore the case $n = 2$ above yields

$$\langle \mathcal{R}_{XV}Y, W\rangle = -\frac{H^f(X,Y)}{f}\langle V, W\rangle. \tag{19.35}$$

The case $n = 3$ gives

$$\langle \mathcal{R}_{XY}Z, V\rangle = 0$$

and hence by a symmetry property of the curvature tensor, $\langle \mathcal{R}_{XY}Z, V\rangle = \langle \mathcal{R}_{ZV}X, Y\rangle = 0$, or, changing notation,

$$\langle \mathcal{R}_{XV}Y, Z\rangle = 0.$$

Thus

$$\mathcal{R}_{VX}Y = \frac{H^f(X,Y)}{f}V. \tag{19.36}$$

We have $\langle \mathcal{R}_{UV}X, W\rangle = -\langle \mathcal{R}_{UV}W, X\rangle = 0$ and by (19.36) and the first Bianchi identity $\langle \mathcal{R}_{UV}X, Y\rangle = (H^f(X,Y)/f) \times (\langle U, V\rangle - \langle V, U\rangle) = 0$ so

$$\mathcal{R}_{UV}X = 0. \tag{19.37}$$

If we use this fact, the symmetry of the curvature tensor and (19.35) we see that

$$\mathcal{R}_{XV}W = \frac{\langle V, W\rangle}{f}\nabla_X \operatorname{grad} f. \tag{19.38}$$

It follows from the case $n = 3$ and $n = 4$ that

$$\mathcal{R}_{XY}Z = \mathcal{R}^{\mathcal{H}}_{XY}Z, \tag{19.39}$$

the basic vector field corresponding to the vector field $\mathcal{R}^B_{X_B Y_B}Z_B$. Hence $\langle \mathcal{R}_{XY}V, Z\rangle = 0$. We also have $\langle \mathcal{R}_{XY}V, W\rangle = \langle \mathcal{R}_{VW}X, Y\rangle = 0$ so

$$\mathcal{R}_{XY}V = 0. \tag{19.40}$$

Ricci curvature of a warped product.

The Ricci curvature, Ric (X, Y) (using O'Neill's convention) is the trace of the map $V \mapsto \mathcal{R}_{XV}Y$ is given in terms of an "orthonormal" frame field $E_1, \ldots, E_n$ by

$$\operatorname{Ric}(X, Y) = \sum \epsilon_i \langle \mathcal{R}_{XE_i}Y, E_i\rangle, \quad \epsilon_i = \langle E_i, E_i\rangle.$$

We will apply this to a frame field whose first dim B vectors lie in Vect B and whose last $d = \dim F$ vectors lie in Vect F. We will assume that $d > 1$ and that $X, Y \in$ Vect B and $U, V \in$ Vect F. We get

$$\text{Ric}(X,Y) = \text{Ric}^B(X,Y) - \frac{d}{f}H^f(X,Y) \qquad (19.41)$$

$$\text{Ric}(X,V) = 0 \qquad (19.42)$$

$$\text{Ric}(V,W) = \text{Ric}^F(V,W) - \langle V, W \rangle f^{\#} \quad \text{where} \qquad (19.43)$$

$$f^{\#} := \frac{\Delta f}{f} + (d-1)\frac{\langle \text{grad } f, \text{grad } f \rangle}{f^2} \qquad (19.44)$$

where Δf is the Laplacian of f.

Geodesics for a warped product

We now compute the equations for a geodesic on $B \times_f F$. Let $\gamma(s) = (\alpha(s), \beta(s))$ be a curve on $B \times_f F$ and suppose temporarily the neither $\alpha'(s) = 0$ nor $\beta'(s) = 0$ in an interval we are studying. So we can embed the tangent vectors along both projected curves in vector fields, X on B and V on F, so that γ is a solution curve to $X + V$ on $B \times_f F$. The condition that γ be a geodesic is then that $\nabla_{X+V}(X+V) = 0$ along γ. But

$$\nabla_{X+V}(X+V) = \nabla_X X + \nabla_X V + \nabla_V X + \nabla_V V$$

$$= \nabla_X^B X + 2\frac{Xf}{f}V - \frac{\langle V, V \rangle}{f}\text{grad } f + \nabla_V^F V.$$

Separating the vertical and horizontal components, and using the fact that $\nabla_X^B X = \alpha''$ along α and $\beta' = V$, $\nabla_V^F V = \beta''$ along β shows that the geodesic equations take the form

$$\alpha'' = \langle \beta', \beta' \rangle_F (f \circ \alpha)^{-1} \text{grad } f \quad \text{on } B \qquad (19.45)$$

$$\beta'' = -\frac{2}{f \circ \alpha}\frac{d(f \circ \alpha)}{ds}\beta' \quad \text{on } F \qquad (19.46)$$

A limiting argument [O-208] shows that these equations hold for all geodesics.

19.3. CURVATURE.

We repeat all the important equations of this subsection:

$$\nabla_X Y = \nabla_X^B Y$$

$$\nabla_X V = \frac{Xf}{f} V = \nabla_V X$$
$$\mathcal{H}\nabla_V W = T_U V$$
$$= -\frac{1}{f}\langle V, W \rangle \operatorname{grad} f$$

vert $\nabla_V W$
$$= \nabla_V^F W$$

geodesic eqns
$$\alpha'' = \langle \beta', \beta' \rangle_F (f \circ \alpha)^{-1} \operatorname{grad} f \text{ on } B$$
$$\beta'' = -\frac{2}{f \circ \alpha} \frac{d(f \circ \alpha)}{ds} \beta' \text{ on } F$$

curvature
$$\mathcal{R}_{XY} Z = \mathcal{R}^{\mathcal{H}}_{XY} Z$$
$$\mathcal{R}_{VX} Y = \frac{H^f(X,Y)}{f} V$$
$$\mathcal{R}_{XV} W = \frac{\langle V, W \rangle}{f} \nabla_X \operatorname{grad} f$$
$$\mathcal{R}_{UV} W = \mathcal{R}^F_{UV} W - \frac{\langle \operatorname{grad} f, \operatorname{grad} f \rangle}{f^2}(\langle U, W \rangle V - \langle V, W \rangle U)$$

Ricci curv
Ric (X, Y) $= \operatorname{Ric}^B(X, Y) - \frac{d}{f}\operatorname{Hess}^B(f)(X, Y))$
Ric (X, V) $= 0$
Ric (V, W) $= \operatorname{Ric}^F(V, W) - \langle V, W \rangle f^{\#}$ where
$f^{\#} := \frac{\Delta f}{f} + (d-1)\frac{\langle \operatorname{grad} f, \operatorname{grad} f \rangle}{f^2}.$

19.3.2 Sectional curvature.

u We return to the general case of a submersion, and recall that the sectional curvature of the plane, $P_{ab} \subset TM_m$, spanned by two independent vectors, $a, b \in TM_m$ is defined as

$$K(P_{ab}) := \frac{\langle \mathcal{R}_{ab} a, b \rangle}{\langle a, a \rangle \langle b, b \rangle - \langle a, b \rangle^2}.$$

We can write the denominator more simply as $||a \wedge b||^2$. In the following formulas, all pairs of vectors are assumed to be independent, with u, v vertical and x, y horizontal, and where x_B denotes $d\pi_m(x)$ and $y_B := d\pi_m(y)$. Substituting into our formulas for the curvature gives

$$K(P_{vw}) = K^{\mathcal{V}}(P_{vw}) - \frac{\langle T_v v, T_w w \rangle - ||T_v w||^2}{||v \wedge w||^2} \tag{19.47}$$

$$K(P_{xv}) = \frac{\langle (\nabla_x T)_v v, x \rangle + ||A_x v||^2 - ||T_v x||^2}{||x||^2 ||v||^2} \tag{19.48}$$

$$K(P_{xy}) = K^B(P_{x_B y_B}) - \frac{3||A_x y||^2}{||x \wedge y||^2}. \tag{19.49}$$

where $K^{\mathcal{V}}$ and K^B denote the sectional curvatures of the fiber and the base.

19.4 Reductive homogeneous spaces.

19.4.1 Homogeneous spaces.

We refer back to Chapter V for the discussion of bi-invariant metrics on Lie groups.

Now suppose that $B = G/H$ where H is a subgroup with Lie algebra, $\mathfrak{h}$ such that $\mathfrak{h}$ has an H invariant complementary subspace, $\mathfrak{m} \subset \mathfrak{g}$. In fact, for simplicity, let us assume that $\mathfrak{g}$ has a non-degenerate bi-invariant scalar product, whose restriction to $\mathfrak{h}$ is non-degenerate, and let $\mathfrak{m} = \mathfrak{h}^\perp$. This defines a G invariant metric on B, and the projection $\to G/H = B$ is a submersion. The left invariant horizontal vector fields are exactly the vector fields $X \in \mathfrak{m}$, and so

$$A_X Y = \frac{1}{2} \mathcal{V}[X, Y], \quad X, Y \in \mathfrak{m}.$$

On the other hand, the fibers are cosets of H, hence totally geodesic since the geodesics are one parameter subgroups. Hence $T = 0$. We can read (19.49) backwards to determine $K_B(P_B)$ as

$$K_B(P_{X_B Y_B}) = K(P_{XY}) + \frac{3}{4} \frac{||\mathcal{V}[X, Y]||^2}{||X \wedge Y||^2}$$

or

$$K_B(P_{X_B Y_B}) = \frac{\frac{1}{4}||\mathcal{H}[X, Y]||^2 + ||\mathcal{V}[X, Y]||^2}{||X \wedge Y||^2}, \quad X, Y \in \mathfrak{m}. \tag{19.50}$$

See O'Neill pp. 313-15 for a slightly more general formulation of this result.

It follows from (19.2) that the geodesics emanating from the point $H \in B = G/H$ are just the curves $(\exp tX)H$, $X \in \mathfrak{m}$.

19.4.2 Normal symmetric spaces.

Formula (19.50) simplifies if all brackets of basic vector fields are vertical. So we assume that $[\mathfrak{m}, \mathfrak{m}] \subset \mathfrak{h}$. Then we get

$$K_B(P_{X_B Y_B}) = \frac{||[X, Y]||^2}{||X \wedge Y||^2} = \frac{\langle [[X, Y], X], Y \rangle}{||X \wedge Y||^2}. \tag{19.51}$$

19.4. REDUCTIVE HOMOGENEOUS SPACES.

For examples where this holds, we need to search for a Lie group G whose Lie algebra $\mathfrak{g}$ has an Ad-invariant non-degenerate scalar product, $\langle\,,\,\rangle$ and a decomposition $\mathfrak{g} = \mathfrak{h} + \mathfrak{m}$ such that

$$\mathfrak{h} \perp \mathfrak{m}$$
$$[\mathfrak{h}, \mathfrak{h}] \subset \mathfrak{h}$$
$$[[\mathfrak{h}, \mathfrak{m}] \subset \mathfrak{m}$$
$$[\mathfrak{m}, \mathfrak{m}] \subset \mathfrak{h}.$$

Let $\theta : \mathfrak{g} \to \mathfrak{g}$ be the linear map determined by

$$\theta X = -X,\quad X \in \mathfrak{m}, \quad \theta U = U, \quad U \in \mathfrak{h}.$$

Then

- θ is an isometry of $\langle\,,\,\rangle$

- $\theta[E, F] = [\theta E, \theta F]\quad \forall E, F \in \mathfrak{g}$

- $\theta^2 = $id.

Conversely, suppose we start with a θ satisfying these conditions. Since $\theta^2 = $ id, we can write $\mathfrak{g}$ as the linear direct sum of the $+1$ and -1 eigenspaces of θ, i.e. define $\mathfrak{h} := \{U|\theta(U) = U\}$ and $\mathfrak{m} := \{X|\theta(X) = -X\}$. Since θ preserves the scalar product, eigenspaces corresponding to different eigenvalues must be orthogonal, and the bracket conditions on $\mathfrak{h}$ and $\mathfrak{m}$ follow automatically from their definition.

One way of finding such a θ is to find a diffeomorphism $\sigma : G \to G$ such that

- G has a bi-invariant metric which is also preserved by σ,

- σ is an automorphism of G, i.e. $\sigma(ab) = \sigma(a)\sigma(b)$,

- $\sigma^2 = $ id.

If we have such a σ, then $\theta := d\sigma_e$ satisfies our requirements. Furthermore, the set of fixed points of σ,

$$F := \{a \in G | \sigma(a) = a\}$$

is clearly a subgroup, which we could take as our subgroup, H. In fact, let F_0 denote the connected component of the identity in F, and let H be any subgroup satisfying $F_0 \subset H \subset F$. Then $M = G/H$ satisfies all our requirements. Such a space is called a normal symmetric space. We construct a large collection of examples of such spaces in the next two subsections.

19.4.3 Orthogonal groups.

We begin by constructing an explicit model for the spaces $\mathbb{R}^{p,q}$ and the orthogonal groups $O(p,q)$. We let $\bullet$ denote the standard Euclidean (positive definite) scalar product on $\mathbb{R}^n$. For any matrix, M, square or rectangular, we let ${}^t M$ denote its transpose. For a given choice of (p,q) with $p+q=n$ we let ϵ denote the diagonal matrix with $+1$ in the first p positions and -1 in the last q positions. Then
$$\langle u, v \rangle := (\epsilon u) \bullet v = u \bullet (\epsilon v)$$
is a scalar product on $\mathbb{R}^n$ of type (p,q).

The condition that a matrix A belong to $O(p,q)$ is then that
$$\epsilon Av \bullet Aw = \epsilon v, w, \quad \forall v, w \in \mathbb{R}^n$$
which is the same as
$$({}^t A \epsilon A v) \bullet w = (\epsilon v) \bullet w \quad \forall v, w \in \mathbb{R}^n$$
which is the same as the condition
$${}^t A \epsilon A v = \epsilon v \quad \forall v \in \mathbb{R}^n.$$
So ${}^t A \epsilon A = \epsilon$ or
$$A \in O(p,q) \Leftrightarrow {}^t A = \epsilon A^{-1} \epsilon. \tag{19.52}$$

Now suppose that $A = \exp tM$, $M \in \mathfrak{g} := o(p,q)$. Then, since the exponential of the transpose of a matrix is the transpose of its exponential, we have
$$\exp s {}^t M = \epsilon \exp(-sM) \epsilon = \exp(-s\epsilon M \epsilon)$$
since $\epsilon^{-1} = \epsilon$. Differentiation at $s=0$ gives
$${}^t M = -\epsilon M \epsilon \tag{19.53}$$
as the condition for a matrix to belong to the Lie algebra $o(p,q)$. If we write M in "block" form
$$M = \begin{pmatrix} a & x \\ y & b \end{pmatrix}$$
then
$${}^t M = \begin{pmatrix} {}^t a & {}^t y \\ {}^t x & {}^t b \end{pmatrix}$$
and the condition to belong to $o(p,q)$ is that
$${}^t a = -a, \quad {}^t b = -b, \quad y = {}^t x$$
so the most general matrix in $o(p,q)$ has the form
$$M = \begin{pmatrix} a & x \\ {}^t x & b \end{pmatrix}, \quad {}^t a = -a, \quad {}^t b = -b. \tag{19.54}$$

19.4. REDUCTIVE HOMOGENEOUS SPACES.

Consider the symmetric bilinear form $X, Y \mapsto \operatorname{tr} XY$, called the "trace form". It is clearly invariant under conjugation, hence, restricted to X, Y both belonging to $o(p,q)$, it is an invariant bilinear form. Let us show that is non-degenerate. Indeed, suppose that

$$X = \begin{pmatrix} a & x \\ {}^t x & b \end{pmatrix}, \quad Y = \begin{pmatrix} c & y \\ {}^t y & d \end{pmatrix}$$

are elements of $o(p,q)$. Then

$$\operatorname{tr} XY = \operatorname{tr}\left(ac + bd + 2x^t y\right).$$

this shows that the subalgebra $h := o(p) \oplus o(q)$ consisting of all "block diagonal" matrices is orthogonal to the subspace m consisting of all matrices with zero entries on the diagonal, i.e. of the form

$$M = \begin{pmatrix} 0 & x \\ {}^t x & 0 \end{pmatrix}.$$

For matrices of the latter form, we have

$$\operatorname{tr} x^t x = \sum_{ij} x_{ij}^2$$

and so is positive definite. On the other hand, since ${}^t a = -a$ and ${}^t b = -b$ we have $a \mapsto \operatorname{tr} a^2 = -\sum_{ij} a_{ij}^2$ is negative definite, and similarly for b. Hence the restriction of the trace form to h is negative definite.

19.4.4 Dual Grassmannians.

Suppose we consider the space $\mathbb{R}^{p+q}$, the positive definite Euclidean space, with orthogonal group $O(p+q)$. Its Lie algebra consists of all anti-symmetric matrices of size $p+q$ and the restriction of the trace form to $o(p+q)$ is negative definite. So we can choose a positive definite invariant scalar product on $g = o(p+q)$ by setting

$$\langle X, Y \rangle := -\frac{1}{2} \operatorname{tr} XY.$$

Let ϵ be as in the preceding subsection, so ϵ is diagonal with p plus 1's and q minus 1's on the diagonal. Notice that ϵ is itself an orthogonal transformation (for the positive definite scalar product on $\mathbb{R}^{p+q}$), and hence conjugation by ϵ is an automorphism of $O(p+q)$ and also of $SO(p+q)$ the subgroup consisting of orthogonal matrices with determinant one.

Let us take $G = SO(p+q)$ and σ to be conjugation by ϵ. So

$$\sigma \begin{pmatrix} a & b \\ c & d \end{pmatrix} = \begin{pmatrix} a & -b \\ -c & d \end{pmatrix}$$

and hence the fixed point subgroup is $F = S(O(p) \times O(q))$. We will take $H = SO(p) \times SO(q)$. The subspace m consists of all matrices of the form

$$X = \begin{pmatrix} 0 & -{}^t x \\ x & 0 \end{pmatrix}$$

and

$$\operatorname{tr} X^2 = -2 \operatorname{tr} {}^t x x$$

so

$$\langle X, X \rangle = \operatorname{tr} {}^t x x,$$

which was the reason for the $\frac{1}{2}$ in our definition of $\langle \, , \, \rangle$.

Our formula for the sectional curvature of a normal symmetric space shows that the sectional curvature of

$$\tilde{G}_{p,q}$$

is non-negative. The special case $p = 1$ the quotient space is the $q-$ dimensional sphere,

$$\tilde{G}_{1,q} = S^q$$

and the x occurring in the above formula is a column vector. Hence $[X, Y]$ where Y corresponds to the column vector y is the operator $= y \otimes^t x - x \otimes^t y \in o(q)$, and

$$\|[X, Y]\|^2 = \|X \wedge Y\|^2,$$

proving that the unit sphere has constant curvature $+1$.

Next let G be the connected component of $O(p, q)$, as described in the preceding subsection, and again take σ to be conjugation by ϵ. This time take

$$\langle X, Y \rangle = \frac{1}{2} \operatorname{tr} XY.$$

The -1 eigenspace, m of σ consists of all matrices of the form

$$\begin{pmatrix} 0 & {}^t x \\ x & 0 \end{pmatrix}$$

and the restriction of $\langle \, , \, \rangle$ to m is positive definite, while the restriction to $H := SO(p) \times SO(q)$ is negative definite. The corresponding symmetric space G/H is denoted by G^*_{pq}. It has negative sectional curvature. In particular, the case $p = 1$ is hyperbolic space, and the same computation as above shows that it has constant sectional curvature equal to -1. This realizes hyperbolic space as the space of timelike lines through the origin in a Lorentz space of one higher dimension.

These two classes of symmetric spaces are dual in the following sense: Suppose that (h, m) and (h^*, m^*) are the Lie algebra data of symmetric spaces G/H and G^*/H^*. Suppose we have

19.5. SCHWARZSCHILD AS A WARPED PRODUCT.

- a Lie algebra isomorphism $\ell : h \to h^*$ such that $\langle \ell U, \ell V \rangle^* = -\langle U, V \rangle$, $\forall U, V \in h$ and

- a linear isometry $i : m \to m^*$ which reverses the bracket:

$$[iX, iY]^* = -[X, Y] \quad \forall X, Y \in m.$$

Then it is immediate from our formula for the sectional curvature that

$$K^*(iX, iY) = -K(X, Y)$$

for any $X, Y \in m$ spanning a non-degenerate plane. We say that the symmetric spaces G/H and G^*/H^* are in duality.

In our case, $H = SO(p) \times SO(q)$ for both $\tilde{G}_{p,q}$ and $G^*_{p,q}$ so we take $\ell =$ id. We define i by

$$i : \begin{pmatrix} 0 & -{}^t x \\ x & 0 \end{pmatrix} \mapsto \begin{pmatrix} 0 & {}^t x \\ x & 0 \end{pmatrix}.$$

It is easy to check that these satisfy our axioms and so $\tilde{G}_{p,q}$ and $G^*_{p,q}$ are dual. For example, the sphere and hyperbolic space are dual in this sense.

19.5 Schwarzschild as a warped product.

In the Schwarzschild model we define

$$P := \{(t, r) | r > 0, r \neq 2M\}$$

with metric

$$-h dt^2 + \frac{1}{h} dr^2, \quad h = h(r) = 1 - \frac{2M}{r}.$$

Then construct the warped product

$$P \times_r S^2$$

where S^2 is the ordinary unit sphere with its standard positive definite metric, call it $d\sigma^2$. So, following O'Neill's conventions, the total metric is of type $(3, 1)$ (timelike = negative square length) given by

$$-h dt^2 + \frac{1}{h} dr^2 + r^2 d\sigma^2.$$

We write $P = P_I \cup P_{II}$ where

$$P_I = \{(t, r) | r > 2M\}, \quad P_{II} = \{(t, r) | r < 2M\}$$

and

$$N = P_I \times_r S^2, \quad B = P_{II} \times_r S^2.$$

N is called the Schwarzschild *exterior* and B is called the *black hole*. In the exterior, ∂_t is timelike. In B, ∂_t is spacelike and ∂_r is timelike.

In either, the vector fields ∂_t, ∂_r are orthogonal and basic. So the base is a surface with orthogonal coordinates.

It is a useful exercise to apply the results about warped products to derive facts about the geometry of the Schwarzschild metric. If you get stuck, see O'Neill, Chapter 13. We derived most of these results earlier in this book, using the Cartan calculus for the curvature computations.

Bibliography

[1] Marcel Berger *A Panoramic View of Riemannian Geometry* Springer (2003) and many other excellent books including
"Arthur L. Besse" *Einstein manifolds* Springer (2008)

[2] Alfred Gray *Tubes* Birkhauser Basel (2004)

[3] Gordon Kane *Modern elementary particle physics* Addison-Wesley(1993)

[4] Yvette Kosmann Schwarzbach *The Noether theorems* Springer (2011)

[5] Thomas Levenson *Einstein in Berlin* Random House (2004)

[6] Lynn Loomis and Shlomo Sternberg *Advanced Calculus* available on my Harvard website

[7] Barrett O'Neill *Semi-Riemannian Geometry with applications to relativity*, Academic Press (1983)

[8] Barrett O'Neill *The geometry of Kerr black holes* A.K. Peters (1995)

[9] Barrett O'Neill *Elementary Differential Geometry*, Revised 2nd Edition, Elsevier, (2006)

[10] Harry Nussbaumer and Lydia Bieri *Discovering the Expanding Universe* Cambridge University Press (2009)

[11] Roger H. Stuewer *The Compton Effect* Science History Publications (1975)

[12] Robert M. Wald *General Relativity* University of Chicago Press (1984)

[13] Steven Weinberg *The quantum theory of fields* 3 volumes Cambridge University Press (1995, 1996, 2003)

Index

acceleration relative to a connection, 88
al Biruni, 189
algebra of exterior differential forms, 52
Ampère's law, 260
associated bundle, 335
associated vector bundle, 337

basic vector fields, 379
bi-invariant metric, 155
Bianchi identity (first), 92
Bianchi identity (second), 172
Bishop frames, 79
bump function, 96
bundle of frames, 318

Cartan connection, 349, 351
Cartan philosophy in Riemannian geometry, 110
Cartan's lemma, 166
Cartan's philosophy in Riemannian geometry, 228
Cartan's theorem on closed geodesics, 202
Chain rule, 55
charge density, 259
Christoffel symbols, 86
Clifford algebra, 366
Clifford superconnections, 375
competing conventions for the curvature, 91
Compton effect, 243
conjugate point, 199
conjugate points, 107
connection forms in a frame field, 162
constituitive equation, 259

contraction of a tensor, 98
contravariant degree, 98
convex open sets, 113
cosmological constant, 145
cosmological constant , 148
covariant degree, 98
covariant differential, 101
covariant differentiation on a principal bundle, 341
covariant divergence of a symmetric tensor field, 279
curvature form(s) in a frame field, 164
curvature identities of the Levi-Civita connection, 135
curvature of a bi-invariant metric, 157
curvature of a linear connection, 91
curvature tensor, 92
curvature, classical expression, 92
cyclic sum, 60

d operator, 53
dark matter, 148
de Rham, 207
densities, 271
derivations, even and odd, 53
deSitter universe, 145
developing a ribbon, 76
dielectric displacement, 259
Differential forms, 52
dirigible, proposed by Meusnier, 44
divergence of a vector field relative to a density, 274
divergence in a semi-Riemannian manifold, 275
divergence theorem, 274
dual Grassmannians, 397

Ehresmann connection, 325
Einstein field equations, 148, 297
Einstein summation convention, 98
electric field stength, 258
energy-momentum, 238
ergosphere, 182
Euclidean frames, 69
exponential map, 107
exponential map and the Jacobi equation, 110
exterior differentiation, 53

Faraday's law of induction, 260
first fundamental form, 34
first structural equation of Cartan, 163
frame fields and coframe fields, 161
frames adapted to a submanifold, 70
free action, 332
Frenet equations, 79
Frenet frames, 79
Friedmann Robertson Walker metrics, 47, 173
Frobenius' theorem, 306, 313

gauge group, 353
gauge theory, history, 347
Gauss map, 26
Gauss's lemma, 190
geodesic, 89
geodesic variation, 106
geodesic, equations for, locally, 89
geodesics, 196
geodesics in orthogonal coordinates, 121
geodesics in the Schwarzschild exterior, 123

Haar measure, 65
Hadamard's theorem, 212
harmonic map, 301
harmonic maps, 299
Hertz's law, 260
Higgs mechanism, 356, 357, 360
Hilbert functional, 295
homogeneous spaces, 394
Hopf-Rinow theorem, 210
Horizontal and basic forms of a fibration, 313

hyperbolic geometry, 168

implicit definition of a surface , 22
inertial coordinates, 109
inertialcoordinates, 37
isometric connection, 117

Jacobi equation, 106
Jacobi fields, 193
Jacobi vector field, 106
Jacobi's identity, 60

Kerr black hole, 178
Kerr-Schild form, 132
Killing fields, 178
Killing form, 158
Koszul's formula, 118

leaves of an integrable system, 308
left action of a group on itself, 63
left invariant metric, 155
Legendre transformation, 291
Leibnitz' rule, 53
Levi-Civita connection, Christoffel symbols of, 120
Levi-Civita's connection, 118
Levi-Civita's theorem, 118, 166
Lie derivative of a density, 273
Lie derivative of differential forms, 55
Lie derivatives of vector fields, 59
Lie superalgebra, 367
linear connection, 85
Linear differential forms, 52
locally symmetric space, 140
London equations, 265
London equations in relativistic form, 267
Lorentz transformations in two dimensions, 232, 234
Lorentz-Fitzgerald contraction, 236

magnetic field, 260
magnetic induction, 260
Magnetoquasistatics, 260
matrix valued differential forms, 62
Maurer-Cartan equation in general, 78
Maurer-Cartan equation(s), 64

Maurer-Cartan form in general, 77, 153
Maurer-Cartan form of a linear Lie group, 64, 65
Maxwell's equations, 262
mean curvature, 29, 42
Meusnier, 43, 45
minimal immersion, 302
minimal surfaces, 42
Minkowksi space in two dimensions, 231
Minkowski space, 241
Morse index, 199
Myer's theorem, 225

Noether's theorem, 179
normal coordinates, 109
normal neighborhood, 108

Ohm's law, 261

parallel transport along a curve, 88
parametrized surfaces, 24
partial velocities, 104
Penrose energy extracton, 183
perihelion advance, 128, 130
polar coordinates in two dimensions, 167
polar map, 111
pregeodesic, 49
principal bundle, 332
principal bundles, 331
Proca equations, 270
Proca Lagrangian, 270
proper time, 235
psychological units, 239, 240
pullback of differential forms, 55

reduction, 311
reduction of a principal bundle, 350
relativistic velocity, 232
rest mass, 238
ribbons, 73
Ricci curvature, 139
Ricci tensor and volume growth, 228
Riemann curvature, symmetry properties of, 136
Riemannian distance, 205

right action of a group on itself, 63
rotating black hole, 178

scalar product induced on exterior powers, 249
Schur's theorem, 173
Schwarzschild metric calculations via Cartan, 170
Schwarzschild solution, 124
second Bianchi identity, 172
second fundamental form, 28
second structural equation of Cartan, 165
sectional curvature, 136
sectional curvature and Riemannian distance, 217
sectional curvature of a semi-Riemannian smersion, 393
semi-Riemannian submersion, 379
soldering, 349
soldering form, 351
source equation, 292
spaces of constant curvature, 138
spontaneous symmetry breaking, 352
star operator on forms, 254
star operator, definition of, 250
starshaped, 112
structure group of a principal bundle, 333
superalgebra, 51
superbundle, 368
superconnection, 369
superconnection form, 374
superconnection, exterior components of, 371
supercurvature, 370
surface of revolution, 47, 124
symmetric space, 214
Synge's formula, 225
Synge's formula for the second variation, 197
Synge's theorem, 201

tangent space, 25
tensor A of a semi-Riemannian submersion, 381

INDEX

tensor T of a semi-Riemannian submersion, 380
tensor analysis, 93
tensor derivation, 99
theorema egregium, 19, 33
torsion of a linear connection, 90
torsion tensor, 91
tubes, 33
twin paradox, 237
two parameter maps, 104

vector field along a curve, 86
volume formula, 30
volume of a thickened hypersurface, 20
von Mangoldt, 212

warped product, 384, 390
wedge product, 52
Weil's formula, 58
Weinberg angle, 160
Weingarten map, 28

A CATALOG OF SELECTED
DOVER BOOKS
IN SCIENCE AND MATHEMATICS

CATALOG OF DOVER BOOKS

Physics

THEORETICAL NUCLEAR PHYSICS, John M. Blatt and Victor F. Weisskopf. An uncommonly clear and cogent investigation and correlation of key aspects of theoretical nuclear physics by leading experts: the nucleus, nuclear forces, nuclear spectroscopy, two-, three- and four-body problems, nuclear reactions, beta-decay and nuclear shell structure. 896pp. 5 3/8 x 8 1/2. 0-486-66827-4

QUANTUM THEORY, David Bohm. This advanced undergraduate-level text presents the quantum theory in terms of qualitative and imaginative concepts, followed by specific applications worked out in mathematical detail. 655pp. 5 3/8 x 8 1/2.
0-486-65969-0

ATOMIC PHYSICS AND HUMAN KNOWLEDGE, Niels Bohr. Articles and speeches by the Nobel Prize–winning physicist, dating from 1934 to 1958, offer philosophical explorations of the relevance of atomic physics to many areas of human endeavor. 1961 edition. 112pp. 5 3/8 x 8 1/2. 0-486-47928-5

COSMOLOGY, Hermann Bondi. A co-developer of the steady-state theory explores his conception of the expanding universe. This historic book was among the first to present cosmology as a separate branch of physics. 1961 edition. 192pp. 5 3/8 x 8 1/2.
0-486-47483-6

LECTURES ON QUANTUM MECHANICS, Paul A. M. Dirac. Four concise, brilliant lectures on mathematical methods in quantum mechanics from Nobel Prize-winning quantum pioneer build on idea of visualizing quantum theory through the use of classical mechanics. 96pp. 5 3/8 x 8 1/2. 0-486-41713-1

THE PRINCIPLE OF RELATIVITY, Albert Einstein and Frances A. Davis. Eleven papers that forged the general and special theories of relativity include seven papers by Einstein, two by Lorentz, and one each by Minkowski and Weyl. 1923 edition. 240pp. 5 3/8 x 8 1/2. 0-486-60081-5

PHYSICS OF WAVES, William C. Elmore and Mark A. Heald. Ideal as a classroom text or for individual study, this unique one-volume overview of classical wave theory covers wave phenomena of acoustics, optics, electromagnetic radiations, and more. 477pp. 5 3/8 x 8 1/2. 0-486-64926-1

THERMODYNAMICS, Enrico Fermi. In this classic of modern science, the Nobel Laureate presents a clear treatment of systems, the First and Second Laws of Thermodynamics, entropy, thermodynamic potentials, and much more. Calculus required. 160pp. 5 3/8 x 8 1/2. 0-486-60361-X

QUANTUM THEORY OF MANY-PARTICLE SYSTEMS, Alexander L. Fetter and John Dirk Walecka. Self-contained treatment of nonrelativistic many-particle systems discusses both formalism and applications in terms of ground-state (zero-temperature) formalism, finite-temperature formalism, canonical transformations, and applications to physical systems. 1971 edition. 640pp. 5 3/8 x 8 1/2. 0-486-42827-3

QUANTUM MECHANICS AND PATH INTEGRALS: Emended Edition, Richard P. Feynman and Albert R. Hibbs. Emended by Daniel F. Styer. The Nobel Prize–winning physicist presents unique insights into his theory and its applications. Feynman starts with fundamentals and advances to the perturbation method, quantum electrodynamics, and statistical mechanics. 1965 edition, emended in 2005. 384pp. 6 1/8 x 9 1/4. 0-486-47722-3

Browse over 9,000 books at www.doverpublications.com

CATALOG OF DOVER BOOKS

Physics

INTRODUCTION TO MODERN OPTICS, Grant R. Fowles. A complete basic undergraduate course in modern optics for students in physics, technology, and engineering. The first half deals with classical physical optics; the second, quantum nature of light. Solutions. 336pp. 5 3/8 x 8 1/2. 0-486-65957-7

THE QUANTUM THEORY OF RADIATION: Third Edition, W. Heitler. The first comprehensive treatment of quantum physics in any language, this classic introduction to basic theory remains highly recommended and widely used, both as a text and as a reference. 1954 edition. 464pp. 5 3/8 x 8 1/2. 0-486-64558-4

QUANTUM FIELD THEORY, Claude Itzykson and Jean-Bernard Zuber. This comprehensive text begins with the standard quantization of electrodynamics and perturbative renormalization, advancing to functional methods, relativistic bound states, broken symmetries, nonabelian gauge fields, and asymptotic behavior. 1980 edition. 752pp. 6 1/2 x 9 1/4. 0-486-44568-2

FOUNDATIONS OF POTENTIAL THERY, Oliver D. Kellogg. Introduction to fundamentals of potential functions covers the force of gravity, fields of force, potentials, harmonic functions, electric images and Green's function, sequences of harmonic functions, fundamental existence theorems, and much more. 400pp. 5 3/8 x 8 1/2. 0-486-60144-7

FUNDAMENTALS OF MATHEMATICAL PHYSICS, Edgar A. Kraut. Indispensable for students of modern physics, this text provides the necessary background in mathematics to study the concepts of electromagnetic theory and quantum mechanics. 1967 edition. 480pp. 6 1/2 x 9 1/4. 0-486-45809-1

GEOMETRY AND LIGHT: The Science of Invisibility, Ulf Leonhardt and Thomas Philbin. Suitable for advanced undergraduate and graduate students of engineering, physics, and mathematics and scientific researchers of all types, this is the first authoritative text on invisibility and the science behind it. More than 100 full-color illustrations, plus exercises with solutions. 2010 edition. 288pp. 7 x 9 1/4. 0-486-47693-6

QUANTUM MECHANICS: New Approaches to Selected Topics, Harry J. Lipkin. Acclaimed as "excellent" (*Nature*) and "very original and refreshing" (*Physics Today*), these studies examine the Mössbauer effect, many-body quantum mechanics, scattering theory, Feynman diagrams, and relativistic quantum mechanics. 1973 edition. 480pp. 5 3/8 x 8 1/2. 0-486-45893-8

THEORY OF HEAT, James Clerk Maxwell. This classic sets forth the fundamentals of thermodynamics and kinetic theory simply enough to be understood by beginners, yet with enough subtlety to appeal to more advanced readers, too. 352pp. 5 3/8 x 8 1/2. 0-486-41735-2

QUANTUM MECHANICS, Albert Messiah. Subjects include formalism and its interpretation, analysis of simple systems, symmetries and invariance, methods of approximation, elements of relativistic quantum mechanics, much more. "Strongly recommended." – *American Journal of Physics*. 1152pp. 5 3/8 x 8 1/2. 0-486-40924-4

RELATIVISTIC QUANTUM FIELDS, Charles Nash. This graduate-level text contains techniques for performing calculations in quantum field theory. It focuses chiefly on the dimensional method and the renormalization group methods. Additional topics include functional integration and differentiation. 1978 edition. 240pp. 5 3/8 x 8 1/2. 0-486-47752-5

Browse over 9,000 books at www.doverpublications.com

CATALOG OF DOVER BOOKS

Physics

MATHEMATICAL TOOLS FOR PHYSICS, James Nearing. Encouraging students' development of intuition, this original work begins with a review of basic mathematics and advances to infinite series, complex algebra, differential equations, Fourier series, and more. 2010 edition. 496pp. 6 1/8 x 9 1/4. 0-486-48212-X

TREATISE ON THERMODYNAMICS, Max Planck. Great classic, still one of the best introductions to thermodynamics. Fundamentals, first and second principles of thermodynamics, applications to special states of equilibrium, more. Numerous worked examples. 1917 edition. 297pp. 5 3/8 x 8. 0-486-66371-X

AN INTRODUCTION TO RELATIVISTIC QUANTUM FIELD THEORY, Silvan S. Schweber. Complete, systematic, and self-contained, this text introduces modern quantum field theory. "Combines thorough knowledge with a high degree of didactic ability and a delightful style." – *Mathematical Reviews*. 1961 edition. 928pp. 5 3/8 x 8 1/2. 0-486-44228-4

THE ELECTROMAGNETIC FIELD, Albert Shadowitz. Comprehensive undergraduate text covers basics of electric and magnetic fields, building up to electromagnetic theory. Related topics include relativity theory. Over 900 problems, some with solutions. 1975 edition. 768pp. 5 5/8 x 8 1/4. 0-486-65660-8

THE PRINCIPLES OF STATISTICAL MECHANICS, Richard C. Tolman. Definitive treatise offers a concise exposition of classical statistical mechanics and a thorough elucidation of quantum statistical mechanics, plus applications of statistical mechanics to thermodynamic behavior. 1930 edition. 704pp. 5 5/8 x 8 1/4.
0-486-63896-0

INTRODUCTION TO THE PHYSICS OF FLUIDS AND SOLIDS, James S. Trefil. This interesting, informative survey by a well-known science author ranges from classical physics and geophysical topics, from the rings of Saturn and the rotation of the galaxy to underground nuclear tests. 1975 edition. 320pp. 5 3/8 x 8 1/2.
0-486-47437-2

STATISTICAL PHYSICS, Gregory H. Wannier. Classic text combines thermodynamics, statistical mechanics, and kinetic theory in one unified presentation. Topics include equilibrium statistics of special systems, kinetic theory, transport coefficients, and fluctuations. Problems with solutions. 1966 edition. 532pp. 5 3/8 x 8 1/2.
0-486-65401-X

SPACE, TIME, MATTER, Hermann Weyl. Excellent introduction probes deeply into Euclidean space, Riemann's space, Einstein's general relativity, gravitational waves and energy, and laws of conservation. "A classic of physics." – *British Journal for Philosophy and Science*. 330pp. 5 3/8 x 8 1/2. 0-486-60267-2

RANDOM VIBRATIONS: Theory and Practice, Paul H. Wirsching, Thomas L. Paez and Keith Ortiz. Comprehensive text and reference covers topics in probability, statistics, and random processes, plus methods for analyzing and controlling random vibrations. Suitable for graduate students and mechanical, structural, and aerospace engineers. 1995 edition. 464pp. 5 3/8 x 8 1/2. 0-486-45015-5

PHYSICS OF SHOCK WAVES AND HIGH-TEMPERATURE HYDRO DYNAMIC PHENOMENA, Ya B. Zel'dovich and Yu P. Raizer. Physical, chemical processes in gases at high temperatures are focus of outstanding text, which combines material from gas dynamics, shock-wave theory, thermodynamics and statistical physics, other fields. 284 illustrations. 1966–1967 edition. 944pp. 6 1/8 x 9 1/4.
0-486-42002-7

Browse over 9,000 books at www.doverpublications.com

CATALOG OF DOVER BOOKS

Mathematics–Bestsellers

HANDBOOK OF MATHEMATICAL FUNCTIONS: with Formulas, Graphs, and Mathematical Tables, Edited by Milton Abramowitz and Irene A. Stegun. A classic resource for working with special functions, standard trig, and exponential logarithmic definitions and extensions, it features 29 sets of tables, some to as high as 20 places. 1046pp. 8 x 10 1/2. 0-486-61272-4

ABSTRACT AND CONCRETE CATEGORIES: The Joy of Cats, Jiri Adamek, Horst Herrlich, and George E. Strecker. This up-to-date introductory treatment employs category theory to explore the theory of structures. Its unique approach stresses concrete categories and presents a systematic view of factorization structures. Numerous examples. 1990 edition, updated 2004. 528pp. 6 1/8 x 9 1/4. 0-486-46934-4

MATHEMATICS: Its Content, Methods and Meaning, A. D. Aleksandrov, A. N. Kolmogorov, and M. A. Lavrent'ev. Major survey offers comprehensive, coherent discussions of analytic geometry, algebra, differential equations, calculus of variations, functions of a complex variable, prime numbers, linear and non-Euclidean geometry, topology, functional analysis, more. 1963 edition. 1120pp. 5 3/8 x 8 1/2. 0-486-40916-3

INTRODUCTION TO VECTORS AND TENSORS: Second Edition--Two Volumes Bound as One, Ray M. Bowen and C.-C. Wang. Convenient single-volume compilation of two texts offers both introduction and in-depth survey. Geared toward engineering and science students rather than mathematicians, it focuses on physics and engineering applications. 1976 edition. 560pp. 6 1/2 x 9 1/4. 0-486-46914-X

AN INTRODUCTION TO ORTHOGONAL POLYNOMIALS, Theodore S. Chihara. Concise introduction covers general elementary theory, including the representation theorem and distribution functions, continued fractions and chain sequences, the recurrence formula, special functions, and some specific systems. 1978 edition. 272pp. 5 3/8 x 8 1/2. 0-486-47929-3

ADVANCED MATHEMATICS FOR ENGINEERS AND SCIENTISTS, Paul DuChateau. This primary text and supplemental reference focuses on linear algebra, calculus, and ordinary differential equations. Additional topics include partial differential equations and approximation methods. Includes solved problems. 1992 edition. 400pp. 7 1/2 x 9 1/4. 0-486-47930-7

PARTIAL DIFFERENTIAL EQUATIONS FOR SCIENTISTS AND ENGINEERS, Stanley J. Farlow. Practical text shows how to formulate and solve partial differential equations. Coverage of diffusion-type problems, hyperbolic-type problems, elliptic-type problems, numerical and approximate methods. Solution guide available upon request. 1982 edition. 414pp. 6 1/8 x 9 1/4. 0-486-67620-X

VARIATIONAL PRINCIPLES AND FREE-BOUNDARY PROBLEMS, Avner Friedman. Advanced graduate-level text examines variational methods in partial differential equations and illustrates their applications to free-boundary problems. Features detailed statements of standard theory of elliptic and parabolic operators. 1982 edition. 720pp. 6 1/8 x 9 1/4. 0-486-47853-X

LINEAR ANALYSIS AND REPRESENTATION THEORY, Steven A. Gaal. Unified treatment covers topics from the theory of operators and operator algebras on Hilbert spaces; integration and representation theory for topological groups; and the theory of Lie algebras, Lie groups, and transform groups. 1973 edition. 704pp. 6 1/8 x 9 1/4. 0-486-47851-3

Browse over 9,000 books at www.doverpublications.com